PERGAMON INTERNATIONAL LIBRARY
of Science, Technology, Engineering and Social Studies

The 1000-volume original paperback library in aid of education,
industrial training and the enjoyment of leisure

Publisher: Robert Maxwell, M.C.

Single Case
Experimental Designs
(PGPS-56)

Pergamon Titles of Related Interest

Barlow/Hayes/Nelson THE SCIENTIST PRACTITIONER: Research
and Accountability in Clinical and Educational Settings
Bellack/Hersen RESEARCH METHODS IN CLINICAL PSYCHOLOGY
Hersen/Bellack BEHAVIORAL ASSESSMENT: A Practical
Handbook, Second Edition
Ollendick/Hersen CHILD BEHAVIORAL ASSESSMENT: Principles
and Procedures

Related Journals*

BEHAVIORAL ASSESSMENT
PERSONALITY AND INDIVIDUAL DIFFERENCES

*Free specimen copies available upon request.

PERGAMON GENERAL PSYCHOLOGY SERIES
EDITORS
Arnold P. Goldstein, *Syracuse University*
Leonard Krasner, *SUNY at Stony Brook*

Single Case Experimental Designs
Strategies for Studying Behavior Change
Second Edition

David H. Barlow
SUNY at Albany

Michel Hersen
University of Pittsburgh School of Medicine

PERGAMON PRESS
New York • Oxford • Toronto • Sydney • Paris • Frankfurt

Pergamon Press Offices:

U.S.A.   Pergamon Press Inc., Maxwell House, Fairview Park,
                             Elmsford, New York 10523, U.S.A.

U.K.   Pergamon Press Ltd., Headington Hill Hall,
                             Oxford OX3 0BW, England

CANADA   Pergamon Press Canada Ltd., Suite 104, 150 Consumers Road,
                             Willowdale, Ontario M2J 1P9, Canada

AUSTRALIA   Pergamon Press (Aust.) Pty. Ltd., P.O. Box 544,
                             Potts Point, NSW 2011, Australia

FRANCE   Pergamon Press SARL, 24 rue des Ecoles,
                             75240 Paris, Cedex 05, France

FEDERAL REPUBLIC   Pergamon Press GmbH, Hammerweg 6,
OF GERMANY   D-6242 Kronberg-Taunus, Federal Republic of Germany

Library of Congress Cataloging in Publication Data

Barlow, David H.

    Single case experimental designs, 2nd ed.

    (Pergamon general psychology series)
    Author's names in reverse order in 1st ed., 1976.
    Includes bibliographies and indexes.
    1. Psychology--Research. 2. Experimental design.
I. Hersen, Michel. II. Title. III. Series. [DNLM:
1. Behavior. 2. Psychology, Experimental. 3. Research
design. BF 76.5 H572s]
BF76.5.B384   1984   150'.724   84-6292
ISBN 0-08-030136-3
ISBN 0-08-030135-5 (soft)

Printed in the United States of America

Contents

TO THE MEMORY OF
Frederic I. Barlow
and
to Members of the Hersen Family
who died in World War II

Preface

In the preface to the first edition of this book we said:

> We do not expect this book to be the final statement on single-case designs. We learned at least as much as we already knew in analyzing the variety of innovative and creative applications of these designs to varying applied problems. The unquestionable appropriateness of these designs in applied settings should ensure additional design innovations in the future.

At the time, this seemed a reasonable statement to make, but we think that few of us involved in applied research anticipated the explosive growth of interest in single-case designs and how many methodological and strategical innovations would subsequently appear. As a result of developments in the 8 years since the first edition, this book can be more accurately described as new than as revised. Fully 5 of the 10 chapters are new or have been completely rewritten. The remaining five chapters have been substantially revised and updated to reflect new guidelines and the current wisdom on experimental strategies involving single-case designs.

Developments in the field have not been restricted to new or modified experimental designs. New thinking has emerged on the analyses of data from these designs, particularly with regard to use of statistical procedures. We were most fortunate in having Alan Kazdin take into account these developments in the revision of his chapter on statistical analyses for single-case experimental designs. Furthermore, the area of techniques of measurement and assessment relevant to single-case designs has changed greatly in the years since the first edition. Don Hartmann, the Editor of *Behavioral Assessment* and one of the leading figures in assessment and single-case designs, has strengthened the book considerably with his lucid chapter. Nevertheless, the primary purpose of the book was, and remains, the provision of a sourcebook of single-case designs, with guidelines for their use in applied settings.

To Sallie Morgan, who is very tired of typing the letters A-B-C over and over again for the past 10 years, we can say that we couldn't have done it without you, or without Mary Newell and Susan Capozzoli. Also, Susan

Cohen made a significant contribution in searching out the seemingly endless articles on single-case designs that have accumulated over the years. And Susan, as well as Janet Klosko and Janet Twomey, deserves credit for compiling for what we hope is a useful index, a task for which they have developed considerable expertise. Finally, this work really is the creation of the community of scientists dedicated to exploring ways to alleviate human suffering and enhance human potential. These intellectual colleagues and forebears are now too numerous to name, but we hope that this book serves our colleagues as well as the next generation.

David H. Barlow
Albany, New York

Michel Hersen
Pittsburgh, Pennsylvania

Epigram

Conversation between Tolman and Allport

TOLMAN: "I know I should be more idiographic in my research, but I just don't know how to be."

ALLPORT: "Let's learn!"

CHAPTER 1

The Single-case in Basic and Applied Research: An Historical Perspective

1.1. INTRODUCTION

The individual is of paramount importance in the clinical science of human behavior change. Until recently, however, this science lacked an adequate methodology for studying behavior change in individuals. This gap in our methodology has retarded the development and evaluation of new procedures in clinical psychology and psychiatry as well as in educational fields.

Historically, the intensive study of the individual held a preeminent place in the fields of psychology and psychiatry. In spite of this background, an adequate experimental methodology for studying the individual was very slow to develop in applied research.* To find out why, it is useful to gain some perspective on the historical development of methodology in the broad area of psychological research.

The purpose of this chapter is to provide such a perspective, beginning with the origins of methodology in the basic sciences of physiology and experimental psychology in the middle of the last century. Because most of this early work was performed on individual organisms, reasons for the development of between-group comparison methodology in basic research (which did not occur until the turn of the century) are outlined. The rapid development of inferential statistics and sampling theory during the early 20th century enabled greater sophistication in the research methodology of experimental psychology. The manner in which this affected research methods in applied areas during the middle of the century is discussed.

---

*In this book applied research refers to experimentation in the area of human behavior change relevant to the disciplines of clinical psychology, psychiatry, social work, and education.

1

In the meantime, applied research was off to a shaky start in the offices of early psychiatrists with a technique known as the case study method. The separate development of applied research is traced from those early beginnings through the grand collaborative group comparison studies proposed in the 1950s. The subsequent disenchantment with this approach in applied research forced a search for alternatives. The rise and fall of the major alternatives—process research and naturalistic studies—is outlined near the end of the chapter. This disenchantment also set the stage for a renewal of interest in the scientific study of the individual. The multiple origins of single-case experimental designs in the laboratories of experimental psychology and the offices of clinicians complete the chapter. Descriptions of single-case designs and guidelines for their use as they are evolving in applied research comprise the remainder of this book.

1.2. BEGINNINGS IN EXPERIMENTAL PHYSIOLOGY AND PSYCHOLOGY

The scientific study of individual human behavior has roots deep in the history of psychology and physiology. When psychology and physiology became sciences, the initial experiments were performed on individual organisms, and the results of these pioneering endeavors remain relevant to the scientific world today. The science of physiology began in the 1830s, with Johannes Müller and Claude Bernard, but an important landmark for applied research was the work of Paul Broca in 1861. At this time, Broca was caring for a man who was hospitalized for an inability to speak intelligibly. Before the man died, Broca examined him carefully; subsequent to death, he performed an autopsy. The finding of a lesion in the third frontal convolution of the cerebral cortex convinced Broca, and eventually the rest of the scientific world, that this was the speech center of the brain. Broca's method was the clinical extension of the newly developed experimental methodology called *extirpation of parts*, introduced to physiology by Marshall Hall and Pierre Flouren in the 1850s. In this method, brain function was mapped out by systematically destroying parts of the brain in animals and noting the effects on behavior.

The importance of this research in the context of the present discussion lies in the demonstration that important findings with wide generality were gleaned from single organisms. This methodology was to have a major impact on the beginnings of experimental psychology.

Boring (1950) fixed the beginnings of experimental psychology in 1860, with the publication of Fechner's *Elemente der Psychophysik*. Fechner is most famous for developing measures of sensation through several psychophysical methods. With these methods, Fechner was able to determine sensory thresholds and just noticeable differences (JNDs) in various sense modalities. What is common to these methods is the repeated measurement of a response at different intensities or different locations of a given stimulus in an individual subject. For example, when stimulating skin with two points in a certain region to determine the minimal separation which the subject reliably recognizes as two stimulations, one may use the method of constant stimuli. In this method the two points repeatedly stimulate two areas of skin at five to seven fixed separations, in random order, ranging from a few millimeters apart to the relatively large separation of 10 mm. During each stimulation, the subject reports whether he or she senses one point or two. After repeated trials, the point at which the subject "notices" two separate points can be determined. It is interesting to note that Fechner was one of the first to apply statistical methods to psychological problems. Fechner noticed that judgments of just noticeable differences in the sensory modalities varied somewhat from trial to trial. To quantify this variation, or "error" in judgment, he borrowed the normal law of error and demonstrated that these errors were normally distributed around a mean, which then became the "true" sensory threshold. This use of descriptive statistics anticipated the application of these procedures to groups of individuals at the turn of the century, when traits or capabilities were also found to be normally distributed around a mean. The emphasis on error, or the average response, raised issues regarding imprecision of measurement that were to be highlighted in between-group comparison approaches (see below and chapter 2). It should be noted, however, that Fechner was concerned with variability *within* the subject, and he continued his remarkable work on series of individuals.

These traditions in methodology were continued by Wilhelm Wundt. Wundt's contributions, and those of his students and followers, most notably Titchener, had an important impact on the field of psychology, but it is the scientific methodology he and his students employed that most interests us.

To Wundt, the subject matter of psychology was immediate experience, such as how a subject experiences light and sound. Since these experiences were private events and could not be directly observed, Wundt created a new method called *introspection*. Mention of the procedure may strike a responsive chord in some modern-day clinicians, but in fact this methodology is quite different from the introspection technique of free association and others, often used in clinical settings to uncover repressed or unconscious material. Nor did introspection bear any relation to armchair dreams or reflections that are so frequent a part of experience. Introspection, as Wundt employed it, was a highly specific and rigorous procedure that was used with individual subjects who were highly trained. This training involved learning to describe experiences in an objective manner, free from emotional or language restraints. For example, the experience of seeing a brightly colored object would be described in terms of shapes and hues without recourse to aesthetic appeal. To illustrate the objectivity of this system, introspection of emotional experiences where scientific calm and objectivity might be disrupted was not allowed. Introspection of this experience was to be done at a later date when the scientific attitude returned. This method, then, became retrospection, and the weaknesses of this approach were accepted by Wundt to preserve objectivity. Like Fechner's psychophysics, which is essentially an introspectionist methodology, the emphasis hinges on the study of a highly trained individual with the clear assumption, after some replication on other individuals, that findings would have generality to the population of individuals. Wundt and his followers comprised a school of psychology known as the Structuralist School, and many topics important to psychology were first studied with this rather primitive but individually oriented form of scientific analysis. The major subject matter, however, continued to be sensation and perception. With Fechner's psychophysical methods, the groundwork for the study of sensation and perception was laid. Perhaps because of these beginnings, a strong tradition of studying individual organisms has ensued in the fields of sensation and perception and physiological psychology. This tradition has not extended to other areas of experimental psychology, such as learning, or to the more clinical areas of investigation that are broadly based on learning principles or theories. This course of events is surprising because the efforts to study principles of learning comprise one of the more famous examples of the scientific study of the single-case. This effort was made by Hermann Ebbinghaus, one of the towering figures in the development of psychology. With a belief in the scientific approach to psychology, and heavily influenced by Fechner's methods (Boring, 1950), Ebbinghaus established principles of human learning that remain basic to work in this area.

Basic to Ebbinghaus's experiments was the invention of a new instrument to measure learning and forgetting—the nonsense syllable. With a long list of nonsense syllables and himself as the subject, he investigated the effects of different variables (such as the amount of material to be remembered) on the efficiency of memory. Perhaps his best known discovery was the retention curve, which illustrated the process of forgetting over time. Chaplin and Kraweic (1960) noted that he "worked so carefully that the results of his experiments have never been seriously questioned" (p. 180). But what is most relevant and remarkable about his work is his emphasis on repeated measures of performance in one individual over time (see chapter 4). As Boring (1950) pointed out, Ebbinghaus made repetition the basis for the experimental measurement of memory. It would be some 70 years before a new approach, called the *experimental analysis of behavior*, was to employ repeated measurement in individuals to study complex animal and human behaviors.

One of the best known scientists in the fields of physiology and psychology during these early years was Pavlov (Pavlov, 1928). Although Pavlov considered himself a physiologist, his work on principles of association and learning was his greatest contribution, and, along with his basic methodology, is so well known that summaries are not required. What is often overlooked, however, is that Pavlov's basic findings were gleaned from single organisms and strengthened by replication on other organisms. In terms of scientific yield, the study of the individual organism reached an early peak with Pavlov, and Skinner would later cite this approach as an important link and a strong bond between himself and Pavlov (Skinner, 1966a).

1.3. ORIGINS OF THE GROUP COMPARISON APPROACH

Important research in experimental psychology and physiology using single cases did not stop with these efforts, but the turn of the century witnessed a new development which would have a marked effect on basic and, at a later date, applied research. This development was the discovery and measurement of individual differences. The study of individual differences can be traced to Adolphe Quetelet, a Belgian astronomer, who discovered that human traits (e.g., height) followed the normal curve (Stilson, 1966). Quetelet interpreted these findings to mean that nature strove to produce the "average" man but, due to various reasons, failed, resulting in errors or variances in traits that grouped around the average. As one moved further from this average, fewer examples of the trait were evident, following the well-known normal distribution. This approach, in turn, had its origins in Darwin's observations on individual variation within a species. Quetelet viewed these variations or errors as unfortunate since he viewed the average man, which he termed *l'homme moyen*, as a cherished goal rather than a descriptive fact of central tendency. If nature were "striving" to produce the average man, but failed due to various accidents, then the average, in this view, was obviously the ideal. Where nature failed, however, man could pick up the pieces, account for the errors, and estimate the average man through statistical techniques. The influence of this finding on psychological research was enormous, as it paved the way for the application of sophisticated statistical procedures to psychological problems. Quetelet would probably be distressed to learn, however, that his concept of the average individual would come under attack during the 20th century by those who observed that there is no average individual (e.g., Dunlap, 1932; Sidman, 1960).

This viewpoint notwithstanding, the study of individual differences and the statistical approach to psychology became prominent during the first half of the 20th century and changed the face of psychological research. With a push from the American functional school of psychology and a developing interest in the measurement and testing of intelligence, the foundation for comparing groups of individuals was laid.

Galton and Pearson expanded the study of individual differences at the turn of the century and developed many of the descriptive statistics still in use today, most notably the notion of correlation, which led to factor analysis, and significant advances in construction of intelligence tests first introduced by Binet in 1905. At about this time, Pearson, along with Galton and Weldon, founded the journal *Biometricka* with the purpose of advancing quantitative research in biology and psychology. Many of the newly devised statistical tests were first published there. Pearson was highly enthusiastic about the statistical approach and seemed to believe, at times, that inaccurate data could be made to yield accurate conclusions if the proper statistics were applied (Boring, 1950). Although this view was rejected by more conservative colleagues, it points up a confidence in the power of statistical procedures that reappears from time to time in the execution of psychological research (e.g., D. A. Shapiro & Shapiro, 1983; M. L. Smith & Glass, 1977; G. T. Wilson & Rachman, 1983).

One of the best known psychologists to adopt this approach was James McKeen Cattell. Cattell, along with Farrand, devised a number of simple mental tests that were administered to freshmen at Columbia University to determine the range of individual differences. Cattell also devised the order of merit method, whereby a number of judges would rank items or people on a given quality, and the average response of the judges constituted the rank of that item vis-à-vis other items. In this way, Cattell had 10 scientists rate a number of eminent colleagues. The scientist with the highest score (on the average) achieved the top rank.

It may seem ironic at first glance that a concern with individual differences led to an emphasis on groups and averages, but differences among individuals, or intersubject variability, and the distribution of these differences necessitate a comparison among individuals and a concern for a description of a group or population as a whole. In this context observations from a single organism are irrelevant. Darwin, after all, was concerned with survival of a species and not the survival of individual organisms.

The invention of many of the descriptive statistics and some crude statistical tests of comparison made it easier to compare performance in large groups of subjects. From 1900 to 1930, much of the research in experimental psychology, particularly learning, took advantage of these statistics to compare groups of subjects (usually rats) on various performance tests (e.g., see Birney & Teevan, 1961). Crude statistics that could attribute differences between groups to something other than chance began to appear, such as the critical ratio test (Walker & Lev, 1953). The idea that the variability or error among organisms could be accounted for or averaged out in large groups was a commonsense notion emanating from the new emphasis on variability among organisms. The fact that this research resulted in an average finding from the hypothetical average rat drew some isolated criticism. For instance, in 1932, while reviewing research in experimental psychology, Dunlap pointed out that there was no average rat, and Lewin (1933) noted that ". . . the only situations which should be grouped for statistical treatment are those which have for the individual rats or for the individual children the same psychological structure and only for such period of time as this structure exists" (p. 328). The new emphasis on variability and averages, however, would have pleased Quetelet, whose slogan could have been "Average is Beautiful."

The influence of inferential statistics

During the 1930s, the work of R. A. Fisher, which subsequently exerted considerable influence on psychological research, first appeared. Most of the sophisticated statistical procedures in use today for comparing groups were invented by Fisher. It would be difficult to pick up psychological or psychiatric journals concerned with behavior change and not find research data analyzed by the ubiquitous analysis of variance. It is interesting, however, to consider the origin of these tests. Early in his career, Fisher, who was a mathematician interested in genetics, made an important decision. Faced with pursuing a career at a biometrics laboratory, he chose instead a relatively obscure agricultural station on the grounds that this position would offer him more opportunity for independent research. This personal decision at the very least changed the language of experimental design in psychological research, introducing agricultural terms to describe relevant designs and variables (e.g., split plot analysis of variance). While Fisher's statistical innovations were one of the more important developments of the century for psychology, the philosophy underlying the use of these procedures is clearly in line with Quetelet's notion of the importance of the average. As a good agronomist, Fisher was concerned with the yield from a given area of land under various soil treatments, plant varieties, or other agricultural variables. Much as in the study of individual differences, the fate of the individual plant is irrelevant in the context of the yield from the group of plants in that area. Agricultural variables are important to the farm and society if the yield is better *on the average* than a similar plot treated differently. The implications of this philosophy for applied research will be discussed in chapter 2.

The work of Fisher was not limited to the invention of sophisticated statistical tests. An equally important contribution was the consideration of the problem of induction or inference. Essentially, this issue concerns generality of findings. If some data are obtained from a group or a plot of land, this information is not very valuable if it is relevant only to that particular group or plot of land because similar data must be collected from each new plot. Fisher (1925) worked out the properties of statistical tests, which made it possible to estimate the relevance of data from one small group with certain characteristics to the universe of individuals with those characteristics. In other words, inference is made from the sample to the population. This work and the subsequent developments in the field of sampling theory made it possible to talk in terms of psychological principles with broad generality and applicability—a primary goal in any science. This type of estimation, however, was based on appropriate statistics, averages, and intersubject variability in the sample, which further reinforced the group comparison approach in basic research.

As the science of psychology grew out of its infancy, its methodology was largely determined by the lure of broad generality of findings made possible through the brilliant work of Fisher and his followers. Because of the emphasis on averages and intersubject variability required by this design in order to make general statements, the intensive study of the single organism, so popular in the early history of psychology, fell out of favor. By the 1950s, when investigators began to consider the possibility of doing serious research in applied settings, the group comparison approach was so entrenched that anyone studying single organisms was considered something of an oddity by no less an authority than Underwood (1957). The zeitgeist in psychological research was group comparison and statistical estimation. While an occasional paper was published during the 1950s defending the study of the single-case (S. J. Beck, 1953; Rosenzweig, 1951), or at least pointing out its place in psychological research (duMas, 1955), very little basic research was carried out on single-cases. A notable exception was the work of B. F. Skinner and his students and colleagues, who were busy developing an approach known as the experimental analysis of behavior, or operant conditioning. This work, however, did not have a large impact on methodology in other areas of psychology during the 1950s, and applied research was just beginning. Against this background, it is not surprising that applied researchers in the 1950s employed the group comparison approach, despite the fact that the origins of the study of clinically relevant phenomena were quite different from the origin of more basic research described above.

1.4. DEVELOPMENT OF APPLIED RESEARCH: THE CASE STUDY METHOD

As the sciences of physiology and psychology were developing during the late 19th and 20th centuries, people were suffering from emotional and behavioral problems and were receiving treatment. Occasionally, patients recovered, and therapists would carefully document their procedures and communicate them to colleagues. Hypotheses attributing success or failure to various assumed causes emanated from these cases, and these hypotheses gradually grew into theories of psychotherapy. Theories proliferated, and procedures based on observations of cases and inferences from these theories grew in number. As Paul (1969) noted, those theories or procedures that could be communicated clearly or that presented new and exciting principles tended to attract followers to the organization, and schools of psychotherapy were formed. At the heart of this process is the *case study* method of investigation (Bolger, 1965). This method (and its extensions) was, with few exceptions, the sole methodology of clinical investigation through the first half of the 20th century.

The case study method, of course, is the clinical base for the experimental study of single-cases and, as such, it retains an important function in present-day applied research (Barlow, 1980; Barlow, Hayes, & Nelson, 1983; Kazdin, 1981) (see section 1.7). Unfortunately, during this period clinicians were unaware, for the most part, of the basic principles of applied research, such as definition of variables and manipulation of independent variables. Thus it is noteworthy from an historical point of view that several case studies reported during this period came tantalizingly close to providing the basic scientific ingredients of experimental single-case research. The most famous of these, of course, is the J. B. Watson and Rayner (1920) study of an analogue of clinical phobia in a young boy, where a prototype of a *withdrawal* design was attempted (see chapter 5). These investigators unfortunately suffered the fate of many modern-day clinical researchers in that the subject moved away before the "reversal" was complete.

Anytime that a treatment produced demonstrable effects on an observable behavior disorder, the potential for scientific investigation was there. An excellent example, among many, was Breuer's classic description of the treatment of hysterical symptoms in Anna O. through psychoanalysis in 1895 (Breuer & Freud, 1957). In a series of treatment sessions, Breuer dealt with one symptom at a time through hypnosis and subsequent "talking through," where each symptom was traced back to its hypothetical causation in circumstances surrounding the death of her father. One at a time, these behaviors disappeared, but only when treatment was administered to each respective behavior. This process of treating one behavior at a time fulfills the basic requirement for a multiple baseline experimental design described in chapter 7, and the clearly observable success indicated that Breuer's treatment was effective. Of course, Breuer did not define his independent variables, in that there were several components to his treatment (e.g., hypnosis, interpretation); but, in the manner of a good scientist as well as a good clinician, Breuer admitted that he did not know which component or components of his treatment were responsible for success. He noted at least two possibilities, the suggestion inherent in the hypnosis or the interpretation. He then described events discovered through his talking therapy as possibly having etiological significance and wondered about the reliability of the girl's report as he hypothesized various etiologies for the symptoms. However, he did not, at the time, firmly link successful treatment with the necessity of discovering the etiology of the behavior disorder. One wonders if the early development of clinical techniques, including psychoanalysis, would have been different if careful observers like Breuer had been cognizant of the experimental implications of their clinical work. Of course, this small leap from uncontrolled case study to scientific investigation of the single case did not occur because of a lack of awareness of basic scientific principles in early clinicians. The result was an accumulation of successful individuals' case studies, with clinicians from varying schools claiming that their techniques were indispensable to success. In many cases their claims were grossly exaggerated. Brill noted in 1909 on psychoanalysis that "The results obtained by the treatment are unquestionably very gratifying. They surpass those obtained by simpler methods in two chief respects; namely, in permanence and in the prophylactic value they have for the future" (Brill, 1909). Much later, in 1935, Kessel and Hyman observed, "this patient was saved from an inferno and we are convinced that this could have been achieved by no other method" (Kessel & Hyman, 1933). From an early behavioral standpoint, Max (1935) noted the electrical aversion therapy produced "95 percent relief" from the compulsion of homosexuality.

These kinds of statements did little to endear the case study method to serious applied researchers when they began to appear in the 1940s and 1950s. In fact, the case study method, if anything, deteriorated somewhat over the years in terms of the amount and nature of publicly observable data available in these reports. Frank (1961) noted the difficulty in even collecting data from a therapeutic hour in the 1930s due to lack of necessary equipment, reluctance to take detailed notes, and concern about confidentiality. The advent of the phonograph record at this time made it possible at least to collect raw data from those clinicians who would cooperate, but this method did not lead to any fruitful new ideas on research. With the advent of serious applied research in the 1950s, investigators tended to reject reports from uncontrolled case studies due to an inability to evaluate the effects of treatment. Given the extraordinary claims by clinicians after successful case studies, this attitude is understandable. However, from the viewpoint of single-case experimental designs, this rejection of the careful observation of behavior change in a case report had the effect of throwing out the baby with the bathwater.

Percentage of success in treated groups

A further development in applied research was the reporting of collections of case studies in terms of percentage of success. Many of these reports have been cited by Eysenck (1952). However, reporting of results in this manner probably did more harm than good to the evaluation of clinical treatment. As Paul (1969) noted, independent and dependent variables were no better defined than in most case reports, and techniques tended to be fixed and "school" oriented. Because all procedures achieved some success, practitioners within these schools concentrated on the positive results, explained away the failures, and decided that the overall results confirmed that their procedures, as applied, were responsible for the success. Due to the strong and overriding theories central to each school, the successes obtained were attributed to theoretical constructs underlying the procedure. This precluded a careful analysis of elements in the procedure or the therapeutic intervention that many have been responsible for certain changes in a given case and had the effect of reinforcing the application of a global, ill-defined treatment from whatever theoretical orientation, to global definitions of behavior disorders, such as *neurosis*. This, in turn, led to statements such as "psychotherapy works with neurotics." Although applied researchers later rejected these efforts as unscientific, one carryover from this approach was the notion of the average response to treatment; that is, if a global treatment is successful on the average with a group of "neurotics," then this treatment will probably be successful with any individual neurotic who requests treatment.

Intuitively, of course, descriptions of results from 50 cases provide a more convincing demonstration of the effectiveness of a given technique than separate descriptions of 50 individual cases. A modification of this approach utilizing updated strategies and procedures and with the focus on individual responses has been termed *clinical replication*. This strategy can make a substantial contribution to the applied research process (see chapter 10). The major difficulty with this approach, however, particularly as it was practiced in early years, is that the category in which these clients are classified most always becomes unmanageably heterogeneous. The neurotics described in Eysenck's (1952) paper may have less in common than any group of people one would choose randomly. When cases are described individually, however, a clinician stands a better chance of gleaning some important information, since specific problems and specific procedures are usually described in more detail. When one lumps cases together in broadly defined categories, individual case descriptions are lost and the ensuing report of percentage success becomes meaningless. This unavoidable heterogeneity in any group of patients is an important consideration that will be discussed in more detail in this chapter and in chapter 2.

Group comparison approach in applied research

By the late 1940s, clinical psychology and, to a lesser extent, psychiatry began to produce the type of clinician who was also aware of basic research strategies. These scientists were quick to point out the drawbacks of both the case study and reports of percentages of success in groups in evaluating the effects of psychotherapy. They noted that any adequate test of psychotherapy would have to include a more precise definition of terms, particularly *outcome criteria* or *dependent variables* (e.g., Knight, 1941). Most of these applied researchers were trained as psychologists, and in psychology a new emphasis was placed on the "scientist-practitioner" model (Barlow et al., 1983). Thus, the source of research methodology in the newly developing areas of applied research came from experimental psychology. By this time, the predominant methodology in experimental psychology was the between-subjects group design.

The group design also was a logical extension of the earlier clinical reports of percentage success in a large group of patients, because the most obvious criticism of this endeavor is the absence of a control group of untreated patients. The appearance of Eysenck's (1952) notorious article comparing percentage success of psychotherapy in large groups to rates of "spontaneous" remission gleaned from discharge rates at state hospitals and insurance company records had two effects. *First*, it reinforced the growing conviction that the effects of psychotherapy could not be evaluated from case reports or "percentage success groups" and sparked a new flurry of interest in evaluating psychotherapy through the scientific method. *Second*, the emphasis on comparison between groups and quasi-control groups in Eysenck's review strengthened the notion that the logical way to evaluate psychotherapy was through the prevailing methodology in experimental psychology—the between-groups comparison designs.

This approach to applied research did not suddenly begin in the 1950s, although interest certainly increased at this time. Scattered examples of research with clinically relevant problems can be found in earlier decades. One interesting example is a study reported by Kantorovich (1928), who applied aversion therapy to one group of twenty alcoholics in Russia and compared results to a control group receiving hypnosis or medication. The success of this treatment (and the direct derivation from Pavlov's work) most likely ensured a prominent place for aversion therapy in Russian treatment programs for alcoholics. Some of the larger group comparison studies typical of the 1950s also began before Eysenck's celebrated paper. One of the best known is the Cambridge-Somerville youth study, which was reported in 1951 (Powers & Witmer, 1951) but was actually begun in 1937. Although this was an early study, it is quite representative of the later group comparison studies in that many of the difficulties in execution and analysis of results were repeated again and again as these studies accumulated.

The major difficulty, of course, was that these studies did not prove that psychotherapy worked. In the Cambridge-Somerville study, despite the advantages of a well-designed experiment, the discouraging finding was that "counseling" for delinquents or potential delinquents had no significant effect when compared to a well-matched control group.

When this finding was repeated in subsequent studies (e.g., Barron & Leary, 1955), the controversy over Eysenck's assertion on the ineffectiveness of psychotherapy became heated. Most clinicians rejected the findings outright because they were convinced that psychotherapy was useful, while scientists such as Eysenck hardened their convictions that psychotherapy was at best ineffective and at worst some kind of great hoax perpetrated on unsuspecting clients. This controversy, in turn, left serious applied researchers groping for answers to difficult methodological questions on how to even approach the issue of evaluating effectiveness in psychotherapy. As a result, major conferences on research in psychotherapy were called to discuss these questions (e.g., Rubenstein & Parloff, 1959). It was not until Bergin reexamined these studies in a very important article (Bergin, 1966; see also Bergin & Lambert, 1978) that some of the discrepancies between clinical evidence from uncontrolled case studies and experimental evidence from between-subject group comparison designs were clarified. Bergin noted that some clients *were* improving in these studies, but others were getting worse. When subjected to statistical averaging of results, these effects canceled each other out, yielding an overall result of no effect when compared to the control group. Furthermore, Bergin pointed out that these therapeutic effects had been described in the original articles, but only as afterthoughts to the major statistical findings of no effect. Reviewers such as Eysenck, approaching the results from a methodological point of view, concentrated on the statistical findings. These studies did not, however, prove that psychotherapy was ineffective for a given individual. What these results demonstrated is that people, particularly clients with emotional or behavioral disorders, are quite different from each other. Thus attempts to apply an ill-defined and global treatment such as psychotherapy to a heterogeneous group of clients classified under a vague diagnostic category such as neurosis are incapable of answering the more basic question on the effectiveness of a specific treatment for a specific individual.

The conclusion that psychotherapy was ineffective was premature, based on this reanalysis, but the overriding conclusion from Bergin's review was that "Is psychotherapy effective?" was the wrong question to ask in the first place, even when appropriate between-group experimental designs were employed. During the 1960s, scientists (e.g., Paul 1967) began to realize that any test of a global treatment such as psychotherapy would not be fruitful and that clinical researchers must start defining the independent variables more precisely and must ask the question: "What specific treatment is effective with a specific type of client under what circumstances?"

1.5. LIMITATIONS OF THE GROUP COMPARISON APPROACH

The clearer definition of variables and the call for experimental questions that were precise enough to be answered were major advances in applied research. The extensive review of psychotherapy research by Bergin and Strupp (1972), however, demonstrated that even under these more favorable conditions, the application of the group comparison design to applied problems posed many difficulties. These difficulties, or objections, which tend to limit the usefulness of a group comparison approach in applied research, can be classified under five headings: (1) ethical objections, (2) practical problems in collecting large numbers of patients, (3) averaging of results over the group, (4) generality of findings, and (5) intersubject variability.

Ethical objections

An oft-cited issue, usually voiced by clinicians, is the ethical problem inherent in withholding treatment from a no-treatment control group. This notion, of course, is based on the assumption that the therapeutic intervention, in fact, works, in which case there would be little need to test it at all. Despite the seeming illogic of this ethical objection, in practice many clinicians and other professional personnel react with distaste to withholding some treatment, however inadequate, from a group of clients who are undergoing significant human suffering. This attitude is reinforced by scattered examples of experiments where control groups did endure substantial harm during the course of the research, particularly in some pharmacological experiments.

Practical problems

On a more practical level, the collection of large numbers of clients homogeneous for a particular behavior disorder is often a very difficult task. In basic research in experimental psychology most subjects are animals (or college sophomores), where matching of relevant behaviors or background variables such as personality characteristics is feasible. When dealing with severe behavior disorders, however, obtaining sufficient clients suitably matched to constitute the required groups in the study is often impossible. As Isaac Marks, who is well known for his applied research with large groups, noted:

> Having selected the technique to be studied, another difficulty arises in assembling a homogeneous sample of patients. In uncommon disorders this is only possible in centers to which large numbers of patients are regularly referred,

from these a tiny number are suitable for inclusion in the homogeneous sample one wishes to study. Selection of the sample can be so time consuming that it severely limits research possibilities. Consider the clinician who wishes to assemble a series of obsessive-compulsive patients to be assigned at random into one of two treatment conditions. He will need at least 20 such cases for a start, but obsessive-compulsive neuroses (not personality) make up only 0.5-3 percent of the psychiatric outpatients in Britain and the USA. This means the clinician will need a starting population of about 2000 cases to sift from before he can find his sample, and even then this assumes that all his colleagues are referring every suitable patient to him. In practice, at a large center such as the Maudsley Hospital, it would take up to two years to accumulate a series of obsessive compulsives for study (Bergin & Strupp, 1972, p. 130).

To Marks's credit, he has successfully undertaken this arduous venture on several occasions (Marks, 1972, 1981), but the practical difficulties in executing this type of research in settings other than the enormous clinical facility at the Maudsley are apparent.

Even if this approach is possible in some large clinical settings, or in state hospital settings where one might study various aspects of schizophrenia, the related economic considerations are also inhibiting. Activities such as gathering and analyzing data, following patients, paying experimental therapists, and on and on require large commitments of research funds, which are often unavailable.

Recognizing the practical limitations on conducting group comparison studies in one setting, Bergin and Strupp set an initial goal in their review of the state of psychotherapy research of exploring the feasibility of large collaborative studies among various research centers. One advantage, at least, was the potential to pool adequate numbers of patients to provide the necessary matching of groups. Their reluctant conclusion was that this type of large collaborative study was not feasible due to differing individual styles among researchers and the extraordinary problems involved in administering such an endeavor (Bergin & Strupp, 1972). Since that time there has been the occasional attempt to conduct large collaborative studies, most notably the recent National Institute of Mental Health study testing the effectiveness of cognitive behavioral treatment of depression (NIMH, 1980). But the extreme expense and many of the administrative problems foreseen by Bergin and Strupp (1972) seem to ensure that these efforts will be few and far between (Barlow et al., 1983).

Averaging of results

A third difficulty noted by many applied researchers is the obscuring of individual clinical outcome in group averages. This issue was cogently raised by Sidman (1960) and Chassan (1967, 1979) and repeatedly finds its way into the informal discussions with leading researchers conducted by Bergin and Strupp and published in their book, *Changing Frontiers in the Science of Psychotherapy* (1972). Bergin's (1966) review of large-outcome studies where some clients improved and others worsened highlighted this problem. As noted earlier, a move away from tests of global treatments of ill-defined variables with the implicit question "Is psychotherapy effective?" was a step in the right direction. But even when specific questions on effects of therapy in homogeneous groups are approached from the group comparison point of view, the problem of obscuring important findings remains because of the enormous complexities of any individual patient included in a given treatment group. The fact that patients are seldom truly "homogeneous" has been described by Kiesler (1966) in his discussion of the patient uniformity myth. To take Marks's example, 10 patients, homogeneous for obsessive-compulsive neurosis, may bring entirely different histories, personality variables, and environmental situations to the treatment setting and will respond in varying ways to treatment. That is, some patients will improve and others will not. The average response, however, will not represent the performance of any individual in the group. In relation to this problem, Bergin (Bergin & Strupp, 1972) noted that he consulted a prominent statistician about a therapy research project who dissuaded him from employing the usual inferential statistics applied to the group as a whole and suggested instead that individual curves or descriptive analyses of small groups of highly homogeneous patients might be more fruitful.

Generality of findings

Averaging and the complexity of individual patients also bring up some related problems. Because results from group studies do not reflect changes in individual patients, these findings are not readily translatable or generalizable to the practicing clinician since, as Chassan (1967) pointed out, the clinician cannot determine which particular patient characteristics are correlated with improvement. In ignorance of the responses of individual patients to treatment, the clinician does not know to what extent a given patient is similar to patients who improved or perhaps deteriorated within the context of an overall group improvement. Furthermore, as groups become more homogeneous, which most researchers agree is a necessary condition to answer specific questions about effects of therapy, one loses the ability to make inferential statements to the population of patients with a particular disorder because the individual complexities in the population will not have been adequately sampled. Thus it becomes difficult to generalize findings at all beyond the specific group of patients in the experiment. These issues of averaging and generality of findings will be discussed in greater detail in chapter 2.

Intersubject variability

A final issue bothersome to clinicians and applied researchers is variability. Between-subject group comparison designs consider only variability between subjects as a method of dealing with the enormous differences among individuals in a group. Progress is usually assessed only once (in a posttest). This large intersubject variability is often responsible for the "weak" effect obtained in these studies, where some clients show considerable improvement and others deteriorate, and the average improvement is statistically significant but clinically weak. Ignored in these studies is within-subject variability or the clinical course of a specific patient during treatment, which is of great practical interest to clinicians. This issue will also be discussed more fully in chapter 2.

1.6. ALTERNATIVES TO THE GROUP COMPARISON APPROACH

Many of these practical and methodological difficulties seemed overwhelming to clinicians and applied researchers. Some investigators wondered if serious, meaningful research on evaluation of psychotherapy was even possible (e.g., Hyman & Berger 1966), and the gap between clinician and scientist widened. One difficulty here was the restriction placed on the type of methodology and experimental design applicable to applied research. For many scientists, a group comparison design was the only methodology capable of yielding important information in psychotherapy studies. In view of the dearth of alternatives available and against the background of case study and "percentage success" efforts, these high standards were understandable and correct. Since there were no clearly acceptable scientific alternatives, however, applied researchers failed to distinguish between those situations where group comparison designs were practical, desirable, and necessary (see section 2.9) and situations where the development of alternative methodology was required. During the 1950s and 1960s, several alternatives were tested.

Many applied researchers reacted to the difficulties of the group comparison approach with a "flight into process" where components of the therapeutic process, such as relationship variables, were carefully studied (Hoch & Zubin, 1964). A second approach, favored by many clinicians, was the "naturalistic study," which was very close to actual clinical practice but had dubious scientific underpinnings. As Kiesler (1971) noted, these approaches are quite closely related because both are based on *correlational* methods, where dependent variables are correlated with therapist or patient variables either within therapy or at some point after therapy. This is distinguished from the *experimental* approach, where independent variables are systematically manipulated.

Naturalistic studies

The advantage of the naturalistic study for most clinicians was that it did little to disrupt the typical activities engaged in by clinicians in day-to-day practice. Unlike with the experimental group comparison design, clinicians were not restricted by precise definitions of an independent variable (treatment, time limitation, or random assignment of patients to groups). Kiesler (1971) noted that naturalistic studies involve ". . . live, unaltered, minimally controlled, unmanipulated 'natural' psychotherapy sequences—so-called experiments of nature" (p. 54). Naturally this approach had great appeal to clinicians for it dealt directly with their activities and, in doing so, promised to consider the complexities inherent in treatment. Typically, measures of multiple therapist and patient behaviors are taken, so that all relevant variables (based on a given clinician's conceptualization of which variables are relevant) may be examined for interrelationships with every other variable.

Perhaps the best known example of this type of study is the project at the Menninger Foundation (Kernberg, 1973). Begun in 1954, this was truly a mammoth undertaking involving 38 investigators, 10 consultants, three different project leaders, and 18 years of planning and data collection. Forty-two patients were studied in this project. This group was broadly defined, although overtly psychotic patients were excluded. Assignment of patient to therapist and to differing modes of psychoanalytic treatment was not random but based on clinical judgments of which therapist or mode of treatment was most suitable for the patient. In other words, the procedures were those normally in effect in a clinical setting. In addition, other treatments, such as pharmacological or organic interventions, were administered to certain patients as needed. Against this background, the investigators measured multiple patient characteristics (such as various components of ego strength) and correlated these variables, measured periodically throughout treatment by referring to detailed records of treatment sessions, with multiple therapeutic activities and modes of treatment. As one would expect, the results are enormously complex and contain many seemingly contradictory findings. At least one observer (Malan, 1973) noted that the most important finding is that purely supportive treatment is ineffective with borderline psychotics, but working through of the transference relationship under hospitalization with this group is effective. Notwithstanding the global definition of treatment and the broad diagnostic categories (borderline psychotic) also present in early group comparison studies, this report was generally hailed as an extremely important breakthrough in psychotherapy research. Methodologists, however, were not so sure. While admitting the benefits of a clearer definition of psychoanalytic terms emanating from the project, May (1973) wondered about the power and significance of the conclusions. Most of this criticism concerns the purported strength of the naturalistic study—that is, the lack of control over factors in the naturalistic setting. If subjects are assigned to treatments based on certain characteristics, were these characteristics responsible for improvement rather than the treatment? What is the contribution of additional treatments received by certain patients? Did nurses and other therapists possibly react differently to patients in one group or another? What was the contribution of "spontaneous remission"?

In its pure state, the naturalistic study does not advance much beyond the uncontrolled case study in the power to isolate the effectiveness of a given treatment, as severe critics of the procedure point out (e.g., Bergin & Strupp, 1972), but this process is an improvement over case studies or reports of "percentage success" in groups because measures of relevant variables are constructed and administered, sometimes repeatedly. However, to increase confidence in any correlational findings from naturalistic studies, it would seem necessary to undermine the stated strengths of the study—that is, the "unaltered, minimally controlled, unmanipulated" condition prevailing in the typical naturalistic project—by randomly assigning patients, limiting access to additional confounding modes of treatment, and observing deviation of therapists from prescribed treatment forms. But if this were done, the study would no longer be naturalistic.

A further problem is obvious from the example of the Menninger project. The practical difficulties in executing this type of study seem very little less than those inherent in the large group comparison approach. The one exception is that the naturalistic study, in retaining close ties to the actual functioning of the clinic, requires less structuring or manipulating of large numbers of patients and therapists. The fact that this project took 18 years to complete makes one consider the significant administrative problem inherent in maintaining a research effort for this length of time. This factor is most likely responsible for the admission from one prominent member of the Menninger team, Robert S. Wallerstein, that he would not undertake such a project again (Bergin & Strupp, 1972). Most seem to have heeded his advice because few, if any, naturalistic studies have appeared in recent years.

Correlational studies, of course, do not have to be quite so "naturalistic" as the Menninger study (Kazdin, 1980a; Kendall & Butcher, 1982). Kiesler (1971) reviewed a number of studies without experimental manipulation that contain adequate definitions of variables and experimental attempts to rule out obvious confounding factors. Under such conditions, and if practically feasible, correlational studies may expose heretofore unrecognized relationships among variables in the psychotherapeutic process. But the fact remains that correlational studies by their nature are incapable of determining causal relationships on the effects of treatment. As Kiesler pointed out, the most common error in these studies is the tendency to conclude that a relationship between two variables indicates that one variable is causing the other. For instance, the conclusion in the Menninger study that working through transference relationships is an effective treatment for borderline psychotics (assuming other confounding factors were controlled or randomized) is open to several different interpretations. One might alternatively conclude that certain behaviors subsumed under the classification *borderline psychotic* caused the therapist to behave in such a way that transference variables changed or that a third variable, such as increased therapeutic attention during this more directive approach, was responsible for changes.

Process research

The second alternative to between-group comparison research was the process approach so often referred to in the APA conferences on psychotherapy research (e.g., Strupp & Luborsky, 1962). Hoch and Zubin's (1964) popular phrase "flight into process" was an accurate description of the reaction of many clinical investigators to the practical and methodological difficulties of the large group studies. Typically, process research has concerned itself with what goes on *during* therapy between an individual patient and therapist instead of the final outcome of any therapeutic effort. In the late 1950s and early 1960s, a large number of studies appeared on such topics as relation of therapist behavior to certain patient behaviors in a given interview situation (e.g., Rogers, Gendlin, Kiesler, & Truax, 1967). As such, process research held much appeal for clinicians and scientists alike. Clinicians were pleased by the focus on the individual and the resulting ability to study actual clinical processes. In some studies repeated measures during therapy gave clinicians an idea of the patient's course during treatment. Scientists were intrigued by the potential of defining variables more precisely within one interview without concerning themselves with the complexities involved before or after the point of study. The increased interest in process research, however, led to an unfortunate distinction between process and outcome studies (see Kiesler, 1966). This distinction was well stated by Luborsky (1959), who noted that process research was concerned with how changes took place in a given interchange between patient and therapist, whereas outcome research was concerned with what change took place as a result of treatment. As Paul (1969) and Kiesler (1966) pointed out, the dichotomization of process and outcome led to an unnecessary polarity in the manner in which measures of behavior change were taken. Process research collected data on patient changes at one or more points during the course of therapy, usually without regard for outcome, while outcome research was concerned only with pre-post measures outside of the therapeutic situation. Kiesler noted that this was unnecessary because measures of change within treatment can be continued throughout treatment until an "outcome" point is reached. He also quoted Chassan (1962) on the desirability of determining what transpired between the beginning and end of therapy in addition to outcome. Thus the major concern of the process researchers, perhaps as a result of this imposed distinction, continued to be changes in patient behavior at points within the therapeutic endeavor. The discovery of meaningful clinical changes as a result of these processes was left to the prevailing experimental strategy of the group comparison approach. This reluctance to relate process variables to outcome and the resulting inability of this approach to evaluate the effects of psychotherapy led to a decline of process research. Matarazzo noted that in the 1960s the number of people interested in process studies of psychotherapy had declined and their students were nowhere to be seen (Bergin & Strupp, 1972). Because process and outcome were dichotomized in this manner, the notion eventually evolved that changes during treatment are not relevant or legitimate to the important question of outcome. Largely overlooked at this time was the work of M. B. Shapiro (e.g., 1961) at the Maudsley Hospital in London, begun in the 1950s. Shapiro was repeatedly administering measures of change to individual cases during therapy and also continuing these measures to an end point, thereby relating "process" changes to "outcome" and closing the artificial gap which Kiesler was to describe so cogently some years later.

1.7. THE SCIENTIST-PRACTITIONER SPLIT

The state of affairs of clinical practice and research in the 1960s satisfied few people. Clinical procedures were largely judged as unproven (Bergin & Strupp, 1972; Eysenck, 1965), and the prevailing naturalistic research was unacceptable to most scientists concerned with precise definition of variables and cause-effect relationships. On the other hand, the elegantly designed and scientifically rigorous group comparison design was seen as impractical and incapable of dealing with the complexities and idiosyncrasies of individuals by most clinicians. Somewhere in between was process research, which dealt mostly with individuals but was correlational rather than experimental. In addition, the method was viewed as incapable of evaluating the clinical effects of treatment because the focus was on changes within treatment rather than on outcome.

These developments were a major contribution to the well-known and oft-cited scientist-practitioner split (e.g., Joint Commission on Mental Illness and Health, 1961). The notion of an applied science of behavior change growing out of the optimism of the 1950s did not meet expectations, and many clinician-scientists stated flatly that applied research had no effect on their clinical practice. Prominent among them was Matarazzo, who noted, "Even after 15 years, few of my research findings affect my practice. Psychological science *per se* doesn't guide me one bit. I still read avidly but this is of little direct practical help. My clinical experience is the only thing that has helped me in my practice to date. . . ." (Bergin & Strupp, 1972, p. 340). This opinion was echoed by one of the most productive and best known researchers of the 1950s, Carl Rogers, who as early as the 1958 APA conference on psychotherapy noted that research had no impact on his clinical practice and by 1969 advocated abandoning formal research in psychotherapy altogether (Bergin & Strupp, 1972). Because this view prevailed among prominent clinicians who were well acquainted with research methodology, it follows that clinicians without research training or expertise were largely unaffected by the promise or substance of scientific evaluation of behavior change procedures. L. H. Cohen (1976, 1979) confirmed this state of affairs when he summarized a series of surveys indicating that 40% of mental health professionals think that no research exists that is relevant to practice, and the remainder believe that less than 20% of research articles have any applicability to professional settings.

Although the methodological difficulties outlined above were only one contribution to the scientist-practitioner split (see Barlow et al., 1963, for a detailed analysis), the concern and pessimism voiced by leading researchers in the field during Bergin and Strupp's comprehensive series of interviews led these commentators to reevaluate the state of the field. Voicing dissatisfaction with the large-scale group comparison design, Bergin and Strupp concluded:

> Among researchers as well as statisticians, there is a growing disaffection from traditional experimental designs and statistical procedures which are held inappropriate to the subject matter under study. This judgment applies with particular force to research in the area of therapeutic change, and our emphasis on the value of experimental case studies underscores this point. We strongly agree that most of the standard experimental designs and statistical procedures have exerted and are continuing to exert, a constricting effect on fruitful inquiry, and they serve to perpetuate an unwarranted overemphasis on methodology. More accurately, the exaggerated importance accorded experimental and statistical dicta cannot be blamed on the techniques proper—after all, they are merely tools— but their veneration mirrors a prevailing philosophy among behavioral scientists which subordinates problems to methodology. The insidious effects of this trend are tellingly illustrated by the typical graduate student who is often more interested in the details of a factorial design than in the problem he sets out to study; worse, the selection of a problem is dictated by the experimental design. Needless to say, the student's approach faithfully reflects the convictions and teachings of his mentors. With respect to inquiry in the area of psychotherapy, the kinds of effects we need to demonstrate at this point in time should be significant enough so that they are readily observable by inspection or descriptive statistics. If this cannot be done, no fixation upon statistical and mathematical niceties will generate fruitful insights, which obviously can come only from the researcher's understanding of the subject matter and the descriptive data under scrutiny. (1972, p. 440)

1.8. A RETURN TO THE INDIVIDUAL

Bergin and Strupp were harsh in their comments on group comparison design and failed to specify those situations where between-group methodology may be practical and desirable (see chapter 2). However, their conclusions on alternative directions, outlined in a paper appropriately titled "New Directions in Psychotherapy Research" (Bergin & Strupp, 1970), had radical and far-reaching implications for the conduct of applied research. Essentially, Bergin and Strupp advised against investing further effort in process and outcome studies and proposed the experimental single-case approach for the purpose of isolating mechanisms of change in the therapeutic process. Isolation of these mechanisms of change would then be followed by construction of new procedures based on a combination of variables whose effectiveness was demonstrated in single-case experiments. As the authors noted, "As a general paradigm of inquiry, the individual experimental case study and the experimental analogue approaches appear to be the primary strategies which will move us forward in our understanding of the mechanisms of change at this point" (Bergin & Strupp, 1970, p. 19). The hope was also expressed that this approach would tend to bring research and practice closer together.

With the recommendations emerging from Bergin and Strupp's comprehensive analysis, the philosophy underlying applied research methodology had come full circle in a little over 100 years. The disillusionment with large-scale between-group comparisons observed by Bergin and Strupp and their subsequent advocacy of the intensive study of the individual is an historical repetition of a similar position taken in the middle of the last century. At that time, the noted physiologist, Claude Bernard, in *An Introduction to the Study of Experimental Medicine* (1957), attempted to dissuade colleagues who believed that physiological processes were too complex for experimental inquiry within a single organism. In support of this argument, he noted that the site of processes of change is in the individual organism, and group averages and variance might be misleading. In one of the more famous anecdotes in science, Bernard castigated a colleague interested in studying the properties of urine in 1865. This colleague had proposed collecting specimens from urinals in a centrally located train station to determine properties of the average European urine. Bernard pointed out that this would yield little information about the urine of any one individual. Following Bernard's persuasive reasoning, the intensive scientific study of the individual in physiology flourished.

But methodology in physiology and experimental psychology is not directly applicable to the complexities present in applied research. Although the splendid isolation of Pavlov's laboratories allowed discovery of important psychological processes without recourse to sophisticated experimental design, it is unlikely that the same results would have obtained with a household pet in its natural environment. Yet these are precisely the conditions under which most applied researchers must work.

The plea of applied researchers for appropriate methodology grounded in the scientific method to investigate complex problems in individuals is never more evident than in the writings of Gordon Allport. Allport argued most eloquently that the science of psychology should attend to the uniqueness of the individual (e.g., Allport, 1961, 1962). In terms commonly used in the 1950s, Allport became the champion of the idiographic (individual) approach, which he considered superior to the nomothetic (general or group) approach.

> Why should we not start with individual behavior as a source of hunches (as we have in the past) and then seek our generalization (also as we have in the past) but finally come back to the individual not for the mechanical application of laws (as we do now) but for a fuller and more accurate assessment then we are now able to give? I suspect that the reason our present assessments are now so often feeble and sometimes even ridiculous, is because we do not take this final step. We stop with our wobbly laws of generality and seldom confront them with the concrete person. (Allport, 1962, p. 407)

Due to the lack of a practical, applied methodology with which to study the individual, however, most of Allport's own research was nomothetic. The increase in the intensive study of the individual in applied research led to a search for appropriate methodology, and several individuals or groups began developing ideas during the 1950s and 1960s.

The role of the case study

One result of the search for appropriate methodology was a reexamination of the role of the uncontrolled case study so strongly rejected by scientists in the 1950s. Recognizing its inherent limitations as an evaluation tool, many clinical investigators (e.g., Barlow, 1980; Kazdin, 1981; Lazarus & Davison, 1971) suggested that the case study could make important contributions to an experimental effort. One of the more important functions of the case study is the generation of new hypotheses, which later may be subjected to more rigorous experimental scrutiny. As Dukes (1965) observed, the case study can occasionally be used to shed some light on extremely rare phenomena or cast doubt on well-established theoretical assumptions. Carefully analyzing threats to internal validity when drawing causal inferences from case studies, Kazdin (1981) concluded that under certain very specific conditions data from case studies can approach data from single-case experimental manipulations. Case studies may also make other important contributions to science (Barlow et al., 1983; see also chapter 10). Nevertheless, the case study generally is not capable of isolating therapeutic mechanisms of change (Hersen & Barlow, 1976; Kazdin, 1981; Leitenberg, 1973), and the inability of many scientists and clinicians to discriminate the critical difference between the uncontrolled case study and the experimental study of an individual case has most likely retarded the implementation of single-case experimental designs (see chapter 5).

The representative case

During this period, other theorists and methodologists were attempting to formulate viable approaches to the experimental study of single cases. Shontz (1965) proposed the study of the representative case as an alternative to traditional approaches in experimental personality research. Essentially, Shontz was concerned with validating previously established personality constructs or measurement instruments on individuals who appear to possess the necessary behavior appropriate for the research problem. Shontz's favorite example was a study of the contribution of psychodynamic factors to epilepsy described by Bowdlear (1955). After reviewing the literature on the presumed psychodynamics in epilepsy, Bowdlear chose a patient who closely approximated the diagnostic and descriptive characteristics of epilepsy presented in the literature (i.e., the representative case). Through a series of questions, Bowdlear then correlated seizures with a certain psychodynamic concept in this patient—acting out dependency. Since this case was "representative," Bowdlear assumed some generalization to other similar cases.

Shontz's contribution was not methodological, because the experiments he cites were largely correlational and in the tradition of process research. Shontz also failed to recognize the value of the single-case study in isolating effective therapeutic variables or building new procedures, as suggested later by Bergin and Strupp (1972). Rather, he proposed the use of a single-case in a deductive manner to test previously established hypotheses and measurement instruments in an individual who is known to be so stable in certain personality characteristics that he or she is "representative" of these characteristics. Conceptually, Shontz moved beyond Allport, however, in noting that this approach was not truly idiographic in that he was not proposing to investigate a subject as a self-contained universe with its own laws. To overcome this objectionable aspect of single-case research, he proposed replication on subjects who differed in some significant way from the first subject. If the general hypothesis were repeatedly confirmed, this would begin to establish a generally applicable law of behavior. If the hypothesis were sometimes confirmed and sometimes rejected, he noted that ". . . the investigator will be in a position either to modify his thinking or to state more clearly the conditions under which the hypothesis does and does not provide a useful model of psychological events" (Shontz, 1965, p. 258). With this statement, Shontz anticipated the applied application of the methodology of direct and systematic replication in basic research (see chapter 10) suggested by Sidman (1960).

Shapiro's methodology in the clinic

One of the most important contributions to the search for a methodology came from the pioneering work of M. B. Shapiro in London. As early as 1951, Shapiro was advocating a scientific approach to the study of individual phenomena, an advocacy that continued through the 1960s (e.g., M. B. Shapiro, 1961, 1966, 1970).

Unlike Allport, however, Shapiro went beyond the point of noting the advantages of applied research with single-cases and began the difficult task of constructing an adequate methodology. One important contribution by Shapiro was the utilization of carefully constructed measures of clinically relevant responses administered repeatedly over time in an individual. Typically, Shapiro would examine fluctuations in these measures and hypothesize on the controlling effects of therapeutic or environmental influences. As such, Shapiro was one of the first to formally investigate questions more relevant to psychopathology than behavior change or psychotherapy *per se* using the individual case. Questions concerning classification and the identification of factors maintaining the disorder and even speculations regarding etiology were all addressed by Shapiro. Many of these studies were correlational in nature, or what Shapiro refers to as simple or complex descriptive studies (1966). As such, these efforts bear a striking resemblance to process studies mentioned above, in that the effect of a therapeutic or potential-maintaining variable was correlated with a target response. Shapiro attempted to go beyond this correlational approach, however, by defining and manipulating independent variables within single-cases. One good example in the area of behavior change is the systematic alteration of two therapeutic approaches in a case of paranoid delusions (M. B. Shapiro & Ravenette, 1959). In a prototype of what was later to be called the A-B-A design, the authors measured paranoid delusions by asking the patient to rate the "intensity" of a number of paranoid ideas on a scale of 1 to 5. The sum of the score across 18 different delusions then represented the patient's paranoid "score." Treatments consisted of "control" discussion concerning guilt feelings about situations in the patient's life, unrelated to any paranoid ideation, and rational discussion aimed at exposing the falseness of the patient's paranoid beliefs. The experimental sequence consisted of 4 days of "guilt" discussion followed by 8 days of rational discussion and a return to 4 days of "guilt" discussion. The authors observed an overall decline in paranoid scores during this experiment, which they rightly noted as correlational and thus potentially due to a variety of causes. Close examination of the data revealed, however, that on weekends when no discussions were held, the patient worsened during the guilt control phase and improved during the rational discussion phase. These fluctuations around the regression line were statistically significant. This effect, of course, is weak and of dubious importance because overall improvement in paranoid scores was not functionally related to treatment. Furthermore, several guidelines for a true experimental analysis of the treatment were violated. Examples of experimental error include the absence of baseline measurement to determine the pretreatment course of the paranoid beliefs and the simultaneous withdrawal of one treatment and introduction of a second treatment (see chapter 3). The importance of the case and other early work from M. B. Shapiro, however, is not the knowledge gained from any one experiment, but the beginnings of the development of a scientifically based methodology for evaluating effects of treatment within a single-case. To the extent that Shapiro's correlational studies were similar to process research, he broke the semantic barrier which held that process criteria were unrelated to outcome. He demonstrated clearly that repeated measures within an individual could be extended to a logical end point and that this end point *was* the outcome of treatment. His more important contribution from our point of view, however, was the demonstration that independent variables in applied research could be defined and systematically manipulated within a single-case, thereby fulfilling the requirements of a "true" experimental approach to the evaluation of therapeutic technique (Underwood, 1957). In addition, his demonstration of the applicability of the study of the individual case to the discovery of issues relevant to psychopathology was extremely important. This approach is only now enjoying more systematic application by some of our creative clinical scientists (e.g., Turkat & Maisto, in press).

Quasi-experimental designs

In the area of research dealing with broad-based educational or social change, most often termed evaluation research, Campbell and Stanley (1963) and Cook and Campbell (1979) proposed a series of important methodological innovations that they termed quasi-experimental designs. Education research, of course, is more often concerned with broad-based effects of programs rather than individual behavioral change. But these designs, many of which are applicable to either groups or individuals, are also directly relevant in our context. The two designs most appropriate for analysis of change in the individual are termed the *term series design* and the *equivalent term series design*. From the perspective of applied clinical research, the time series design is similar to M. B. Shapiro's effort to extend process observation throughout the course of a given treatment to a logical end point or outcome. This design goes beyond observations within treatment, however, to include observations from repeated measures in a period preceding and following a given intervention. Thus one can observe changes from a baseline as a result of a given intervention. While the inclusion of a baseline is a distinct methodological improvement, this design is basically correlational in nature and is unable to isolate effects of therapeutic mechanisms or establish cause-effect relationships. Basically, this design is the A-B design described in chapter 5. The equivalent time series design, however, involves experimental manipulation of independent variables through alteration of treatments, as in the M. B. Shapiro and Ravenette study (1959), or introduction and withdrawal of one treatment in an A-B-A fashion. Approaching the study of the individual from a different perspective than Shapiro, Campbell and Stanley arrived at similar conclusions on the possibility of manipulation of independent variables and establishment of cause-effect relationships in the study of a single-case.

What was perhaps the more important contribution of these methodologists, however, was the description of various limitations of these designs in their ability to rule out alternative plausible hypotheses (internal validity) or the extent to which one can generalize conclusions obtained from the designs (external validity) (see chapter 2).

Chassan and intensive designs

It remained for Chassan (1967, 1979) to pull together many of the methodological advances in single-case research to that point in a book that made clear distinctions between the advantages and disadvantages of what he termed extensive (group) design and intensive (single-case) design. Drawing on long experience in applied research, Chassan outlined the desirability and applicability of single-case designs evolving out of applied research in the 1950s and early 1960s. While most of his own experience in single-case design concerned the evaluation of pharmacologic agents for behavior disorders, Chassan also illustrated the uses of single-case designs in psychotherapy research, particularly psychoanalysis. As a statistician rather than a practicing clinician, he emphasized the various statistical procedures capable of establishing relationships between therapeutic intervention and dependent variables within the single-case. He concentrated on the correlation type of design using trend analysis but made occasional use of a prototype of the A-B-A design (e.g., Bellak & Chassan, 1964), which, in this case, extended the work of M. B. Shapiro to evaluation of drug effects but, in retrospect, contained some of the same methodological faults. Nevertheless, the sophisticated theorizing in the book on thorny issues in single-case research, such as generality of findings from a single-case, provided the most comprehensive treatment of these issues to this time. Many of Chassan's ideas on this subject will appear repeatedly in later sections of this book.

1.9. THE EXPERIMENTAL ANALYSIS OF BEHAVIOR

While innovative applied researchers such as Chassan and M. B. Shapiro made methodological advances in the experimental study of the single-case, their advances did not have a major impact on the conduct of applied research outside of their own settings. As late as 1965, Shapiro noted in an invited address to the Eastern Psychological Association that a large majority of research in prominent clinical psychology journals involved between-group comparisons with little and, in some cases, no reference to the individual approach that he advocated. He hoped that his address might presage the beginning of a new emphasis on this method. In retrospect, there are several possible reasons for the lack of impact. *First*, as Leitenberg (1973) was later to point out, many of the measures used by M. B. Shapiro in applied research were indirect and subjective (e.g., questionnaires), precluding the observation of direct behavioral effects that gained importance with the rise of behavior therapy (see chapter 4). *Second*, Shapiro and Chassan, in studies of psychotherapy, did not produce the strong, clinically relevant changes that would impress clinicians, perhaps due to inadequate or weak independent variables or treatments, such as instructions within interview procedures. *Finally*, the advent of the work of Shapiro and Chassan was associated with the general disillusionment during this period concerning the possibilities of research in psychotherapy. Nevertheless, Chassan and Shapiro demonstrated that meaningful applied research was possible and even desirable in the area of psychotherapy. These investigators, along with several of Shapiro's students (e.g., Davidson & Costello, 1969; Inglis, 1966; Yates, 1970), had an important influence on the development and acceptance of more sophisticated methodology, which was beginning to appear in the 1960s.

It is significant that it was the rediscovery of the study of the single-case in basic research, coupled with a new approach to problems in the applied area, that marked the beginnings of a new emphasis on the experimental study of the single-case in applied research. One indication of the broad influence of this combination of events was the emergence of a journal in 1968 (*Journal of Applied Behavior Analysis*) devoted to single-case methodology in applied research and the appearance of this experimental approach in increasing numbers in the major psychological and psychiatric journals. The methodology in basic research was termed the *experimental analysis of behavior*; the new approach to applied problems became known as *behavior modification* or *behavior therapy*.

Some observers have gone so far as to define behavior therapy in terms of single-case methodology (Yates, 1970; 1975) but, as Leitenberg (1973) pointed out, this definition is without empirical support because behavior therapy is a clinical approach employing a number of methodological strategies (see

Kazdin, 1978, and Krasner, 1971a, for a history of behavior therapy). The relevance of the experimental analysis of behavior to applied research is the development of sophisticated methodology enabling intensive study of individual subjects. In rejecting a between-subject approach as the only useful scientific methodology, Skinner (1938, 1953) reflected the thoughts of the early physiologists such as Claude Bernard and emphasized repeated objective measurement in a single subject over a long period of time under highly controlled conditions. As Skinner noted (1966b), ". . . instead of studying a thousand rats for one hour each, or a hundred rats for ten hours each, the investigator is likely to study one rat for a thousand hours" (p. 21), a procedure that clearly recognizes the individuality of an organism. Thus, Skinner and his colleagues in the animal laboratories developed and refined the single-case methodology that became the foundation of a new applied science. Culminating in the definitive methodological treatise by Sidman (1960), entitled *Tactics of Scientific Research*, the assumption and conditions of a true experimental analysis of behavior were outlined. Examples of fine-grain analyses of behavior and the use of withdrawal, reversal, and multi-element experimental designs in the experimental laboratories began to appear in more applied journals in the 1960s, as researchers adapted these strategies to the investigation of applied problems.

It is unlikely, however, that this approach would have had a significant impact on applied clinical research without the growing popularity of behavior therapy. The fact that M. B. Shapiro and Chassan were employing rudimentary prototypes of withdrawal designs (independent of influences from the laboratories of operant conditioning) without marked effect on applied research would seem to support this contention. In fact, even earlier, F. C. Thorne (1947) described clearly the principle of single-case research, including A-B-A withdrawal designs, and recommended that clinical research proceed in this manner, without apparent effect (Barlow et al., 1983). The growth of the behavior therapy approach to applied problems, however, provided a vehicle for the introduction of the methodology on a scale that attracted attention from investigators in applied areas. Behavior therapy, as the application of the principles of general-experimental and social psychology to the clinic, also emphasized direct measurement of clinically relevant target behaviors and experimental evaluation of independent variables or "treatments." Since many of these "principles of learning" utilized in behavior therapy originally emanated from operant conditioning, it was a small step for behavior therapists to also borrow the operant methodology to validate the effectiveness of these same principles in applied settings. The initial success of this approach (e.g., Ullmann & Krasner, 1965) led to similar evaluations of additional behavior therapy techniques that did not derive directly from the operant laboratories (e.g., Agras et al., 1971; Barlow, Leitenberg, & Agras, 1969). During this period, methodology originally intended for the animal laboratory was adapted more fully to the investigation of applied problems and "applied behavior analysis" became an important supplementary and, in some cases, alternative methodological approach to between-subjects experimental designs.

The early pleas to return to the individual as the cornerstone of an applied science of behavior have been heeded. The last several years have witnessed the crumbling of barriers that precluded publication of single-case research in any leading journal devoted to the study of behavioral problems. Since the first edition of this book, a proliferation of important books has appeared devoted, for example, to strategies for evaluating data from single-case designs (Kratochwill, 1978b), to the application of these methods in social work (Jayaratne & Levy, 1979), or to the philosophy underlying this approach to applied research (J. M. Johnston & Pennypacker, 1980). Other excellent books have appeared concentrating specifically on descriptions of design alternatives (Kazdin, 1982b), and major handbooks on research are not complete without a description of this approach (e.g., Kendall & Butcher, 1982).

More importantly, the field has not stood still. From their more recent origins in evaluating the application of operant principles to behavior disorders, single-case designs are now fully incorporated into the armamentarium of applied researchers generally interested in behavior change beyond the subject matter of the core mental health professions or education. Professions such as rehabilitation medicine are turning increasingly to this approach as appropriate to the subject matter at hand (e.g., Schindele, 1981), and the field is progressing. New design alternatives have appeared only recently, and strategies involved in more traditional approaches have been clarified and refined. We believe that the recent methodological developments and the demonstrated effectiveness of this methodology provide a base for the establishment of a true science of human behavior with a focus on the paramount importance of the individual. A description of this methodology is the purpose of this book.

CHAPTER 2

General Issues in a Single-Case Approach

2.1. INTRODUCTION

Two issues basic to any science are variability and generality of findings. These issues are handled somewhat differently from one area of science to another, depending on the subject matter. The first section of this chapter concerns variability.

In applied research, where individual behavior is the primary concern, it is our contention that the search for sources of variability in individuals must occur if we are to develop a truly effective clinical science of human behavior change. After a brief discussion of basic assumptions concerning sources of variability in behavior, specific techniques and procedures for dealing with behavioral variability in individuals are outlined. Chief among these are repeated measurement procedures that allow careful monitoring of day-to-day variability in individual behavior, and rapidly changing, improvised experimental designs that facilitate an immediate search for sources of variability in an individual. Several examples of the use of this procedure to track down sources of intersubject or intrasubject variability are presented.

The second section of this chapter deals with generality of findings. Historically, this has been a thorny issue in applied research. The seeming limitations in establishing wide generality from results in a single-case are obvious, yet establishment of generality from results in large groups has also proved elusive. After a discussion of important types of generality of findings, the shortcomings of attempting to generalize from group results in applied research are discussed. Traditionally, the major problems have been an inability to draw a truly random sample from human behavior disorders and the difficulty of generalizing from groups to an individual. Applied researchers attempted to solve the problem by making groups as homogeneous as possible so that results would be applicable to an individual who showed the characteristics of the homogeneous group. An alternative method of establishing generality of findings is the replication of single-case experiments. The relative merits of establishing generality of findings from homogeneous groups and replication of single-case experiments are discussed at the end of this section.

Finally, some research questions that cannot be answered through experimentation on single-cases are listed, and strategies for combining some strengths of single-case and between-subject research approaches are suggested.

2.2. VARIABILITY

The notion that behavior is a function of a multiplicity of factors finds wide agreement among scientists and professional investigators. Most scientists also agree that as one moves up the phylogenetic scale, the sources of variability in behavior become greater. In response to this, many scientists choose to work with lower life forms in the hope that laws of behavior will emerge more readily and be generalizable to the infinitely more complex area of human behavior. Applied researchers do not have this luxury. The task of the investigator in the area of human behavior disorders is to discover functional relations among treatments and specific behavior disorders over and above the welter of environmental and biological variables impinging on the patient at any given time. Given these complexities, it is small wonder that most treatments, when tested, produce small effects or, in Bergin and Strupp's terms, *weak results* (Bergin & Strupp, 1972).

Variability in basic research

Even in basic research, behavioral variability is enormous. In attempting to deal with this problem, many experimental psychologists assumed that variability was intrinsic to the organism rather than imposed by experimental or environmental factors (Sidman, 1960). If variability were an intrinsic component of behavior, then procedures had to be found to deal with this issue before meaningful research could be conducted. The solution involved experimental designs and confidence level statistics that would elucidate functional relations among independent and dependent variables over and above the intrinsic variability. Sidman (1960) noted that this is not the case in some other sciences, such as physics. Physics assumes that variability is imposed by error of measurement or other identifiable factors. Experimental efforts are then directed to discovering and eliminating as many sources of variability as possible so that functional relations can be determined with more precision. Sidman proposed that basic researchers in psychology also adopt this strategy. Rather than assuming that variability is intrinsic to the organism, one should make every effort to discover sources of behavioral variability among organisms such that laws of behavior could be studied with the precision and specificity found in physics. This precision, of course, would require close attention to the behavior of the individual organism. If one rat behaves differently from three other rats in an experimental condition, the proper tactic is to find out why. If the experimenter succeeds, the factors that produce that variability can be eliminated and a "cleaner" test of the effects of the original independent variable can be made. Sidman recognized that behavioral variability may never be entirely eliminated, but that isolation of as many sources of variability as possible would enable an investigator to estimate how much variability actually is intrinsic.

Variability in applied research

Applied researchers, by and large, have not been concerned with this argument. Every practitioner is aware of multiple social or biological factors that are imposed on his or her data. If asked, many investigators might also assume some intrinsic variability in clients attributable to capriciousness in nature; but most are more concerned with the effect of uncontrollable but potentially observable events in the environment. For example, the sudden appearance of a significant relative or the loss of a job during treatment of depression may affect the course of depression to a far greater degree than the particular intervention procedure. Menstruation may cause marked changes in behavioral measures of anxiety. Even more disturbing are the multiple unidentifiable sources of variability that cause broad fluctuation in a patient's clinical course. Most applied researchers assume this variability is imposed rather than intrinsic, but they may not know where to begin to factor out the sources.

The solution, as in basic research, has been to accept broad variability as an unavoidable evil, to employ experimental design and statistics that hopefully control variability, and to look for functional relations that supersede the "error."

As Sidman observed when discussing these tactics in basic research:

The rationale for statistical immobilization of unwanted variables is based on the assumed random nature of such variables. In a large group of subjects, the reasoning goes, the uncontrolled factor will change the behavior of some subjects in one direction and will affect the remaining subjects in the opposite way. When the data are averaged over all the subjects, the effects of the uncontrolled variables are presumed to add algebraically to zero. The composite data are then regarded as though they were representative of one ideal subject who had never been exposed to the uncontrolled variables at all (1960, p. 162).

Although one may question this strategy in basic research, as Sidman has, the amount of control an experimenter has over the behavioral history and current environmental variables impinging on the laboratory animal makes this strategy at least feasible. In applied research, when control over behavioral histories or even current environmental events is limited or nonexistent, there is far less probability of discovering a treatment that is effective over and above these uncontrolled variables. This, of course, was the major cause of the inability of early group comparison studies to demonstrate that the treatment under consideration was effective. As noted in chapter 1, some clients were improving while others were worsening, despite the presence of the treatment. Presumably, this variability was not intrinsic but due to current life circumstances of the clients.

Clinical vs. statistical significance

The experimental designs and statistics gleaned from the laboratories of experimental psychology have an added disadvantage in applied research. The purpose of research in any basic science is to discover functional relations among dependent and independent variables. Once discovered, these functional relationships become principles that add to our knowledge of behavior. In applied research, however, the discovery of functional relations is not sufficient. The purpose of applied research is to effect *meaningful* clinical or socially relevant behavioral changes. For example, if depression were reliably measurable on a 0–100 scale, with 100 representing severe depression, a treatment that improved *each patient* in a group of depressives from 80 to 75 would be statistically significant if all depressives in the control group remained at 80. This statistical significance, however, would be of little use to the practicing clinician because a score of 75 could still be in the suicidal range. An improvement of 40 or 50 points might be necessary before the clinician would consider the change clinically important. Elsewhere, we have referred to the issue as *statistical versus clinical significance* (Barlow & Hersen, 1973), and this issue has been raised repeatedly during the last decade (e.g., Garfield & Bergin, 1978). In this simplified example, statisticians might observe that this issue is easily correctable by setting a different criterion level for "effectiveness." In the jungle of applied research, however, when any effect superseding the enormous "error" or variance in a group of heterogeneous clients is remarkable, the clinician and even the researcher will often overlook this issue and consider a treatment that is statistically significant to also be clinically effective.

As Chassan (1960, 1979) pointed out, statistical significance can underestimate clinical effectiveness as well as overestimate it. This unfortunate circumstance occurs when a treatment is quite effective with a few members of the experimental group while the remaining members do not improve or deteriorate somewhat. Statistically, then, the experimental group does not differ from the control group, whose members are relatively unchanged. When broad divergence such as this occurs among clients in response to an intervention, statistical treatments will average out the clinical effects along with changes due to unwanted sources of variability. In fact, this type of intersubject variability is the rule rather than the exception. Bergin (1966) clearly illustrated the years that were lost to applied research because clinical investigators overlooked the marked effectiveness of these treatments on *some* clients (see also, Bergin & Lambert, 1978; Strupp & Hadley, 1979). The issue of clinical versus statistical significance is, of course, not restricted to between-group comparisons but is something applied researchers must consider whenever statistical tests are applied to clinical data (see chapter 9).

Nevertheless, the advantages of attempting to eliminate the enormous intersubject variability in applied research through statistical methods have intuitive appeal for both researchers and clinicians who want quick answers to pressing clinical or social questions. In fact, to the clinician who might observe one severely depressive patient inexplicably get better while another equally depressed patient commits suicide, this variability may well seem to be intrinsic to the nature of the disorder rather than imposed by definable social or biological factors.

Highlighting variability in the individual

In any case, whether variability in applied research is intrinsic to some degree or not, the alternative to the treatment of intersubject variability by statistical means is to highlight variability and begin the arduous task of determining sources of variability in the individual. To the applied researcher, this task is staggering. In realistic terms he or she must look at each individual who differs from other clients in terms of response to treatment and attempt to determine why. Since the complexities of human environments, both external and internal, are enormous, the possible causes of these differences number in the millions.

With the complexities involved in this search, one may legitimately question where to begin. Since intersubject variability begins with one client differing in response from some other clients, a logical starting point is the individual. If one is to concentrate on individual variability, however, the manner in which one observes this variability must also change. If one depressed patient deteriorates during treatment while others improve or remain stable, it is difficult to speculate on reasons for this deterioration if the only data available are observations before and after treatment. It would be much to the advantage of the clinical researcher to have followed this one patient's course *during* treatment so that the beginning of deterioration could be pinpointed. In this hypothetical case the patient may have begun to improve until a point midway in treatment, when deterioration began. Perhaps a disruption in family life occurred or the patient missed a treatment session, while other patients whose improvement continued did not experience these events. It would then be possible to speculate on these or other factors that were correlated with such change. In single-case research the investigator could adjust to the variability with immediate alteration in experimental design to test out hypothesized sources of these changes.*

Repeated measures

The basis of this search for sources of variability is repeated measurement of the dependent variable or problem behavior. If this tactic has a familiar ring to practitioners, it is no accident, for this is precisely the strategy every practitioner uses daily. It is no secret to clinicians or other behavior change agents in applied settings that behavioral improvement from an initial observation to some end point sandwiches marked variability in the behavior between these points. A major activity of clinicians is observing this variability and making appropriate changes in treatment strategies or environmental circumstances, where possible, to eliminate these fluctuations from a general improving trend. Because measures in the clinic seldom go beyond gross observation, and treatment consists of a combination of factors, it is difficult for clinicians to pinpoint potential sources of variability, but they speculate; with increased clinical experience, effective clinicians may guess rightly more often than wrongly. In some cases, weekly observation may go on for years. As Chassan (1967) pointed out:

> The existence of variability as a basic phenomenon in the study of individual psychopathology implies that a single observation of a patient state, in general, can offer only a minimum of information about the patient state. While such information is literally better than no information, it provides no more data than does any other statistical sample of one (1967, p. 182)

He then quoted Wolstein (1954) from a psychoanalytic point of view, who comments on diagnostic categories:

> These terms are "ad hoc" definitions which move the focus of inquiry away from repetitive patterns with observable frequencies to fixed momentary states. But this notion of the momentary present is specious and deceptive; it is neither fixed nor momentary nor immediately present, but an inferred condition (p. 39).

---

*For an excellent discussion of the concept of variability and the relationship of measurement to variability see J. M. Johnston and Pennypacker (1981).

The relation of this strategy to process research, described in chapter 1, is obvious. But the search for sources of individual variability cannot be restricted to repeated measures of one small segment of a client's course somewhere between the beginning and the end of treatment, as in process research. With the multitude of events impinging on the organism, significant behavior fluctuation may occur at any time—from the beginning of an intervention until well after completion of treatment. The necessity of repeated, frequent measures to begin the search for sources of individual variability is apparent. Procedures for repeated measures of a variety of behavior problems are described in chapter 4.

Rapidly changing designs

If one is committed to determining sources of variability in individuals, repeated measurement alone is insufficient. In a typical case, no one event is clearly associated with behavioral fluctuation, and repeated observation will permit only a temporal correlation of several events with the behavioral fluctuation. In the clinic this temporal correlation provides differing degrees of evidence on an intuitive level concerning causality. For instance, if a claustrophobic became trapped in an elevator on the way to the therapist's office and suddenly worsened, the clinician could make a reasonable inference that this event caused the fluctuation. Usually, of course, sources of variability are not so clear, and the applied researcher must guess from among several correlated events. However, it would add little to science if an investigator merely reported at the end of an experiment that fluctuation in behaviors were observed and were correlated with several events. The task confronting the applied researcher at this point is to devise experimental designs to isolate the cause of the change or the lack of change. One advantage of single-case experimental designs is that the investigator can begin an immediate search for the cause of an experimental behavior trend by altering the experimental design on the spot. This feature, when properly employed, can provide immediate information on hypothesized sources of variability. In Skinner's words:

> A prior design in which variables are distributed, for example, in a Latin square, may be a severe handicap. When effects on behavior can be immediately observed, it is more efficient to explore relevant variables by manipulating them in an improvised and rapidly changing design. Similar practices have been responsible for the greater part of modern science (Honig, 1966, p. 21).

More recently, this feature of single-case designs has been termed *response guided experimentation* (Edgington, 1983, 1984).

2.3. EXPERIMENTAL ANALYSIS OF SOURCES OF VARIABILITY THROUGH IMPROVISED DESIGNS

In single-case designs there are at least three patterns of variability highlighted by repeated measurement. In the first pattern a subject may not respond to a treatment previously demonstrated as effective with other subjects. In a second pattern a subject may improve when no treatment is in effect, as in a baseline phase. This "spontaneous" improvement is often considered to be the result of "placebo" effects. These two patterns of intersubject variability are quite common in applied research. In a third pattern the variability is intrasubject in that marked cyclical patterns emerge in the measures that supersede the effect of any independent variable. Using improvised and rapidly changing designs, it is possible to follow Skinner's suggestion and begin an immediate search for sources of this variability. Examples of these efforts are provided next.

Subject fails to improve

One experiment from our laboratories illustrates the use of an "improvised and rapidly changing design" to determine why one subject did not improve with a treatment that had been successful with other subjects. The purpose of this experiment was to explore the effects of a classical conditioning procedure on increasing heterosexual arousal in homosexuals desiring this additional arousal pattern (Herman, Barlow, & Agras, 1974a). In this study, heterosexual arousal as measured by penile circumference change to slides of nude females was the major dependent variable. Measures of homosexual arousal and reports of heterosexual urges and fantasies were also recorded. The design is a basic A-B-A-B with a baseline procedure, making it technically an A-B-C-B-C, where A is baseline; B is a control phase, backward conditioning; and C is the treatment phase, or classical conditioning. In classical conditioning the client viewed two slides for one minute each. One slide depicted a female, which became the CS. A male slide, to which the client became aroused routinely, became the UCS. During classical conditioning, the client viewed the CS (female slide) for one minute, followed immediately by the UCS (male slide) for 1 minute in the typical classical conditioning paradigm. During the B, or control phase, however, the order of presentation was reversed (UCS-CS), resulting in a backward conditioning paradigm which, of course, should not produce any learning.

During Experiment 1 (see Figure 2-1), no increases in heterosexual arousal were noted during baseline or backward conditioning. A sharp rise occurred, however, during classical conditioning. This was followed by a downward trend in heterosexual arousal during a return to the backward conditioning

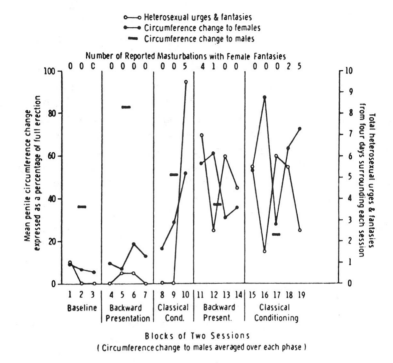

FIGURE 2-1. Mean penile circumference change to male and female slides expressed as a percentage of full erection and total heterosexual urges and fantasies collected from 4 days surrounding each session. Data are presented in blocks of two sessions (circumference change to males averaged over each phase). Reported incidence of masturbation accompanied by female fantasy is indicated for each blocked point. (Figure 1, p. 36, from: Herman, S. H., Barlow, D. H., and Agras, W. W. [1974]. An experimental analysis of classical conditioning as a method of increasing heterosexual arousal in homosexuals. *Behavior Therapy*, 5, 33–47. Copyright 1974 by Association for the Advancement of Behavior Therapy. Reproduced by permission.)

control phase, and further increases in arousal during a second classical conditioning phase, suggesting that the classical conditioning procedure was producing the observed increase.

In attempting to replicate this finding on a second client (see Figure 2-2), some variation in responding was noted. Again, no increase in heterosexual arousal occurred during baseline or backward conditioning phases; but none occurred during the first classical conditioning phase either, even though the number of UCS slides was increased from one to three. At this point, it was noted that his response latency to the male slide was approximately 30 seconds. Thus the classical conditioning procedure was adjusted slightly, such

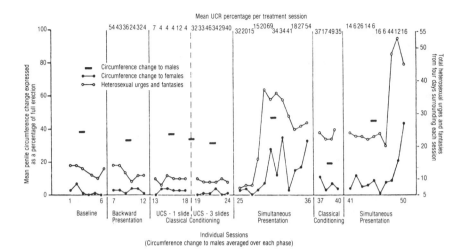

FIGURE 2-2. Mean penile circumference change to male and female slides expressed as a percentage of full erection and total heterosexual urges and fantasies collected from 4 days surrounding each session. Data are presented for individual sessions with circumference change to males averaged over each phase. Mean UCR percentage is indicated for each treatment session. (Figure 2, p. 40, from: Herman, S. H., Barlow, D. H., and Agras, W. S. [1974]. An experimental analysis of classical conditioning as a method of increasing heterosexual arousal in homosexuals. *Behavior Therapy*, **5**, 33–47. Copyright 1974 by Association for the Advancement of Behavior Therapy. Reproduced by permission.)

that 30 seconds of viewing the female slide alone was followed by 30 seconds of viewing both the male and female slides simultaneously (side by side), followed by 30 seconds of the male slide alone. This adjustment (labeled simultaneous presentation) produced increases in heterosexual arousal in the separate measurement sessions, which reversed during a return to the original classical conditioning procedure and increased once again during the second phase, in which the slides were presented simultaneously. The experiment suggested that classical conditioning was also effective with this client but only after a sensitive temporal adjustment was made.

Merely observing the "outcome" of the 2 subjects at the end of a fixed point in time would have produced the type of intersubject variability so common in outcome studies of therapeutic techniques. That is, one subject would have improved with the initial classical conditioning procedure whereas one subject would have remained unchanged. If this pattern continued over additional subjects, the result would be the typical weak effect (Bergin & Strupp, 1972) with large intersubject variability. Highlighting the variability through repeated measurement in the individual and improvising a new experimental design as soon as a variation in response was noted (in this case no response) allowed an immediate search for the cause of this unrespon-siveness. It should also be noted that this research tactic resulted in immediate clinical benefit to the patient, providing a practical illustration of the merging of scientist and practitioner roles in the applied researcher.

Subject improves "spontaneously"

A second source of variability quite common in single-case research is the presence of "spontaneous" improvement in the absence of the therapeutic variable to be tested. This effect is illustrated in a second experiment on increasing heterosexual arousal in homosexuals (Herman, Barlow, & Agras, 1974b).

In this study, the original purpose was to determine the effectiveness of orgasmic reconditioning, or pairing masturbation with heterosexual cues, in producing heterosexual arousal. The heterosexual cues chosen were movies of a female assuming provocative sexual positions. The initial phase consisted of measurements of arousal patterns without any "treatment," which served as a baseline of sexual arousal. Before pairing masturbation with this movie, a control phase was administered where all elements of the treatment were present with the exception of masturbation. That is, the subject was in-structed that this was "treatment" and that looking at movies would help him learn heterosexual arousal. Although no increase in heterosexual arousal was expected during this phase, this procedure was experimentally necessary to isolate the pairing of masturbation with the cues in the next phase as the effective treatment. The effects of masturbation were never tested in this experiment, however, since the first subject demonstrated unexpected but substantial increases in heterosexual arousal during the "control" phase, in which he simply viewed the erotic movie (see Figure 2-3). Once again it became necessary to improvise a new experimental design at the end of this control phase, in an attempt to determine the cause of this unexpected increase. On the hunch that the erotic heterosexual movie was responsible for these gains rather than other therapeutic variables such as expectancy, a second erotic movie without heterosexual content was introduced, in this case a homosexual movie. Heterosexual arousal dropped in this condition and increased once again when the heterosexual movie was introduced. This experiment, and subsequent replication, demonstrated that the erotic hetero-sexual movie was responsible for improvement. Determination of the effects of masturbation was delayed for future experimentation.

Subject displays cyclical variability

A third pattern of variability, highlighted by repeated measurement in individual cases, is observed when behavior varies in a cyclical pattern. The behavior may follow a regular pattern (i.e., weekly) or may be irregular. A

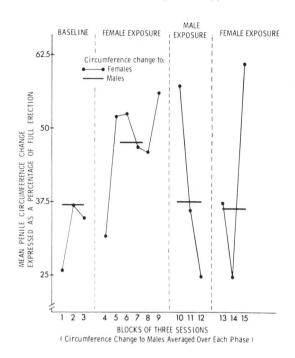

FIGURE 2-3. Mean penile circumference change expressed as a percentage of full erection to nude female (averaged over blocks of three sessions) and nude male (averaged over each phase) slides. (Figure 1, p. 338, from: Herman, S. H., Barlow, D. H., and Agras, W. S. [1974]. An experimental analysis of exposure to "explicit" heterosexual stimuli as an effective variable in changing arousal patterns of homosexuals. *Behaviour Research and Therapy*, **12**, 335-345. Copyright 1974 by Pergamon. Reproduced by permission.)

common temporal pattern, of course, is the behavioral or emotional fluctuation noted during menstruation. Of more concern to the clinician is the marked fluctuation occurring in most behavioral disorders over a period of time. In most instances the fluctuation cannot be readily correlated with specific, observable environmental or psychological events, due to the extent of the behavioral or emotional fluctuation and the number of potential variables that may be affecting the behavior. As noted in the beginning of this chapter, experimental clinicians can often make educated guesses, but the technique of repeated measurement can illustrate relationships that might not be readily observable.

A good example of this method is found in an early case of severe, daily asthmatic attacks reported by Metcalfe (1956). In the course of assessment, Metcalfe had the patient record in diary form asthmatic attacks as well as all activities during the day, such as games, shopping expeditions, meetings with her mother, and other social visits. These daily recordings revealed that asthmatic attacks most often followed meetings with the patient's mother, particularly if these meetings occurred in the home of the mother. After this relationship was demonstrated, the patient experienced a change in her life circumstances which resulted in moving some distance away from her mother. During the ensuing 20 months, only nine attacks were recorded despite the fact that these attacks had occurred daily for a period of 2 years prior to intervention. What is more remarkable is that eight of the attacks followed her now infrequent visits to her mother.

Once again, the procedure of repeated measurement highlighted individual fluctuation, allowing a search for correlated events that bore potential causal relationships to the behavior disorder. It should be noted that no experimental analysis was undertaken in this case to isolate the mother as the cause of asthmatic attacks. However, the dramatic reduction of high-frequency attacks after decreased contact with the mother provided reasonably strong evidence about the contributory effects of visits to the mother, in an A-B fashion. What is more convincing, however, is the reoccurrence of the attacks at widely spaced intervals after visits to the mother during the 20-month follow-up. This series of naturally occurring events approximates a contrived A-B-A-B. . . design and effectively isolates the mother's role in the patient's asthmatic attacks (see chapter 5).

Searching for "hidden" sources of variability

In the preceding case functional relations become obvious without experimental investigation, due to the overriding effects of one variable on the behavior in question and a series of fortuitous events (from an experimental point of view) during follow-up. Seldom in applied research is one variable so predominant. The more usual case is one where marked fluctuations in behavior occur that cannot be correlated with any one variable. In these cases, close examination of repeated measures of the target behavior and correlated internal or external events does not produce an obvious relationship. Most likely, many events may be correlated at one time or another with deterioration or improvement in a client. At this point, it becomes necessary to employ sophisticated experimental designs if one is to search for the source of variability. The experienced applied researcher must first choose the most likely variables for investigation from among the many impinging on the client at any one time. In the case described above, not only visits to the mother but visits to other relatives as well as stressful situations at work might all have contributed to the variance. The task of the clinical investigator is to tease out the relevant variables by manipulating one variable, such as visits to mother, while holding other variables constant. Once the contribution of visits to mother to behavioral fluctuation has been determined, the investigator must go on to the next variable, and so on.

In many cases, behavior is a function of an interaction of events. These events may be naturally occurring environmental variables or perhaps a combination of treatment variables which, when combined, affect behavior differently from each variable in isolation. For example, when testing out a variety of treatments for anorexia nervosa (Agras, Barlow, Chapin, Abel, & Leitenberg, 1974), it was discovered that size of meals served to the patients seemed related to caloric intake. An improvised design at this point in the experiment demonstrated that size of meals was related to caloric intake only if feedback and reinforcement were present. This discovery led to inclusion of this procedure in a recommended treatment package for anorexia nervosa. Experimental designs to determine the effects of combinations of variables will be discussed in section 6.6 of chapter 6.

2.4. BEHAVIOR TRENDS AND INTRASUBJECT AVERAGING

When testing the effects of specific interventions on behavior disorders, the investigator is less interested in small day-to-day fluctuations that are a part of so much behavior. In these cases the investigator must make a judgment on how much behavioral variability to ignore when looking for functional relations among overall trends in behavior and treatment in question. To the investigator interested in determining all sources of variability in individual behavior, this is a very difficult choice. For applied researchers, the choice is often determined by the practical considerations of discovering a therapeutic variable that "works" for a specific behavior problem in an individual. The necessity of determining the effects of a given treatment may constrain the applied researcher from improvising designs in midexperiment to search for a source of each and every fluctuation that appears.

In correlational designs, where one simply introduces a variable and observes the "trend," statistics have been devised to determine the significance of the trend over and above the behavioral fluctuation (Campbell & Stanley, 1966; Cook & Campbell, 1979; see also chapter 9). In experimental designs such as A-B-A-B, where one is looking for cause-effect relationships, investigators will occasionally resort to averaging two or more data points within phases. This intrasubject averaging, which is sometimes called *blocking*, will usually make trends in behavior more visible, so that the clinician can judge the magnitude and clinical relevance of the effect. This procedure is dangerous, however, if the investigator is under some illusion that the variability has somehow disappeared or is unimportant to an understanding of the controlling effects of the behavior in question. This method is simply a procedure to make large and clinically significant changes resulting from introduction and withdrawal of treatment more apparent. To illustrate the procedure, the

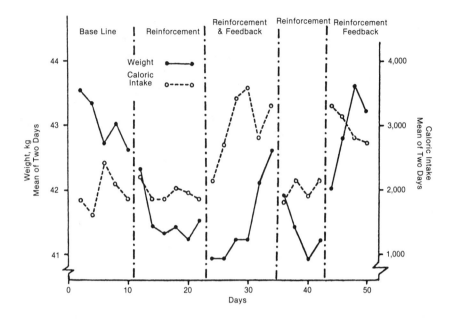

FIGURE 2-4. Data from an experiment examining the effect of feedback on the eating behavior of a patient with anorexia nervosa (Patient 4). (Figure 3, p. 283, from: Agras, W. S., Barlow, D. H., Chapin, H. N., Abel, G. G., and Leitenberg, H. [1974]. Behavior modification of anorexia nervosa. *Archives of General Psychiatry,* **30**, 279–286. Copyright 1974 by American Medical Association. Reproduced by permission.)

original data on caloric intake in a subject with anorexia nervosa will be presented for comparison with published data (Agras et al., 1974). The data as published are presented in Figure 2-4. After the baseline phase, material reinforcers such as cigarettes were administered contingent on weight gain in a phase labeled *reinforcement*. In the next phase, informational *feedback* was added to reinforcement. Feedback consisted of presenting the subject with daily weight counts of caloric intake after each meal and counts of number of mouthfuls eaten. The data indicate that caloric intake was relatively stable during the reinforcement phase but increased sharply when feedback was added to reinforcement. Six data points are presented in each of the reinforcement and reinforcement-feedback phases. Each data point represents the mean of 2 days. With this method of data presentation, caloric intake during reinforcement looks quite stable.

In fact, there was a good deal of day-to-day variability in caloric intake during this phase. If one examines the day-to-day data, caloric intake ranged from 1,450 to 3,150 over the 12-day phase (see Figure 2-5). Since the variabil-

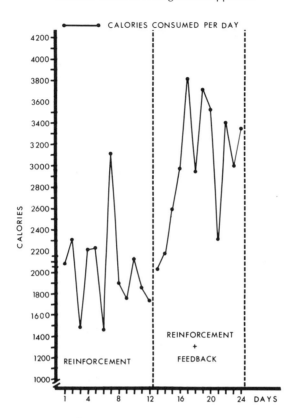

FIGURE 2-5. Caloric intake presented on a daily basis during reinforcement and reinforcement and feedback phases for the patient whose data is presented in Figure 2-4. (Replotted from Figure 3, p. 283, from: Agras, W. S., Barlow, D. H., Chapin, H. N., Abel, G. G., and Leitenberg, H. [1974]. Behavior modification of anorexia nervosa. *Archives of General Psychiatry*, **30**, 279–286. Copyright 1974 by American Medical Association. Reproduced by permission.)

ity assumed a pattern of roughly one day of high caloric intake followed by a day of low intake, the average of 2 days presents a stable pattern. When feedback was added during the next 12-day phase, the day-to-day variability remained, but the range was displaced upward, from 2,150 to 3,800 calories per day. Once again, this pattern of variability was approximately one day of high caloric intake followed by a low value. In fact, this pattern obtained throughout the experiment.

In this experiment, feedback was clearly a potent therapeutic procedure over and above the variability, whether one examines the data day-by-day or in blocks of 2 days. The averaged data, however, present a clear picture of the effect of the variable over time. Since the major purpose of the experiment was to demonstrate the effects of various therapeutic variables with anorexics, we chose to present the data in this way. It was not our intention, however, to ignore the daily variability. The fairly regular pattern of change suggests several environmental or metabolic factors that may account for these changes. If one were interested in more basic research on eating patterns in anorexics, one would have to explore possible sources of this variability in a finer analysis than we chose to undertake here.

It is possible, of course, that feedback might not have produced the clear and clinically relevant increase noted in these data. If feedback resulted in a small increase in caloric intake that was clearly visible only when data were averaged, one would have to resort to statistical tests to determine if the increase could be attributed to the therapeutic variable over and above the day-to-day variability (see chapter 9). Once again, however, one may question the clinical relevance of the therapeutic procedure if the improvement in behavior is so small that the investigator must use statistics to determine if change actually occurred. If this situation obtained, the preferred strategy might be to improvise on the experimental design and augment the therapeutic procedure such that more relevant and substantial changes were produced. The issue of clincial versus statistical significance, which was discussed in some detail above, is a recurring one in single-case research. In the last analysis, however, this is always reduced to judgments by therapists, educators, etc. on the magnitude of change that is relevant to the setting. In most cases, these magnitudes are greater than changes that are merely statistically significant.

The above example notwithstanding, the conservative and preferred approach of data presentation in single-case research is to present all of the data so that other investigators may examine the intrasubject variability firsthand and draw their own conclusions on the relevance of this variability to the problem.

Large intrasubject variability is a common feature during repeated measurements of target behaviors in a single-case, particularly in the beginning of an experiment, when the subject may be accommodating to intrusive measures. How much variability the researcher is willing to tolerate before introducing an independent variable (therapeutic procedure) is largely a question of judgment on the part of the investigator. Similar procedural problems arise when introduction of the independent variable itself results in increased variability. Here the experimenter must consider alteration in length of phases to determine if variability will decrease over time (as it often does), clarifying the effects of the independent variable. These procedural questions will be discussed in some detail in chapter 3.

2.5. RELATION OF VARIABILITY
TO GENERALITY OF FINDINGS

The search for sources of variability within individuals and the use of improvised and fast-changing experimental designs appear to be contrary to one of the most cherished goals of any science—the establishment of generality of findings. Studying the idiosyncrasies of one subject would seem, on the surface, to confirm Underwood's (1957) observation that intensive study of individuals will lead to discovery of laws that are applicable only to that individual. In fact, the identification of sources of variability in this manner leads to increases in generality of findings.

If one assumes that behavior is lawful, then identifying sources of variability in one subject should give us important leads in sources of variability in other similar subjects undergoing the same treatments. As Sidman (1960) pointed out,

> Tracking down sources of variability is then a primary technique for establishing generality. Generality and variability are basically antithetical concepts. If there are major undiscovered sources of variability in a given set of data, any attempt to achieve subject or principle generality is likely to fail. Every time we discover and achieve control of a factor that contributes to variability, we increase the likelihood that our data will be reproducible with new subjects and in different situations. Experience has taught us that precision of control leads to more extensive generalization of data (p. 152).

And again,

> It is unrealistic to expect that a given variable will have the same effects upon all subjects under all conditions. As we identify and control a greater number of the conditions that determine the effects of a given experimental operation, in effect we decrease the variability that may be expected as a consequence of the operation. It then becomes possible to produce the same results in a greater number of subjects. Such generality could never be achieved if we simply accepted inter-subject variability and gave equal status to all deviant subjects in an investigation (p. 190).

In other words, the more we learn about the effects of a treatment on different individuals, in different settings, and so on, the easier it will be to determine if that treatment will be effective with the next individual walking into the office. But if we ignore differences among individuals and simply average them into a group mean, it will be more difficult to estimate the effects on the next individual, or "generalize" the results. In applied research, when intersubject and intrasubject variability are enormous, and putative sources of the variability are difficult to control, the establishment of generality is a difficult task indeed. But the establishment of a science of human behavior change depends heavily on procedures to establish generality of findings. This important issue will be discussed in the next section.

2.6. GENERALITY OF FINDINGS

Types of generality

Generalization means many things. In applied research, generalization usually refers to the process in which behavioral or attitudinal changes in the treatment setting "generalize" to other aspects of the client's life. In educational research this can mean generalization of behavioral changes from the classroom to the home. Generalization of this type can be determined by observing behavioral changes outside of the treatment setting.

There are at least three additional types of generality in behavior change research, however, that are more relevant to the present discussion. The first is generality of findings across subjects or clients; that is, if a treatment effects certain behavior changes in one subject, will the same treatment also work in other subjects with similar characteristics? As we shall see below, this is a large question because subjects can be "similar" in many different ways. For instance, subjects may be similar in that they have the same diagnostic labels or behavioral disorders (e.g., schizophrenia or phobia). In addition, subjects may be of similar age (e.g., between 14 and 16) or come from similar socioeconomic backgrounds.

Generality across behavior change agents is a second type. For instance, will a therapeutic technique that is effective when applied by one behavior change agent also be effective when applied to the same problem by different agents? A common example is the classroom. If a young, attractive, female teacher successfully uses reinforcement principles to control disruptive behavior in her classroom, will an older female teacher who is more stern also be able to apply successfully the same principles to similar problems in her class? Will an experienced therapist be able to treat a middle-aged claustrophobic more effectively than a naive therapist who uses exactly the same procedure?

A third type of generality concerns the variety of settings in which clients are found. The question here is will a given treatment or intervention applied by the same or similar therapist, to similar clients, work as well in one setting as another? For example, would reinforcement principles that work in the classroom also work in a summer camp setting, or would desensitization of an agoraphobic in an urban office building be more difficult than in a rural setting?

These questions are very important to clinicians who are concerned with which treatments are most effective with a given client in a given setting. Typically, clinicians have looked to the applied researcher to answer these questions.

Problems in generalizing from a single-case

The most obvious limitation in studying a single-case is that one does not know if the results from this case would be relevant to other cases. Even if one isolates the active therapeutic variable in a given client through a rigorous single-case experimental design, critics note that there is little basis for inferring that this therapeutic procedure would be equally effective when applied to clients with similar behavior disorders (client generality) or that different therapists using this technique would achieve the same results (therapist generality). Finally, one does not know if the technique would work in a different setting (setting generality). This issue, more than any other, has retarded the development of single-case methodology in applied research and has caused many authorities on research to deny the utility of studying a single-case for any other purpose than the generation of hypotheses (e.g., Kiesler, 1971). Conversely, in the search for generality of applied research findings, the group comparison approach appeared to be the logical answer (Underwood, 1957).

In the specific area of individual human behavior, however, there are issues that limit the usefulness of a group approach in establishing generality of findings. On the other hand, the newly developing procedures of direct, systematic, and clinical replication offer an alternative, in some instances, for establishing generality of findings relevant to individuals. The purpose of this section is to outline the major issues, assumptions, and goals of generality of findings as related to behavior change in an individual and to describe the advantages and disadvantages of the various procedures to establishing generality of findings.

2.7. LIMITATIONS OF GROUP DESIGNS IN ESTABLISHING GENERALITY OF FINDINGS

In chapter 1, section 1.5, several limitations of group designs in applied research noted by Bergin and Strupp (1972) were outlined. One of the limitations referred to difficulties in generalizing results from a group to an individual. In this category, two problems stand out. The first is inferring that results from a relatively homogeneous group are representative of a given population. The second is generalizing from the average response of a heterogeneous group to a particular individual. These two problems will be discussed in turn.

Random sampling and inference in applied research

After the brilliant work of R. A. Fisher, early applied researchers were most concerned with drawing a truly random sample of a given population, so that results would be generalizable to this population. For instance, if one wished to draw some conclusion on the effects of a given treatment for schizophrenia, one would have to draw a random sample of all schizophrenics.

In reference to the three types of generality mentioned above, this means that the clients under study (e.g., schizophrenics) must be a random sample of all schizophrenics, not only for behavioral components of the disorder, such as loose associations or withdrawn behavior, but also for other patient characteristics such as age, sex, and socioeconomic status. These conditions must be fulfilled before one can infer that a treatment that demonstrates a statistically significant effect would also be effective for other schizophrenics outside of the study. As Edgington (1967) pointed out, "In the absence of random samples hypothesis testing is still possible, but the significance statements are restricted to the effect of the experimental treatments on the subjects actually used in the experiment, generalization to other individuals being based on logical nonstatistical considerations" (p. 195). If one wishes to make statements about effectiveness of a treatment across therapists or settings, random samples of therapists and settings must also be included in the study.

Random sampling of characteristics in the animal laboratories of experimental psychology is feasible, at least across subjects, since most relevant characteristics such as genetic and environmental determinants of individual behavior can be controlled. In clinical or educational research, however, it is extremely difficult to sample adequately the population of a particular syndrome. One reason for this is the vagueness of many diagnostic categories (e.g., schizophrenia). In order to sample the population of schizophrenics one must be able to pinpoint the various behavioral characteristics that make up this diagnosis and ensure that any sample adequately represents these behaviors. But the relative unreliability of this diagnostic category, despite improvements in recent years (Spitzer, Forman, & Nee, 1979), makes it very difficult to determine the adequacy of a given sample. In addition, the therapeutic emphasis may differ from setting to setting. In one center, bizarre behavior and hallucinations may be emphasized. In another center, a thought disorder may be the primary target of assessment (Neale & Oltmanns, 1980; Wallace, Boone, Donahoe, & Foy, in press).

A second problem that arises when one is attempting an adequate sample of a population is the availability of clients who have the needed behavior or characteristics to fill out the sample (see chapter 1, section 1.5). In laboratory animal research this is not a problem because subjects with specified characteristics or genetic backgrounds can be ordered or produced in the laboratories. In applied research, however, one must study what is available, and this may result in a heavy weighting on certain client characteristics and inadequate sampling of other characteristics. Results of a treatment applied to this sample cannot be generalized to the population. For example, techniques to control disruptive behavior in the classroom will be less than generalizable if they are tested in a class where students are from predominantly middle-class suburbs and inner-city students are underrepresented.

Even in the great snake phobic epidemic of the 1960s, where the behavior in question was circumscribed and clearly defined, the clients to whom various treatments were applied were almost uniformly female college sophomores whose fear was neither too great (they could not finish the experiment on time) nor too little (they would finish it too quickly). Most investigators admitted that the purpose of these experiments was not to generalize treatment results to clinical populations, but to test theoretical assumptions and generate hypotheses. The fact remains, however, that these results cannot even be generalized beyond female college sophomores to the population of snake fearers, where age, sex, and amount of fear would all be relevant.

It should be noted that all examples above refer to generality of findings across clients with similar behavior and background characteristics. Most studies at least consider the importance of generality of findings along this dimension, although few have been successful. What is perhaps more important is the failure of most studies to consider the generality problem in the other two dimensions—namely, setting generality and behavior change agent (therapist) generality. Several investigators (e.g., Kazdin, 1973b, 1980b; McNamara & MacDonough, 1972) have suggested that this information may be more important than client generality. For example, Paul (1969) noted after a survey of group studies that the results of systematic desensitization seemed to be a function of the qualifications of the therapist rather than differences among clients. Furthermore, in regard to setting generality, Brunswick (1956) suggested that, "In fact, proper sampling of situations and problems may be in the end more important than proper sampling of subjects considering the fact that individuals are probably on the whole much more alike than are situations among one another" (p. 39). Because of these problems, many sophisticated investigators specializing in research methodology have accepted the impracticability of random sampling in this context and have sought other methods for establishing generality (e.g., Kraemer, 1981).

The failure to be able to make statistically inferential statements, even about populations of clients based on most clinical research studies, does not mean that no statements about generality can be made. As Edgington (1966) pointed out, one can make statements at least on generality of findings to similar clients based on logical non-statistical considerations. Edgington referred to this as *logical generalization*, and this issue, along with generality to settings and therapists, will be discussed below in relation to the establishment of generality of findings from a single-case.

Problems in generalizing from the group to the individual

The above discussion might be construed as a plea for more adequate sampling procedures involving larger numbers of clients seen in many different settings by a variety of therapists—in other words, the notion of the "grand collaborative study," which emerged from the conferences on research in psychotherapy in the 1960s (e.g., Bergin & Strupp, 1972; Strupp & Luborsky, 1962). On the contrary, one of the pitfalls of a truly random sample in applied research is that the more adequate the sample, in that all relevant population characteristics are represented, the less relevance will this finding have for a specific individual. The major issue here is that the better the sample, the more heterogeneous the group. The average response of this group, then, will be less likely to represent a given individual in the group. Thus, if one were establishing a random sample of severe depressives, one should include clients of various ages, and racial, and socioeconomic backgrounds. In addition, clients with various combinations of the behavior and thinking or perceptual disorder associated with severe depression must be included. It would be desirable to include some patients with severe agitation, others demonstrating psychomotor retardation, still others with varying degrees and types of depressive delusions, and those with somatic correlates such as terminal sleep disturbance. As this sample becomes truly more random and representative, the group becomes more heterogeneous. The specific effects of a given treatment on an individual with a certain combination of problems becomes lost in the group average. For instance, a certain treatment might alleviate severe agitation and terminal sleep disturbance but have a deleterious effect on psychomotor retardation and depressive delusions. If one were to analyze the results, one could infer that the treatment, on the average, is better than no treatment for the population of patients with severe depression. For the individual clinician, this finding is not very helpful and could actually be dangerous if the clinician's patient had psychomotor retardation and depressive delusions.

Most studies, however, do not pretend to draw a truly random sample of patients with a given diagnosis or behavior disorder. Even the most recent, excellent, example of a general collaborative study on treatments for depression where random sampling was perhaps feasible did not attempt random sampling (NIMH, 1980). Most studies choose clients or patients on the basis of availability after deciding on inclusion and exclusion criteria and then randomly assign these subjects into two or more groups that are matched on relevant characteristics. Typically, the treatment is administered to one group while the other group becomes the no-treatment control. This arrangement, which has characterized much clinical and educational research, suffers for two reasons; (1) To the extent that the "available" clients are not a random sample, one cannot generalize to the population; and (2) to the extent that the group is heterogeneous on any of a number of characteristics, one cannot make statements about the individual. The only statement that can be made concerns the average response of a group with that particular makeup which, unfortunately, is unlikely to be duplicated again. As Bergin (1966) noted, it was even difficult to say anything important about individuals within the group based on the average response because his analysis demonstrated that some were improving and some deteriorating (see Strupp & Hadley, 1979). The result, as Chassan (1967, 1979) eloquently pointed out, was that the behavior change agent did not know which treatment or aspect of treatment was effective that was statistically better than no treatment but that actually might make a particular patient worse.

Improving generality of findings to the individual through homogeneous groups: Logical generalization

What Bergin and Strupp (1972) and others (e.g., Kiesler, 1971; Paul, 1967) recognized was that if anything important was going to be said about the individual, after experimenting with a group, then the group would have to be homogeneous for relevant client characteristics. For example, in a study of a group of agoraphobics, they should all be in one age-group with a relatively homogeneous amount of fear and approximately equal background (personality) variables. Naturally, clients in the control group must also be homogeneous for these characteristics.

Although this approach sacrifices random sampling and the ability to make inferential statements about the population of agoraphobics, one can begin to say something about agoraphobics with the same or similar characteristics as those in the study through the process of logical generalization (Edgington, 1967, 1980a). That is, if a study shows that a given treatment is successful with a homogeneous group of 20- to 30-year-old female agoraphobics with certain personality characteristics, then a clinician can be relatively confident that a 25-year-old female agoraphobic with those personality characteristics will respond well to that same treatment. (Recently some experts have suggested that one should not assemble groups that are too homogeneous, for even the ability to generalize on more logical grounds might be greatly restricted [Kraemer, 1981].)

The process of logical generalization depends on similarities between the patients in the homogeneous group and the individual in question in the clinician's office. Which features of a case are important for extending logical generalization and which features can be ignored (e.g., hair color) will depend on the judgment of the clinician and the state of knowledge at the time. But if one can generalize in logical fashion from a patient whose results or characteristics are well specified as part of a homogeneous group, then one can also logically generalize from a single individual whose response and biographical characteristics are specified. In fact, the rationale has enabled applied researchers to generalize the results of single-case experiments for years (Dukes, 1965; Shontz, 1965). To increase the base for generalization from a single-case experiment, one simply repeats the same experiment several times on similar patients, thereby providing the clinician with results from a number of patients.

2.8. HOMOGENEOUS GROUPS VERSUS REPLICATION OF A SINGLE-CASE EXPERIMENT

Because the issue of generalization from single-case experiments in applied research is a major source of controversy (Agras, Kazdin, & Wilson, 1979; Kazdin, 1980b, 1982b; Underwood, 1957), the sections to follow will describe our views of the relative merits of replication studies versus generalization from homogeneous groups.

As a basis for comparison, it is useful to compare the single-case approach with Paul's (1967, 1969) incisive analysis of the power of various experimental designs using groups of clients. Within the context of the power of these various designs to establish cause-effect relationships, Paul reviewed the several procedures commonly used in applied research. These procedures range from case studies with and without measurement, from which cause-effect relationships can seldom if ever be extracted, through series of cases typically reporting percentage of success with no control group. Finally, Paul cited the two major between-group experimental designs capable of establishing functional relationships between treatments and the average response of clients in the group. The first is what Paul referred to as the nonfactorial design with no-treatment control, in other words the comparison of an experimental (treatment) group with a no-treatment control group. The second design is the powerful factorial design, which not only establishes cause-effect relations between treatments and clients but also specifies what type of clients under what conditions improve with a given treatment; in other words, client-treatment interactions. The single-case replication strategy paralleling the nonfactorial design with no-treatment control is *direct replication*. The replication strategy paralleling the factorial design is called *systematic replication*.

Direct replication and treatment/no-treatment control group design

When Paul's article was written (1967), applied research employing single-case designs, usually of the A-B-A variety, was just beginning to appear (e.g., Ullmann & Krasner, 1965). Paul quickly recognized the validity or power of this design, noting that "The level of product for this design approaches that of the nonfactorial group design with no-treatment controls" (p. 117). When Paul spoke of level of product here he was referring, in Campbell and Stanley's (1963) terms, to internal validity, that is, the power of the design to isolate the independent variable (treatment) as responsible for experimental effects—and to external validity or the ability to generalize findings across relevant domains such as client, therapist, and setting. We would agree with Paul's notions that the level of product of a single-case experimental design only "approaches" that of treatment/no-treatment group designs, but for somewhat different reasons. It is our contention that the single-case A-B-A design approaches rather than equals the nonfactorial group design with no-treatment controls only because the number of clients is considerably less in a single-case design ($N = 1$) than in a group design, where 8, 10, or more clients are not uncommon. It is our further contention that, in terms of external validity or generality of findings, a series of single-case designs in similar clients in which the original experiment is directly replicated three or four times can far surpass the experimental group/no-treatment control group design. Some of the reasons for this assertion are outlined next.

Results generated from an experimental group/no-treatment control group study as well as a direct replication series of single-case experimental designs yield some information on generality of findings across clients but cannot address the question of generality across different therapists or settings. Typically, the group study employs one therapist in one setting who applies a given treatment to a group of clients. Measures are taken on a pre-post basis. Premeasures and postmeasures are also taken from a matched group of clients in the control group who do not receive the intervening treatment. For example, 10 depressive patients homogeneous on behavioral and emotional aspects of their depression, as well as personality characteristics, would be compared to a matched group of patients who did not receive treatment. Logical generalization to other patients (but not to other therapists or settings) would depend on the degree of homogeneity among the depressives in both groups. As noted above, the less homogeneous the depression in the experiment, the greater the difficulty for the practicing clinician in determining if that treatment is effective for his or her particular patient. A solution to this problem would be to specify in some detail the characteristics of each patient in the treatment group and present individual data on each patient. The clinician could then observe those patients that are most like his or her particular client and determine if these experimental patients improved more than the average response in the control group. For example, after describing in detail the case history and presenting symptomatology of 10 depressives, one could administer a pretest measuring severity of depression to the 10 depressives and a matched control group of 10 depressives. After treatment of the 10 depressives in the experimental group, the posttest would be administered. When results are presented, the improvement (or lack of improvement) of each patient in the treatment group could be presented either graphically or in numerical form along with the means and standard deviations for the control group. After the usual procedure to determine statistical significance, the clinician could examine the *amount* of improvement of each patient in the experimental group to determine (1) if the improvement were *clinically* relevant, and (2) if the improvement exceeded any drift toward improvement in the control group. To the extent that some patients in the treatment group were similar to the clinician's patient, the clinician could begin to determine, through logical generalization, whether the treatment might be effective with his or her patient.

However, a series of single-case designs where the original experiment is replicated on a number of patients also enables one to determine generality of findings across patients (but not across therapists or settings). For example, in the same hypothetical group of depressives, the treatment could be administered in an A-B-A-B design, where A represents baseline measurement and B represents the treatment. The comparison here is still between treatment and no treatment. As results accumulate across patients, generality of findings is established, and the results are readily translatable to the practicing clinician, since he or she can quickly determine which patient with which characteristics improved and which patient did not improve. To the extent that therapist and treatment are alike across patients, this is the clinical prototype of a direct replication series (Sidman, 1960), and it represents the most common replication tactic in the experimental single-case approach to date.

Given these results, other attributes of the single-case design provide added strength in generalizing results to other clients. The first attribute is flexibility (noted in section 2.3). If a particular procedure works well in one case but works less well or fails when attempts are made to replicate this in a second or third case, slight alterations in the procedure can be made immediately. In many cases, reasons for the inability to replicate the findings can be ascertained immediately, assuming that procedural deficiencies were, in fact, responsible for the lack of generality. An example of this result was outlined in section 2.3, describing intersubject variability. In this example, one patient improved with treatment, but a second did not. Use of an improvised experimental design at this point allowed identification of the reason for failure. This finding should increase generality of findings by enabling immediate application of the altered procedure to another patient with a similar response pattern. This is an example of Sidman's (1960) assertion that "tracking down sources of variability is then a primary technique of establishing generality" (see also Kazdin, 1973b; Leitenberg, 1973; Skinner, 1966b). If alterations in the procedure do not produce clinical improvement, either differences in background, *personality* characteristics, or differences within the behavior disorder itself can be noted, suggesting further hypotheses on procedural changes that can be tested on this type of client at a later date.

Finally, using the client as his or her own control in successive replications provides an added degree of strength in generalizing the effect of treatment across differing clients. In group or single-case designs employing no-treatment controls or attention-placebo controls, it is possible and even quite likely that certain environmental events in a no-treatment control group or phase will produce considerable improvements (e.g., placebo effects). In a nonfactorial group design, where treated clients show more improvement than clients in a no-treatment control, one can conclude that the treatment is effective and then proceed in generalizing results to other clients in clinical situations. However, the *degree* of the contribution of nonspecific environmental factors to the improvement of *each individual* client is difficult to judge. In a single-case design (for example, the A-B-A-B or true withdrawal design), the influence of environmental factors on each individual client can be estimated by observing the degree of deterioration when treatment is withdrawn. If environmental or other factors are operating during treatment, improvement will continue during the withdrawal phase, perhaps at a slower rate, necessitating further experimental inquiry. Even in a nonfactorial group design with powerful effects, the contribution of this factor to *individual* clients is difficult to ascertain.

Systematic and clinical replication and factorial designs

Direct replication series and nonfactorial designs with no-treatment controls come to grips with only one aspect of generality of findings—generality across clients. These designs are not capable of simultaneously answering questions on generality of findings across therapists, settings, or clients that differ in some substantial degree from the original homogeneous group. For example, one might ask, if the treatment works for 25-year-old female agoraphobics with certain personality characteristics, will it also work for a 40-year-old female agoraphobic with different personality characteristics?

In the therapist domain, the obvious question concerns the effectiveness of treatment as related to that particular therapist. If the therapist in the hypothetical study were an older, more experienced therapist, would the treatment work as well with a young therapist? Finally, even if several therapists in one setting were successful, could therapists in another setting and geographical area attain similar results?

To answer all of these questions would require literally hundreds of experimental group/no-treatment control group studies where each of the factors relevant to generalization was varied one at a time (e.g., type of therapist, type of client). Even if this were feasible, however, the results could not always be attributed to the factor in question as replication after replication ensued, because other sources of variance due to faulty random assignment of clients to the group could appear.

In reviewing the status and goals of psychotherapy research, many clinical investigators (e.g., Kazdin, 1980b, 1982b; Kiesler, 1971; Paul, 1967) proposed the application of one of the most sophisticated experimental designs in the armamentarium of the psychological researcher—the factorial design—as an answer to the above problem. In this design, relevant factors in all three areas of generality of concern to the clinician can be examined. The power of this design is in the specificity of the conclusion.

For example, the effects of two antidepressant pharmacological agents and a placebo might be evaluated in two different settings (the inpatient ward of a general hospital and an outpatient community mental health center) on two groups of depressives (one group with moderate to severe depression and a second group with mild depression). A therapist in the psychiatric ward setting would administer each treatment to one half of each group of depressives—the moderate to severe group and the mild group. All depressives would be matched as closely as possible on background variables such as age, sex, and personality characteristics. The same therapist could then travel to the community mental health center and carry out the same procedure. Thus we have a 2 × 2 × 2 factorial design. Possible conclusions from this study are numerous, but results might be so specific as to indicate that antidepressants do work but only with moderate to severe depressives and only if hospitalized in a psychiatric ward. It would not be possible to draw conclusions on the importance of a particular type of therapist because this factor was not systematically varied. Of course, the usual shortcomings of group designs are also present here because results would be presented in terms of group averages and intersubject variability. However, to the extent that subjects in each experimental cell were homogeneous and to the extent that improvement was large and clinically important rather than merely statistically significant, then results would certainly be a valuable contribution. The clinical practitioner would be able to examine the characteristics of those subjects in the improved group and conclude that under similar conditions (i.e., an inpatient psychiatric unit) his or her moderate to severe depressive patient would be likely to improve, assuming, of course, that this patient resembled those in the study. Here again, the process of logical generalization rather than statistical inference from a sample to a population is the active mechanism.

Thus, while the factorial design can be effective in specifying generality of findings across all important domains in applied research (within the limits discussed above), one major problem remains: Applied researchers seldom do this kind of study. As noted in chapter 1, section 1.5, the major reasons for this are practical. The enormous investment of money and time necessary to collect large numbers of homogeneous patients has severely inhibited this type of endeavor. And often, even in several different settings, the necessary number of patients to complete a study is just not available unless one is willing to wait years. Added to this are procedural difficulties in recruiting and paying therapists, ensuring adequate experimental controls such as double-blind procedures within a large setting, and overcoming resistance to assigning a large number of patients to placebo or control conditions, as well as coping with the laborious task of recording and analyzing large amounts of data (Barlow & Hersen, 1973; Bergin & Strupp, 1972).

In addition, the arguments raised in the last section on inflexibility of the group design are also applicable here. If one patient does not improve or reacts in an unusual way to the therapeutic procedure, administration of the procedure must continue for the specified number of sessions. The unsuccessful or aberrant results are then, of course, averaged into the group results from that experimental cell, thus precluding an immediate analysis of the intersubject variability, which will lead to increased generality.

Systematic and clinical replication procedures involve exploring the effects of different settings, therapists, or clients on a procedure previously demonstrated as successful in a direct replication series. In other words, to borrow the example from the factorial design, a single-case design may demonstrate that a treatment for severe depression works on an inpatient unit. Several direct replications then establish generality among homogeneous patients. The next task is to replicate the procedure once again, in different settings with different therapists or with patients with different background characteristics. Thus the goals of systematic and clinical replication in terms of generality of findings are similar to those of the factorial study.

At first glance, it does not appear as if replication techniques within single-case methodology would prove any more practical in answering questions concerning generality of findings across therapists, settings, and types of behavior disorder. While direct replication can begin to provide answers to questions on generality of findings across similar clients, the large questions of setting and therapist generality would also seem to require significant collaboration among diverse investigators, long-range planning, and a large investment of money and time—the very factors that were noted by Bergin and Strupp (1972) to preclude these important replication effects. The surprising fact concerning this particular method of replication, however, is that these issues are not interfering with the establishment of generality of findings, since systematic and clinical replication is in progress in a number of areas of applied research. In view of the fact that systematic and clinical replication has the same advantages of logical generalization as direct replication, the information yielded by the procedure has direct applicability to the clinic. Examples from these ongoing systematic replication and clinical series and procedures and guidelines for replication will be described in chapter 10.

2.9. APPLIED RESEARCH QUESTIONS REQUIRING ALTERNATIVE DESIGNS

It was observed in chapter 1 that applied researchers during the 1950s and 1960s often considered single-case versus between-group comparison research as an either-or proposition. Most investigators in this period chose one methodology or the other and eschewed the alternative. Much of this polemic characterized the idiographic-nomothetic dichotomy in the 1950s (Allport, 1961). This type of argument, of course, prevented many investigators from asking the obvious question: Under what condition is one type of design more appropriate than another? As single-case designs have become more sophisticated, the number of questions answered by this strategy has increased. But there are many instances in which single-case designs either cannot answer the relevant applied research question or are less applicable. The purpose of this book, of course, is to make a case for the relevance of single-case experimental designs and to cover those issues, areas, and examples where a single-case approach is appropriate and important. We would be remiss, however, in ignoring those areas where alternative experimental designs offer a better answer.

Actuarial questions

There are several related questions or issues that require experimental strategies involving groups. Baer (1971) referred to one as *actuarial*, although he might have said *political*. The fact is, after a treatment has been found effective, society wants to know the magnitude of its effects. This information is often best conveyed in terms of percentage of people who improved compared to an untreated group. If one can say that a treatment works in 75 out of 100 cases where only 15 out of 100 would improve without treatment, this is the kind of information that is readily understood by society. In a systematic replication series, the results would be stated differently. Here the investigator would say that under certain conditions the treatment works, while under other conditions it does not work, and other therapeutic variables must be added. While this statement might be adequate for the practicing clinician or educator, little information on the magnitude of effect is conveyed. Because society supports research and, ultimately, benefits from it, this actuarial approach is not trivial. As Baer (1971) pointed out, this problem ". . . is similar to that of any insurance company, we merely need to know how often a behavioral analysis changes the relevant behavior of society toward the behavior, just as the insurance company needs to know how often age predicts death rates" (p. 366). It should be noted, however, that a study such as this cannot answer why a treatment works; it is simply capable of communicating the size of the effect. But if the treatment package is the result of a series of single-case designs, then one should already know why it works, and demonstration of the magnitude of effect is all that is needed.

Several cautions should be noted when proceeding in this manner. First, the cost and practical limitation of running a large-group study do not allow unlimited replication of this effort, if it can be done at all. Thus one should have a well-developed treatment package that has been thoroughly tested in single-case experimental designs and replications before embarking on this effort. Preferably, the investigator should be well into a systematic replicaton series in order to have some idea of the client, setting, or therapeutic variables that predict success. Groups can then be constructed in a homogeneous fashion. Premature application of the group comparison design, where a treatment or the conditions under which it is effective have not been adequately worked out, can only produce the characteristic weak effect with large intersubject variability that is so prevalent in group comparison studies to date (Bergin & Strupp, 1972). Of course, well-developed clinical replication series, where a comprehensive treatment package is replicated across many individuals with a given problem, can also specify size or effect and the percentage of clinical success. But the information from the comparison group would be missing.

Modification of group behavior

A related issue on the appropriateness of group design arises when the applied researcher is not concerned with the fate of the individual but rather with the effectiveness of a given procedure on a well-defined group. A particularly good example is the classroom. If the problem is a mild but annoying one, such as disruptive behavior in the classroom, the researcher and school administrator may be more interested in quickly determining what procedure is effective in remedying this problem for the classroom as a whole. The goal in this case is changing behavior of a well-defined group rather than individuals within that group. It may not be important that two or three children remain somewhat out of order if the classroom is substantially more quiet. A particularly good example is an experiment on the modification of classroom noise reported in chapter 7, Figure 7-5 (C. W. Wilson & Hopkins, 1973). A similar approach might be desirable with any coexisting group of people, such as a ward in a state hospital where the control of disruptive behavior would allow more efficient execution of individual therapeutic programs (see chapter 5, Figure 5-17) (Ayllon & Azrin, 1965). This stands in obvious contrast to a series of patients with severe clinical problems who do not coexist in some geographical location but are seen sequentially and assigned to a group only for experimental consideration. In this case, the applied researcher would be ill-disposed to ignore the significant human suffering of those individuals who did not improve or perhaps deteriorated.

When group behavior is the target, however, and a comparison of treated and untreated classrooms, for example, is desirable, one is not limited to between-subject designs in these instances because within-subject designs are also feasible. There are many examples where A-B-A or multiple baseline designs have been used in classroom research with repeated measures of the average behavior of the group (e.g., Wolf & Risley, 1971; see also chapters 5 and 6).

Once again, it is a good idea to have a treatment that has been adequately worked out on individuals before attempting to modify behavior of a group. If not, the investigator will encounter intolerable intersubject variability that will weaken the effects of the intervention.

2.10. BLURRING THE DISTINCTION BETWEEN DESIGN OPTIONS

The purpose of this book in general and this chapter in particular is to illustrate the underlying rationale for single-case experimental designs. To achieve this goal, the strategies and underlying rationale of more traditional between-group designs have been placed in sharp relief relative to single-case designs, to highlight the differences. This need not be the case. As described throughout this chapter, group designs could be carried out with close attention to individual change and repeated measures across time.

If one were comparing treatment and no treatment, for example, 10 depressed patients could be individually described and repeated measures could be taken of their progress. Amount of change could then be reported in clinically relevant terms. These data could be contrasted with the same reporting of individual data for a no-treatment group. Of course, statistical inferences could be made concerning group differences, based on group averages and intersubject variability within groups, but one would still have the individual data to fall back on. This would be important for purposes of logical generalization, which forms the only rational basis for generalizing results from one group of individual subjects to another individual subject. In our experience as editors of major journals, data from group studies are being reported increasingly in this manner, as investigators alter their underlying rationale for generality of findings from inferential to logical. With individuals carefully described and closely tracked during treatment, the investigator is in a position to speculate on sources of intersubject variability. That is, if one subject improves dramatically while another improves only marginally or perhaps deteriorates during treatment, the investigator can immediately analyze, at least in a *post hoc* fashion, differences between these clients. The investigator would be greatly assisted in making these judgments by repeated measurement within these group studies because the investigator could determine if a specific client was making good progress and then faltered, or simply did not respond at all from the beginning of treatment. Events correlated with a sudden change in the direction of progress could be noted for future reference. All that the investigator would be lacking would be the flexibility inherent in single-case design which would allow a quick change in experimental strategy or an experimental strategy based on the responses of the individual client (Edgington, 1983) to immediately track down the sources of this intersubject variability. Of course, many other factors must be considered when choosing appropriate designs, particularly practical considerations such as time, expense, and availability of subjects.

Once again we would suggest that if one is going to generalize from group studies to the variety of individuals entering a practitioner's office, then it is essential that data from individual clients be described so that the process of logical generalization can be applied in its most powerful form. In view of the inapplicability of making statistical inferences to hypothetical populations, based on random sampling, logical generalization is the *only* method available to us, and we must maximize its strength with thorough description of individuals in the study.

With these cautions in place, and with a full understanding of the rationale and strengths of single-case designs, the investigator can then make a reasoned choice on design options. For example, for comparing two treatments with no treatment, where each treatment should be effective but the relative effectiveness is unknown, one might choose an alternating-treatments design (see chapter 8) or a more traditional between-group comparison design with close attention to individual change. The strengths and advantages of alternating-treatments designs are fully discussed in chapter 8, but if one has a large number of subjects available and a fixed treatment protocol that for one reason or another cannot be altered during treatment, regardless of progress, then one may wish to use a between-group strategy with appropriate attention to individual data. Subsequent experimental strategies could be employed using single-case experimental designs during follow-up to deal with minimal responders or those who do not respond at all or perhaps deteriorate. But sources of intersubject variability *must* be tracked down eventually if we are to advance our science and ensure the generality of our results. Treatment in between-group designs could also be applied in a relatively "pure" form, much as it would be in a clinical setting. Occasionally we will refer to these options in the context of describing the various single-case design options throughout this book.

A further blurring of the distinction occurs when single-case designs are applied to groups of subjects. Section 5.6 and Figure 5-17 describe the application of an A-B-A withdrawal design to a large group of subjects. Similarly, a multiple baseline design applied to a large group is discussed in section 7.2. Data are described in terms of group averages in both experiments. These experimental designs, then, approach the tradition of within-subject designs (Edwards, 1968), where the same group of subjects experiences repeated experimental conditions. Appropriate statistical analyses have long been available for these design options (e.g., Edwards, 1968).

Despite the blurring of experimental traditions that is increasingly taking place, the overriding strength of single-case designs and their replications lies in the use of procedures that are appropriate to studying the subject matter at hand—the individual. It is to a description of these procedures that we now turn.

CHAPTER 3

General Procedures in Single-case Research

3.1. INTRODUCTION

Advantages of the experimental single-case design and general issues involved in this type of research were briefly outlined in chapter 2. In the present chapter a more detailed analysis of general procedures characteristic of all experimental single-case research will be undertaken. Although previous discussion of these procedures has appeared periodically in the psychological and psychiatric literatures (Barlow & Hersen, 1973; Hersen, 1982; Kazdin, 1982b; Kratchowill, 1978b; Levy & Olson, 1979), a more comprehensive analysis, from both a theoretical and an applied framework, is very much needed.

A review of the literature on applied clinical research since the 1960s shows that there is a substantial increase in the number of articles reporting the use of the experimental single-case design strategy. These papers have appeared in a wide variety of educational, psychological, and psychiatric journals. However, many researchers have proceeded without the benefit of carefully thought-out guidelines, and, as a consequence, needless errors in design and practice have resulted. Even in the *Journal of Applied Behavior Analysis*, which is primarily devoted to the experimental analysis model of research, errors in procedure and practice are not uncommon in reported investigations.

In the succeeding sections of this chapter, theoretical and practical applications of repeated measurement, methods for choosing an appropriate baseline, changing one independent variable at a time, reversals and withdrawals, length of phases, and techniques for evaluating effects of "irreversible" procedures will be considered. For heuristic purposes, both correct and incorrect applications of the aforementioned will be examined. Illustrations of actual and hypothetical cases will be provided. In addition, discussions of strategies to assess response maintenance following successful treatment is provided.

3.2. REPEATED MEASUREMENT

Aspects of repeated measurement techniques have already been discussed in chapter 2. However, in this section we will examine some of the issues in greater detail. In the typical psychotherapy outcome study (e.g., Bellack, Hersen, & Himmelhock, 1981), in which the randomly assigned or matched-group design is used, dependent measures (e.g., Beck Depression Inventory scores) usually are obtained only on a pretherapy, posttherapy, and follow-up basis. Occasionally, however, a midtherapy assessment is carried out. Thus possible fluctuations, including upward and downward trends and curvilinear relationships, occurring throughout the course of therapy are omitted from the analysis. However, whether espousing a behavioral, client-centered, existential, or psychoanalytic position, the experienced clinician is undoubtedly cognizant that changes unfortunately *do not* follow a smooth linear function from the beginning of treatment to its ultimate conclusion.

Practical implications and limitations

There are a number of important practical implications and limitations in applying repeated measurement techniques when conducting experimental single-case research (see chapter 2 for general discussion). First of all, the operations involved in obtaining such measurements (whether they be motoric, physiological, or attitudinal) must be clearly specified, observable, public, and replicable in all respects. When measurement techniques require the use of human observers, independent reliability checks must be established (see chapter 4 for specific details). Secondly, measurements taken repeatedly, especially over extended periods of time, must be done under exacting and totally standardized conditions with respect to measurement devices used, personnel involved, time or times of day measurements are recorded, instructions to the subject, and specific environmental conditions (e.g., location) where the measurement sessions occur.

Deviations from any of the aforementioned conditions may well lead to *spurious* effects in the data and might result in erroneous conclusions. This is of particular import at the point where the prevailing condition is experimentally altered (e.g., change from baseline to reinforcement conditions). In the event that an adventitious change in measurement conditions were to coincide with a modification in experimental procedure, resulting differences in the data could not be scientifically attributed to the experimental manipulation, inasmuch as a correlative change may have taken place. Under these circumstances, the conscientious experimenter would either have to renew efforts or experimentally manipulate and evaluate the change in measurement technique.

The importance of maintaining standard measurement conditions bears some illustration. Elkin, Hersen, Eisler, and Williams (1973) examined the separate and combined effects of feedback, reinforcement, and increased food presentation in a male anorexia nervosa patient. With regard to measurement, two dependent variables—caloric intake and weight—were examined daily. Caloric intake was monitored throughout the 42-day study without the subject's knowledge. Three daily meals (each at a specified time) were served to the subject while he dined alone in his room for a 30-minute period. At the conclusion of each of the three daily meals, unknown to the subject, the caloric value of the food remaining on his tray was subtracted from the standard amount presented. Also, the subject was weighed daily at approximately 2:00 P.M., in the same room, on the same scale, with his back turned toward the dial, and, for the most part, by the same experimenter. In this study, consistency of the experimenter was not considered crucial to maintaining accuracy and freedom from bias in measurement. However, maintaining consistency of the time of day weighed was absolutely essential, particularly in terms of the number of meals (two) consumed until that point.

There are certain instances when a change in the experimenter will seriously affect the subject's responses over time. Indeed, this was empirically evaluated by Agras, Leitenberg, Barlow, and Thomson (1969), in an alternating treatment design (see chapter 8). However, in most single-case research, unless explicitly planned, such change may mar the results obtained. For example, when employing the Behavioral Assertiveness Test (Eisler, Miller, & Hersen, 1973) over time repeatedly as a standard behavioral measure of assertiveness, it is clear that the use of different role models to promote responding might result in unexpected interaction with the experimental condition (e.g., feedback or instructions) being manipulated. Even when using more objective measurement tecniques, such as the mechanical strain gauge for recording penile circumference change (Barlow, Becker, Leitenberg, & Agras, 1970) in sexual deviates, extreme care should be exercised with respect to instructions given and to the role of the examiner (male research assistant) involved in the measurement session (cf. Wincze, 1982; Wincze & Lange, 1981). A substitute for the original male experimenter, particularly in the case of a homosexual pedophile in the early stages of his experimental treatment, could conceivably result in spurious correlated changes in penile circumference data.

There are several other important issues to be considered when using repeated measurement techniques in applied clinical research. For example, frequency of measurements obtained per unit of time should be given more careful attention. The experimenter obviously must ensure that a sufficient number of measurements are recorded so that a representative sample is obtained. On the other hand, the experimenter must exercise caution to avoid taking too many measurements in a given period of time, as fatigue on the part of the subject may result. This is of paramount importance when taking measurements that require an active response on the subject's part (e.g., number of erections to sexual stimuli over a specific time period, or repeated modeling of responses during the course of a session in assertive training).

A unique problem related to measurement traditionally faced by investigators working in institutional settings (state hospitals, training schools for the retarded, etc.) involved the major environmental changes that take place at night and on weekends. The astute observer who has worked in these settings is quite familiar with the distinction that is made between the "day" and "night" hospital and the "work week" and the "weekend" hospital. Unless the investigator is in the favored position to exert considerable control over the environment (as were Ayllon and Azrin, 1968, in their studies on token economy), careful attention should be paid to such differences. One possible solution would be to restrict the taking of measurements across similar conditions (e.g., measurements taken only during the day). A second solution would involve plotting separate data for day and night measurements.

A totally different measurement problem is faced by the experimenter who is intent on using self-report data on a repetitive basis (Herson, 1978). When using this type of assessment tecnique, the possibility always exists, even in clinical subjects, that the subject's natural responsivity will not be tapped, but that data in conformity to "experimental demand" (Orne, 1962) are being recorded. The use of alternate forms and the correlation of self-report (attitudinal) measures with motoric and physiological indexes of behavior are some of the methods to ensure validity of responses. This is of particular utility when measures obtained from the different response systems correlate both highly and positively. Discrepancies in verbal and motoric indexes of behavior have been a subject of considerable speculation and study in the behavioral literature, and the reader is referred to the following for a more complete discussion of those issues: Barlow, Mavissakalian, and Schofield (1980); D. C. Cohen (1977); and Hersen (1973).

A final issue, related to repeated measurement, involves the problem of extreme daily variability of a target behavior under study. For example, repetitive time sampling on a random basis within specified time limits is a most useful technique for a variable subject to extreme fluctuations and responsivity to environmental events (see Hersen, Eisler, Alford, & Agras, 1973; J. G. Williams, Barlow, & Agras, 1972). Similar problems in measurement include the area of cyclic variation, an excellent example being the effect of the female's estrus cycle on behavior. Issues related to cyclic variation in terms of extended measurement sessions will be discussed more specifically in section 3.6 of this chapter.

3.3 CHOOSING A BASELINE

In most experimental single-case designs (the exception is the B-A-B design), the initial period of observation involves the repeated measurement of the natural frequency of occurrence of the target behaviors under study. This initial period is defined as the baseline, and it is most frequently designated as the A-phase of study (Barlow, Blanchard, Hayes, & Epstein, 1977; Barlow & Hersen, 1973; Hersen, 1982; Risley & Wolf, 1972; Van Hasselt & Hersen, 1981). It should be noted that this phase was earlier labeled $O_1O_2O_3O_4$ by Campbell and Stanley (1966) in their analysis of quasi-experimental designs for research (time series analysis).

The primary purpose of baseline measurement is to have a standard by which the subsequent efficacy of an experimental intervention can be evaluated. In addition, Risley and Wolf (1972) pointed out that, from a statistical framework, the baseline period functions as a predictor for the level of the target behavior attained in the future. A number of statistical techniques for analyzing time series data have appeared in the literature (Edgington, 1982; Wallace & Elder, 1980); the use of these methods will be discussed in chapter 9.

Baseline stability

When selecting a baseline, its stability and range of variability must be carefully examined. McNamara and MacDonough (1972) have raised an issue that is continuously faced by all of those involved in applied clinical research. They specifically posed the following question: "How long is long enough for a baseline?" (p. 364). Unfortunately, there is no simple response or formula that can be applied to this question, but a number of suggestions have been made. Baer, Wolf, and Risley (1968) recommended that baseline measurement be continued over time "until its stability is clear" (p. 94). McNamara and MacDonough concurred with Wolf and Risley's (1971) recommendation that repeated measurement be applied until a stable pattern emerges. However, there are some practical and ethical limitations to extending initial measurement beyond certain limits. The first involved a problem of logistics.

For the experimenter working in an institutional setting (unless in an extended-care facility), the subject under study will have to be discharged within a designated period of time, whether upon self-demand, familial pressure, or exhaustion of insurance company compensation. Secondly, even in a facility giving extended care to its patients, there is an obvious ethical question as to how long the applied clinical researcher can withhold a treatment application. This assumes even greater magnitude when the target behavior under study results in serious discomfort either to the subject or to others in the environment (see J. M. Johnston, 1972, p. 1036). Finally, although McNamara and MacDonough (1972) argued that "The use of an extended baseline is a most easily implemented procedure which may help to identify regularities in the behavior under study" (p. 361), unexpected effects on behavior may be found as a result of extended measurement through self-recording procedures (Hollon & Bemis, 1981). Such effects have been found when subjects were asked to record their behaviors under repeated measurement conditions. For example, McFall (1970) found that when he asked smokers to monitor their rate of smoking, increases in their actual smoking behavior occurred. By contrast, smokers asked to monitor rate of resistance to smoking did not show parallel changes in their behavior. The problem of self-recorded and self-reported data will be discussed in more detail in chapter 4.

In the context of basic animal research, where the behavioral history of the organism can be determined and controlled, Sidman (1960) has recommended that, for stability, rates of behavior should be within a 5 percent range of variability. Indeed, the "basic science" research is in a position to create baseline data through a variety of interval and ratio scheduling effects. However, even in animal resarch, where scheduling effects are programmed to ensure stability of baseline conditions, there are instances where unexpected variations take place as a consequence of extrinsic variables. When such variability is presumed to be *extrinsic* rather than *intrinsic*, Sidman (1960) has encouraged the researcher to first examine the source of variability through the method of experimental analysis. Then extrinsic sources of variation can be systematically eliminated and controlled.

Sidman acknowledged, however, that the applied clinical researcher, by virtue of his or her subject matter, when control over the behavioral history is nearly impossible, is at a distinct disadvantage. He noted that "The behavioral engineer must continuously take variability as he finds it, and deal with it as an unavoidable fact of life" (Sidman, 1960, p. 192). He also acknowledged that "The behavioral engineeer seldom has the facilities or the time that would be required to eliminate variability he encounters in a given problem" (p. 193). When variability in baseline measurements is extensive in applied clinical research, it might be useful to apply statistical techniques for purposes of comparing one phase to the next. This would certainly appear to be the case when such variability exceeds a 50 percent level. The use of statistics under these circumstances would then meet the kind of criticism that has been leveled at the applied clinical researcher who uses single-case methodology. For example, Bandura (1969) argued that there is no difficulty in interpreting performance changes when differences between phases are large (e.g., the absence of overlapping distributions) and when such differences can be replicated across subjects (see chapter 10). However, he underscored the difficulties in reaching valid conclusions when there is "considerable variability during baseline conditions" (p. 243).

Examples of baselines

With the exception of a brief discussion in Hersen (1982) and in Barlow and Hersen's (1973) paper, which was primarily directed toward a psychiatric readership, the different varieties of baselines commonly encountered in applied clinical research have neither been examined nor presented in logical sequence in the experimental literature. Thus the primary function of this section is to provide and familiarize the interested applied researcher with examples of baseline patterns. For the sake of convenience, hypothetical examples, based on actual patterns reported in the literature, will be illustrated and described. Methods for dealing with each pattern will be outlined, and an attempt to formulate some specific rules (à la cookbook style) will be undertaken.

The issue concerning the ultimate length of the baseline measurement phase was previously discussed in some detail. However, it should be pointed out here that "A minimum of three separate observation points, plotted on the graph, during this baseline phase are required to establish a trend in the data" (Barlow & Hersen, 1973, p. 320). Thus three successively increasing or decreasing points would constitute establishment of either an upward or downward trend in the data. Obviously, in two sets of data in which the same trend is exhibited, differences in the slope of the line will indicate the extent or power of the trend. By contrast, a pattern in which only minor variation is seen would indicate the recording of a stable baseline pattern. An example of such a stable baseline pattern is depicted in Figure 3-1. Mean number of facial tics averaged over three daily 15-minute videotaped sessions are presented for a 6-day period. Visual inspection of these data reveal no apparent upward or downward trend. Indeed, data points are essentially parallel to the abscissa, while variability remains at a minimum. This kind of baseline pattern, which shows a constant rate of behavior, represents the most desirable trend, as it permits an unequivocal departure for analyzing the subsequent efficacy of a treatment intervention. Thus the beneficial or detrimental effects of the following intervention should be clear. In addition, should there be an absence of effects following introduction of a treatment, it will also be apparent. Absence of such effects, then, would graphically appear as a

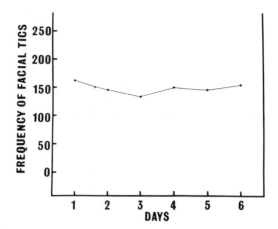

FIGURE 3-1. The stable baseline. Hypothetical data for mean number of facial tics averaged over three daily 15-minute videotaped sessions.

continuation of the steady trend first established during the baseline measurement phase.

A second type of baseline trend that frequently is encountered in applied clinical research is such that the subject's condition under study appears to be worsening (known as the *deteriorating baseline*—Barlow & Hersen, 1973). Once again, using our hypothetical data on facial tics, an example of this kind of baseline trend is presented in Figure 3-2. Examination of this figure shows a steadily increasing linear function, with the number of tics observed augmenting over days. The deteriorating baseline is an acceptable pattern inasmuch as the subsequent application of a successful treatment intervention should lead to a reversed trend in the data (i.e., a decreasing linear function over days). However, should the treatment be ineffective, no change in the slope of the curve would be noted. If, on the other hand, the treatment application leads to further deterioration (i.e., if the treatment is actually detrimental to the patient—see Bergin, 1966), it would be most difficult to assess its effects using the deteriorating baseline. In other words, a differential analysis as to whether a trend in the data was simply a continuation of the baseline pattern or whether application of a detrimental treatment specifically led to its continuation could not be made. Only if there appeared to be a pronounced change in the slope of the curve following introduction of a detrimental treatment could some kind of valid conclusion be reached on the basis of visual inspection. Even then, the withdrawal and reintroduction of the treatment would be required to establish its controlling effects. But from both clinical and ethical considerations, this procedure would be clearly unwarranted.

A baseline pattern that provides difficulty for the applied clinical researcher

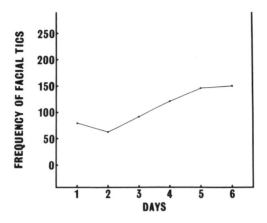

FIGURE 3-2. The increasing baseline (target behavior deteriorating). Hypothetical data for mean number of facial tics averaged over three daily 15-minute videotaped sessions.

is one that reflects steady improvement in the subject's condition during the course of initial observation. An example of this kind of pattern appears in Figure 3-3. Inspection of this figure shows a linear decrease in tic frequency over a 6-day period. The major problem posed by this pattern, from a research standpoint, is that application of a treatment strategy while improvement is already taking place will not allow for an adequate assessment of the intervention. Secondly, should improvement be maintained following initiation of the treatment intervention, the experimenter would be unable to attribute such continued improvement to the treatment unless a marked change in the slope of the curve were to occur. Moreover, removal of the treatment and its subsequent reinstatement would be required to show any controlling effects.

An alternative (and possibly a more desirable) strategy involves the continuation of baseline measurement with the expectation that a plateau will be reached. At that point, a steady pattern will emerge and the effects of treatment can then be easily evaluated. It is also possible that improvement seen during baseline assessment is merely a function of some extrinsic variable (Sidman, 1960) of which the experimenter is currently unaware. Following Sidman's recommendations, it then behooves the methodical experimenter, assuming that time limitations and clinical and ethical considerations permit, to evaluate empirically, through experimental analysis, the possible source (e.g., "placebo" effects) of covariation. The results of this kind of analysis could indeed lead to some interesting hunches, which then might be subjected to further verification through the experimental analysis method (see chapter 2, section 2.3).

The extremely variable baseline presents yet another problem for the

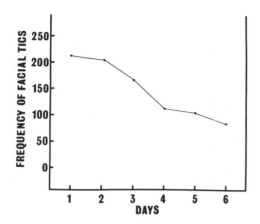

FIGURE 3-3. The decreasing baseline (target behavior improving). Hypothetical data for mean number of facial tics averaged over three daily 15-minute videotaped sessions.

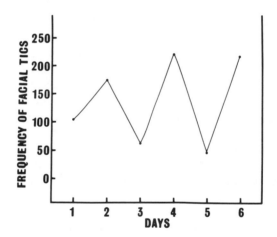

FIGURE 3-4. The variable baseline. Hypothetical data for mean number of facial tics averaged over three 15-minute videotaped sessions.

clinical researcher. Unfortunately, this kind of baseline pattern is frequently obtained during the course of applied clinical research, and various strategies for dealing with it are required. An example of the variable baseline is presented in Figure 3-4. An examination of these data indicate a tic frequency of about 24 to 255 tics per day, with no discernible upward or downward trend clearly in evidence. However, a distinct pattern of alternating low and high trends is present. One possibility (previously discarded in dealing with extreme initial variability) is to simply extend the baseline observation until some semblance of stability is attained, an example of which appears in Figure 3-5.

A second strategy involves the use of inferential statistics when comparing baseline and treatment phases, particularly where there is considerable overlap between succeeding distributions. However, if overlap is that extensive, the statistical model will be equally ineffective in finding differences, as appropriate probability levels will not be reached. Further details regarding graphic presentation and statistical analyses of data will appear in chapter 9.

A final strategy for dealing with the variable baseline is to assess systematically the sources of variability. However, as pointed out by Sidman (1960), the amount of work and time involved in such an analysis is better suited to the "basic scientist" than the applied clinical researcher. There are times when the clinical researcher will have to learn to live with such variability or to select measures that fluctuate to a lesser degree.

Another possible baseline pattern is one in which there is an initial period of deterioration, which is then followed by a trend toward improvement (see Figure 3-6). This type of baseline (increasing-decreasing) poses a number of problems for the experimenter. *First*, when time and conditions permit, an empirical examination of the covariants leading to reversed trends would be of heuristic value. *Second*, while the trend toward improvement is continued in the latter half of the baseline period of observation, application of a treatment will lead to the same difficulties in interpretation that are present in the improving baseline, previously discussed. Therefore, the most useful course of action to pursue involves continuation of measurement procedures until a stable and steady pattern emerges.

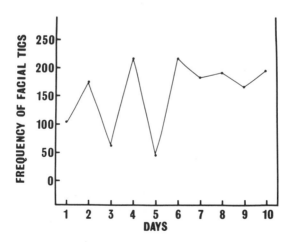

FIGURE 3-5. The variable-stable baseline. Hypothetical data for mean number of facial tics averaged over three daily 15-minute videotaped sessions.

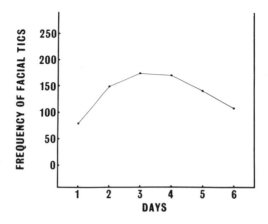

FIGURE 3-6. The increasing-decreasing baseline. Hypothetical data for mean number of facial tics averaged over three daily 15-minute videotaped sessions.

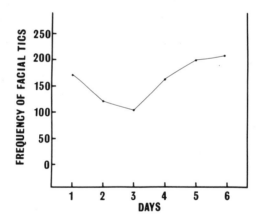

FIGURE 3-7. The decreasing-increasing baseline. Hypothetical data for mean number of facial tics averaged over three daily 15-minute videotaped sessions.

Very similar to the increasing-decreasing pattern is its reciprocal, the decreasing-increasing type of baseline (see Figure 3-7). This kind of baseline pattern often reflects the placebo effects of initially being part of an experiment or being monitored (either self or observed). Although placebo effects are always of interest to the clinical researcher, when he or she is faced with time pressures, the preferred course of action is to continue measurement procedures until a steady pattern in the data is clear. If extended baseline measurement is not feasible, introduction of the treatment, following the worsening of the target behavior under study, is an acceptable procedure, particularly if the controlling effects of the procedure are subsequently demonstrated via its withdrawal and reinstatement.

A final baseline trend, the unstable baseline, also causes difficulty for the applied clinical researcher. A hypothetical example of this type of baseline, obtained under extended measurement conditions, appears in Figure 3-8. Examination of these data reveals not only extreme variability but also the absence of a particular pattern. Therefore, the problems found in the variable baseline are further compounded here by the lack of any trend in the data. This, of course, heightens the difficulty in evaluating these data through the method of experimental analysis. Even the procedure of blocking data usually fails to eliminate all instability on the basis of visual analysis. To date, no completely satisfactory strategy for dealing with the variable baseline has appeared; at best, the kinds of strategies for dealing with the variable baseline are also recommended here.

3.4 CHANGING ONE VARIABLE AT A TIME

A cardinal rule of experimental single-case research is to change *one* variable at a time when proceeding from one phase to the next (Barlow & Hersen, 1973). Barlow and Hersen pointed out that when two variables are simultaneously manipulated, the experimental analysis does not permit conclusions as to which of the two components (or how much of each) contributes to improvements in the target behavior. It should be underscored that the *one*-variable rule holds, regardless of the particular phase (beginning, middle, or end) that is being evaluated. These strictures are most important when

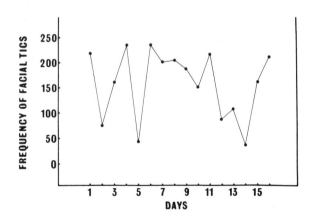

FIGURE 3-8. The unstable baseline. Hypothetical data for mean number of facial tics averaged over three daily 15-minute videotaped sessions.

examining the interactive effects of treatment variables (Barlow & Hersen, 1973; Elkin et al., 1973; Leitenberg, Agras, Thomson, & Wright, 1968). A more complete discussion of interaction designs appears in chapter 6, section 6.5.

Correct and incorrect applications

A frequently committed error during the course of experimental single-case research involves the simultaneous manipulation of two variables so as to assess their presumed interactive effects. A review of the literature suggests that this type of error is often made in the latter phases of experimentation. In order to clarify the issues involved, selected examples of correct and incorrect applications will be presented.

For illustrative purposes, let us asume that baseline measurement in a study consists of the number of social responses (operationally defined) emitted by a chronic schizophrenic during a specific period of observation. Let us further assume that subsequent introduction of a single treatment variable involves application of contingent (token) reinforcement following each social response that is observed on the ward. At this point in our hypothetical example, only one variable (token reinforcement) has been added across the two experimental phases (baseline to the first treatment phase). In accordance with design principles followed in the A-B-A-B design, the third phase would consist of a return to baseline conditions, again changing (removing) only one variable across the second and third phases. Finally, in the fourth phase, token reinforcement would be reinstated (addition of one variable from Phase 3 to 4). Thus, we have a procedurally correct example of the A-B-A-B design (see chapter 5) in which only one variable is altered at a time from phase to phase.

In the following example we will present an inaccurate application of single-case methodology. Using our previously described measurement situation, let us assume that baseline assessment is now followed by a treatment combination comprised of token reinforcement and social reinforcement. At this point, the experiment is labeled A-BC. Phase 3 is a return to baseline conditions (A), while Phase 4 consists of socal reinforcement alone (C). Here we have an example of an A-BC-A-C design, with A = baseline, BC = token and social reinforcement, A = baseline, and C = social reinforcement. In this experiment the researcher is hopeful of teasing the relative effects of token and social reinforcement. However, this a *totally erroneous* assumption on his or her part. From the A-BC-A portion of this experiment, it is feasible only to assess the combined BC effect over baseline (A), assuming that the appropriate trends in the data appear. Evaluation of the individual effects of the two variables (social and token reinforcement) comprising the treatment package is not possible. Moreover, application of the C condition (social reinforcement alone) following the second baseline also does not permit firm conclusions, either with respect to the effects of social reinforcement alone or in contrast to the combined treatment of token and social reinforcement. The experimenter is not in a position to examine the interactive effects of the BC and C phases, as they are not adjacent to one another.

If our experimenter were interested in accurately evaluating the interactive effects of token and social reinforcement, the following extended design would be considered appropriate: A-B-A-B-BC-B-BC. When this experimental strategy is used, the interactive effects of social and token reinforcement can be examined systematically by comparing differences in trends between the adjacent B (token reinforcement) and BC (token and social reinforcement) phases. The subsequent return to B and reintroduction of the combined BC would allow for analysis of the additive and controlling effects of social reinforcement, assuming expected trends in the data occur.

A published example of the correct manipulation of variables across phases appears in Figure 3-9. In this study, Leitenberg et al., (1968) examined the separate and combined effects of feedback and praise on the mean

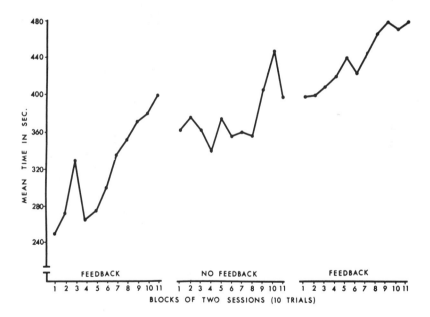

FIGURE 3-9. Time in which a knife was kept exposed by a phobic patient as a function of feedback, feedback *plus* praise, and no feedback or praise conditions. (Figure 2, p. 131, from Leitenberg, H., Agras, W. S., Thomson, L., & Wright, D. E. (1968), Feedback in behavior modification: An experimental analysis in two phobic cases. *Journal of Applied Behavior Analysis*, **1**, 131–137. Copyright 1968 by Society for the Experimental Analysis of Behavior, Inc. Reproduced by permission.)

number of seconds a knife-phobic patient allowed himself to be exposed to a knife. An examination of the seven phases of study reveals the following progression of variables: (1) feedback, (2) feedback and praise, (3) feedback, (4) no feedback and no praise, (5) feedback, (6) feedback and praise, and (7) feedback. A comparison of adjacent phases shows that only one variable was manipulated (added or subtracted) at a time across phases. In a similar design, Elkin et al., (1973) assessed additive and subtractive effects of therapeutic variables in a case of anorexia nervosa. The following progression of variables was used in a six-phase experiment: (1) 3,000 calories—*baseline*, (2) 3,000 calories—*feedback*, (3) 3,000 calories—*feedback and reinforcement*, (4) 4,500 calories—*feedback and reinforcement*, (5) 3,000 calories — *feedback and reinforcement*, (6) 4,500 calories—*feedback and reinforcement*. Again, changes from one phase to the next (italicized) never involved more than the manipulation of a single variable.

Exceptions to the rule

In a number of experimental single-case studies (Barlow et al., 1969; Eisler, Hersen, & Agras, 1973; Pendergrass, 1972; Ramp, Ulrich & Dulaney, 1971) legitimate exceptions to the rule of maintaining a consistent stepwise progression (additive or subtractive) across phases have appeared. In this section the exceptions will be discussed, and examples of published data will be presented and analyzed. For example, Ramp et al. (1971) examined the effects of instructions and delayed time-out in a 9-year-old male elementary school student who proved to be a disciplinary problem. Two target behaviors (intervals out of seat without permission and intervals talking without permission) were selected for study in four separate phases. During baseline, the number of 10-second time intervals in which the subject was out of seat or talking were recorded for 15-minutes sessions. In Phase 2 instructions simply involved the teacher's informing the subject that permission for being out of seat and talking were required (raising his hand). The third phase consisted of a delayed time-out procedure. A red light, mounted on the subject's desk, was illuminated for a 1-3-second period immediately following an instance of out-of-seat or talking behavior. Number of illuminations recorded were cumulated each day, with each classroom violation resulting in a 5-minute detention period in a specially constructed time-out booth while other children participated in gym and recess activities. The results of this study appear in Figure 3-10. Relabeling of the four experimental phases yields an A-B-C-A design. Inspection of the figure shows that the baseline (A) and instructions (B) phases do not differ significantly for either of the two target behaviors under study. Thus although the independent variables differ across these phases, the resulting dependent measures are essentially alike. However,

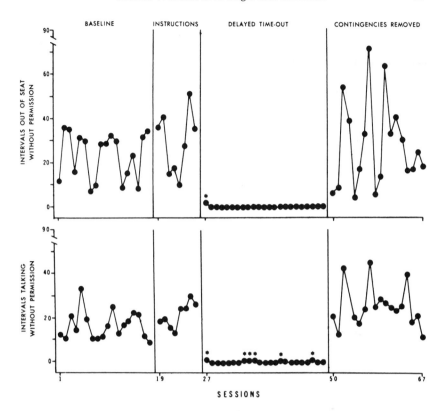

FIGURE 3-10. Each point represents one session and indicates the number of intervals in which the subject was out of his seat (top) or talking without permission (bottom). A total of 90 such intervals was possible within a 15-minute session. Asterisks over points indicate sessions that resulted in time being spent in the booth. (Figure 1, p. 237, from: Ramp, E., Ulrich, R., & Dulaney, S. (1971). Delayed timeout as a procedure for reducing disruptive classroom behavior: A case study. *Journal of Applied Behavior Analysis*, **4**, 235–239. Copyright 1971 by Society for the Experimental Analysis of Behavior, Inc. Reproduced by permission.)

institution of the delayed time-out contingency (C) yielded a marked decrease in classroom violations. Subsequent removal of the time-out contingency in Phase 4 (A) led to a renewed increase in classroom violations.

Since the two initial phases (A and B) yield similar data (instructions did not appear to be effective), equivalence of the baseline and instructions phases are assumed. If one then collapses data across these two phases, an A-C-A design emerges, with some evidence demonstrated for the controlling effects of delayed time-out. In this case the A-C-A design follows the experimental analysis used in the case of the A-B-A design (see chapter 5). However, further confirmation of the controlling effects would require a return to the C condition (delayed time-out). This new design would then be labeled as follows: A = B-C-A-C. It should be noted that without the functional equivalence of the first two phases (A = B) this would essentially be an incorrect experimental procedure. The functional equivalence of different adjacent experimental phases warrants further illustration. An excellent example was provided by Pendergrass (1972), who used an A-B-A = C-B design strategy. In her study, Pendergrass evaluated the effects of time-out and observation of punishment being administered (time-out) to a cosubject in an 8-year-old retarded boy. Two negative high-frequency behaviors were selected as targets for study. They were (1) banging objects on the floor and on others (bang), and (2) the subject's biting of his lips and hand (bite). Only one of the two target behaviors (bang) was directly subjected to treatment effects, but generalization and side effects of treatment on the second behavior (bite) were examined concurrently. Results of the study are presented in Figure 3-11. Time-out following baseline assessment led to a significant decrease in both the punished and unpunished behaviors. A return to baseline conditions in Phase 3 resulted in high levels of both target behaviors. Institution of the "watch" condition (observation of punishment) did not lead to an appreciable decrease, hence the functional equivalence of Phases 3 (A) and 4 (C). In Phase 5 the reinstatement of time-out led to renewed improvement in target behaviors.

In this study the ineffectiveness of the watch condition is functionally

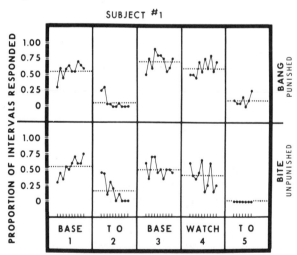

FIGURE 3-11. Proportion of total intervals in which Bang (punished) and Bite (unpunished) responses were recorded for S1 in 47 free-play periods. (Figure 1, p. 88, from: Pendergrass, V. E. (1972). Timeout from positive reinforcement following persistent, high-rate behavior in retardates. *Journal of Applied Behavior Analysis*, **5**, 85–91. Copyright 1972 by Society for Experimental Analysis of Behavior, Inc. Reproduced by permission.)

equivalent to the continuation of the baseline phase (A), despite obvious differences in procedure. With respect to labeling of this design, it is most appropriately designated as follows: A-B-A = C-B (the equal sign between A and C represents their functional equivalence insofar as dependent measures are concerned).

A further exception to the basic rule occurs when the experimenter is interested in the total impact of a treatment package containing two or more components (e.g., instructions, feedback, and reinforcement). In this case, more than one variable is manipulated at a time across adjacent experimental phases. An example of this type of design appeared in a series of analogue studies reported by Eisler, Hersen, and Agras (1973). In one of their studies the combined effects of videotape feedback and focused instructions were examined in an A-BC-A-BC design, with A = baseline and BC = videotape feedback and focused instructions. As is apparent from inspection of Figure 3-12, analysis of these data follows the A-B-A-B design pattern, with the exception that the B phase is represented by a compound treatment variable (BC). However, it should be pointed out that, despite the fact that improvements over baseline appear for both target behaviors (looking and smiling)

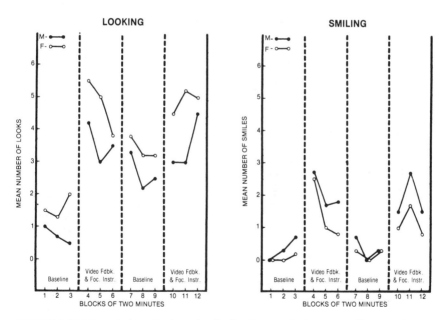

FIGURE 3-12. Mean number of looks and smiles for three couples in 10-second intervals plotted in blocks of 2 minutes for the Videotape Feedback Plus Focused Instructions Design. (Figure 3, p. 556, from: Eisler, R. M., Hersen, M., & Agras, W. S. (1973). Effects of videotape and instructional feedback on nonverbal marital interaction: An analog study. *Behavior Therapy,* **4,** 551–558. Copyright 1973 by Association for the Advancement of Behavior Therapy. Reproduced by permission.)

during videotape feedback and focused instructions conditions, this type of design will obviously allow for no conclusions as to the relative contribution of each treatment component.

A final exception to the one-variable rule appears in a study by Barlow, Leitenberg, and Agras (1969), in which the controlling effects of the noxious scene in covert sensitization were examined in 2 patients (a case of pedophilia and one of homosexuality). In each case an A-BC-B-BC experimental design was used (Barlow & Hersen, 1973). In both cases the four experimental phases were as follows: (1) A = baseline, (2) BC = covert sensitization treatment (verbal description of variant sexual activity and introduction of the nauseous scene), (3) B = verbal description of deviant sexual activity but *no* introduction to the nauseous scene, and (4) BC = covert sensitization (verbal description of sexual activity and introduction of the nauseous scene). For purposes of illustration, data from the pedophilic case appears in Figure 3-13. Examination of the design strategy reveals that covert sensitization treatment (BC) required instigation of both components. Thus initial differences between baseline (A) and acquisition (BC) only suggest efficacy of the total treatment package. When the nauseous scene is removed during extinction (B), the resulting increase in deviant urges and card sort scores similarly *suggests* the controlling effects of the nauseous scene. In reacquisition (BC), where the nauseous scene is reinstated, renewed decreases in the

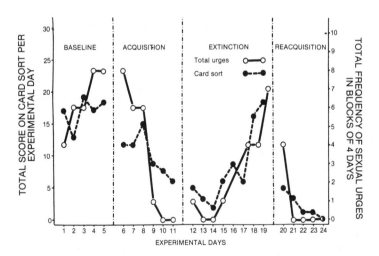

FIGURE 3-13. Total score on card sort per experimental day and total frequency of pedophilic sexual urges in blocks of 4 days surrounding each experimental day. (Lower scores indicate less sexual arousal.). (Figure 1, p. 599, from: Barlow, D. H., Leitenberg, H., & Agras, W. S. (1969). Experimental control of sexual deviation through manipulation of the noxious scene in covert sensitization. *Journal of Abnormal Psychology, 74,* 596–601. Copyright 1969 by the American Psychological Association. Reproduced by permission.)

data confirm its controlling effects. Therefore, despite an initial exception to changing one variable at a time across adjacent phases, a stepwise subtractive and additive progression is maintained in the last two phases, with valid conclusions derived from the ensuing experimental analysis.

Issues in drug evaluation

Issues discussed in the previous section that pertain to changing of variables across adjacent experimental phases and the functional equivalence in data following procedurally different operations are identical when analyzing the effects of drugs on behavior. It is of some interest that experimenters with both a behavior modification bias (e.g., Liberman, Davis, Moon, & Moore, 1973) and those adhering to the psychoanalytic tradition (e.g., Bellak & Chassan, 1964) have used remarkably similar design strategies when investigating drug effects on behavior, either alone or in combination with psychotherapeutic procedures.

Keeping in mind that one-variable rule, the following sequence of experimental phases has appeared in a number of studies: (1) no drug, (2) placebo, (3) active drug, (4) placebo, and (5) active drug. This kind of design, in which a stepwise application of variables appears, permits conclusions with respect to possible placebo effects (no-drug to placebo phase) and those with respect to the controlling influences of active drugs (placebo, active drug, placebo, active drug). Within the experimental analysis framework, Liberman et al. (1973) have labeled this sequence the A-A_1-B-A_1-B design. More specifically, they examined the effects of stelazine on a number of asocial responses emitted by a withdrawn schizophrenic patient. The particular sequence used was as follows: (A) no drug, (A_1) placebo, (B) stelazine, (A_1) placebo, and (B) stelazine. Similarly, within the psychoanalytic framework, Bellak and Chassan (1964) assessed the effects of chlordiazepoxide on variables (primary process, anxiety, confusion, hostility, "sexual flooding," depersonalization, ability to communicate) rated by a therapist during the course of 10 weekly interviews. A double-blind procedure was used in which neither the patient nor the therapist was informed about changes in placebo and active medication conditions. In this study, an A-A_1-B-A_1-B design was employed with the following sequential pattern: (A) no drug, (A_1) placebo, (B) chlordiazepoxide, (A_1) placebo, and (B) chlordiazepoxide.

Once again, pursuing the one variable rule, Liberman et al., (1973) have shown how the combined effects of drugs and behavioral manipulations can be evaluated. Maintaining a constant level of medication (600 mg of chlorpromazine per day), the controlling effects of time-out on delusional behavior (operationally defined) were examined as follows: (1) baseline plus 600 mg of clorpromazine, (2) time-out plus 600 mg of chlorpromazine, and (3) removal of time-out plus 600 mg of chlorpromazine. In this study (AB-

CB-AB) the only variable manipulated across phases was the time contingency.

There are several other important issues related to the investigation of drug effects in single-case experimental designs that merit careful analysis. They include the double-blind evaluation of results, long-term carryover effects of phenothiazines, and length of phases. These will be discussed in some detail in section 3.6 of this chapter and in chapter 7.

3.5. REVERSAL AND WITHDRAWAL

In their survey of the methodological aspects of applied behavior analysis, Baer et al. (1968) stated that there are two types of experimental designs that can be used to show the controlling effects of treatment variables in individuals. These two basic types are commonly referred to as the *reversal* and *multiple-baseline* design strategies. In this section we will concern ourselves only with the reversal design. The prototypic A-B-A design and all of its numerous extensions and permutations (see chapter 5 for details) are usually placed in this category (Barlow et al., 1977; Barlow & Hersen, 1973; Hersen, 1982; Kazdin, 1982b; Van Hasselt & Hersen, 1981).

When speaking of a reversal, one typically refers to the removal (withdrawal) of the treatment variable that is applied after baseline measurement has been concluded. In practice, the reversal involves a withdrawal of the B phase (in the A-B-A design) after behavioral change has been successfully demonstrated. If the treatment (B phase) indeed exerts control over the targeted behavior under study, a decreased or increased trend (depending on which direction indicates deterioration) in the data should follow its removal.

In describing their experimental efforts when using A-B-A designs, applied clinical researchers frequently have referred to both their procedures and resulting data as reversals. This, then, represents a terminological confusion between the independent variable and the dependent variable. However, from either a semantic, logical, or scientific standpoint, it is untenable that both a cause and an effect should be given an identical label. A careful analysis reveals that a reversal involves a *specific technical operation*, and that its result (changes in the target behavior[s]) is simply examined in terms of rates of the data (increased, decreased, or no change) in relation to patterns seen in the previous experimental phase. To summarize, a reversal is an active procedure; the obtained data may or may not reflect a particular trend.

The reversal design

A still finer distinction regarding reversals was made by Leitenberg (1973) in his examination of experimental single-case design strategies. He contended that the reversal design (e.g., A-B-A-B design) is inappropriately labeled, and that the term *withdrawal* (i.e., withdrawal of treatment in the second A phase) is a more accurate description of the actual technical operation. Indeed, a distinction between a withdrawal and a reversal was made, and Leitenberg showed how the latter refers to a specific kind of experimental strategy. It should be underscored that, although ". . . this distinction . . . is typically not made in the behavior modification literature" (Leitenberg, 1973), the point is well taken and should be considered by applied clinical researchers.

To illustrate and clarify this distinction, an excellent example of the reversal design, selected from the child behavior modification literature, will be presented. Allen, Hart, Buell, Harris, and Wolf (1964) were concerned with the contingent effects of reinforcement on the play behavior of a 4½-year-old girl who evidenced social withdrawal with peers in a preschool nursery setting. Two target behaviors were selected for study: (1) percentage of interaction with adults, and (2) percentage of interaction with children. Observations were recorded daily during 2-hour morning sessions. As can be seen in Figure 3-14, baseline data show that about 15 percent of the child's time was spent interacting with children, whereas approximately 45 percent of the time was spent in interactions with adults. The remaining 40 percent involved "isolate" play. Inasmuch as the authors hypothesized that teacher attention fostered interactions with adults, in the second phase of experimentation an effort was made to demonstrate that the same teacher attention, when presented contingently in the form of praise following the child's interaction with other children, would lead to an increase in such interactions. Conversely, isolate play and approaches to adults were ignored. Inspection of Figure 3-14 reveals that contingent reinforcement (praise) increased the percentage of interaction with children and led to a concomitant decrease in interactions with adults. In the third phase a "true" *reversal* of contingencies was put into effect. That is to say, contingent reinforcement (praise) was now administered when the child approached adults, but interaction with other children was ignored. Examination of Phase 3 data reflects the reversal in contingencies. Percentage of time spent with children decreased substantially while percentage of time spent with adults showed a marked increase. Phase 2 contingencies were then reinstated in Phase 4, and the remaining points on the graph are concerned with follow-up measures.

Reversal and withdrawal designs compared

A major difference between the reversal and withdrawal designs is that in the third phase of the reversal design, following instigation of the therapeutic procedure, the same procedure is now applied to an alternative but incompatible behavior. By contrast, in the withdrawal design, the A phase following

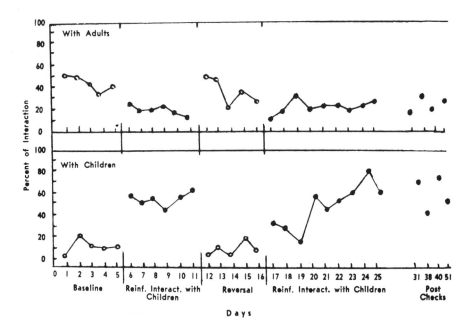

FIGURE 3-14. Daily percentages of time spent in social interaction with adults and with children during approximately 2 hours of each morning session. (Figure 2, p. 515, from: Allen K. E., Hart, B. M., Buell, J. S., Harris, F. R., & Wolf, M. M. (1964). Effects of social reinforcement on isolate behavior of a nursery school child. *Child Development*, **35** 511–518. Copyright 1964. Reproduced by permission of The Society for Research in Child Development, Inc.)

introduction of the treatment variable (e.g., token reinforcement) simply involves its removal and a return to baseline conditions. Leitenberg (1973) argued that "Actually, the reversal design although it can be quite dramatic is somewhat more cumbersome . . ." (pp. 90–91) than the more frequently employed withdrawal design. Moreover, the withdrawal design is much better suited for investigations that do not emanate from the operant (reinforcement) framework (e.g., the investigation of drugs and examination of nonbehavioral therapies).

Withdrawal of treatment

The specific point at which the experimenter removes the treatment variable (second A phase in the A-B-A design) in the withdrawal design is multidetermined. Among the factors to be considered are time limitations imposed by the treatment setting, staff cooperation when working in institutions (J. M. Johnston, 1972), and ethical considerations when removal of treatment can possibly lead to some harm to the subject (e.g., head banging in a retardate) or others in the environment (e.g., physical assaults toward

wardmates in disturbed inpatients). Assuming that these important environmental considerations can be dealt with adequately and judiciously, a variety
of parametric issues must be taken into account before instituting withdrawal
of the treatment variable. One of these issues involved the overall length of
adjacent treatment phases; this will be examined in section 3.6 of this chapter.

In this section we will consider the implementation of treatment withdrawal
in relation to data trends appearing in the first two phases (A and B) of study.
We will illustrate both correct and incorrect applications using hypothetical
data. Let us consider an example in which A refers to baseline measurement
of the frequency of social responses emitted by a withdrawn schizophrenic.
The subsequent treatment phase (B) involves contingent reinforcement in the
form of praise, while the third phase (A) represents the withdrawal of
treatment and a return to original baseline conditions. For purposes of
illustration, we will assume stability of "initial" baseline conditions for each
of the following examples.

In our first example (see Figure 3-15) data during contingent reinforcement
show a clear upward trend. Therefore, institution of withdrawal procedures
at the conclusion of this phase will allow for analysis of the controlling effects
of reinforcement, particularly if the return to baseline results in a downward
trend in the data. Equally acceptable is a baseline pattern (second A phase) in
which there is an immediate loss of treatment effectiveness, which is then
maintained at a low-level stable rate (this pattern is the same as the initial
baseline phase).

In our second example (see Figure 3-16) data during contingent reinforcement show the immediate effects of treatment and are maintained throughout
the phase. After these initial effects, there is no evidence of an increased rate
of responding. However, the withdrawal of contingent reinforcement at the
conclusion of the phase *does* permit analysis of its controlling effects. Data in
the second baseline show no overlap with contingent reinforcement, as there
is a return to the stable but low rate of responding seen in the first baseline (as

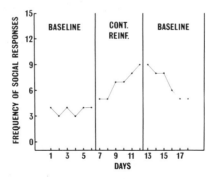

FIGURE 3-15. Increasing treatment phase followed by decreasing baseline. Hypothetical data
for frequency of social responses in a schizophrenic patient per 2-hour period of observation.

in Figure 3-16). Equally acceptable would be a downward trend in the data as depicted in the second baseline in Figure 3-14.

In our third example of a correct withdrawal procedure, examination of Figure 3-17 indicates that contingent reinforcement resulted in an immediate increase in rate, followed by a linear decrease, and then a renewed increase in rate which then stabilized. Although it would be advisable to analyze contributing factors to the decrease and subsequent increase (Sidman, 1960), institution of the withdrawal procedure at the conclusion of the contingent reinforcement phase allows for an analysis of its controlling effects, particularly as a decreased rate was observed in the second baseline.

An example of the incorrect application of treatment withdrawal appears

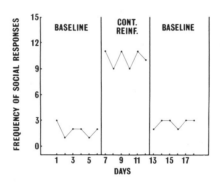

FIGURE 3-16. High-level treatment phase followed by low-level baseline. Hypothetical data for frequency of social responses in a schizophrenic patient per 2-hour period of observation.

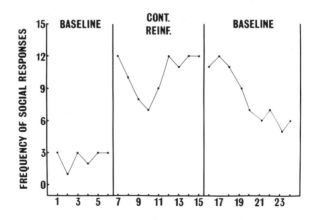

FIGURE 3-17. Decreasing-increasing-stable treatment phase followed by decreasing baseline. Hypothetical data for frequency of social responses in a schizophrenic patient per 2-hour period of observation.

in Figure 3-18. Inspection of the figure reveals that after a stable pattern is obtained in baseline, introduction of contingent reinforcement leads to an immediate and dramatic improvement, which is then followed by a marked decreasing linear function. This trend is in evidence despite the fact that the last data point in contingent reinforcement is clearly above the highest point achieved in baseline. Removal of treatment and a return to baseline conditions on Day 13 similarly result in a decreasing trend in the data. Therefore, no conclusions as to the controlling effects of contingent reinforcement are possible, as it is not clear whether the decreasing trend in the second baseline is a function of the treatment's withdrawal or mere continuation of the trend begun during treatment. Even if withdrawal of treatment were to lead to the stable low-level pattern seen in the first baseline period, the same problems in interpretation would be posed.

When the aforementioned trend appears during the course of experimental treatment, it is recommended that the phase be continued until a more consistent pattern emerges. However, if this strategy is pursued, the equivalent length of adjacent phases is altered (see section 3.6). A second strategy, although admittedly somewhat weak, is to reintroduce treatment in Phase 4 (thus, we have an A-B-A-B design), with the expectation that a reversed trend in the data will reflect improvement. There would then be limited evidence for the treatment's controlling effects.

A similar problem ensues when treatment is withdrawn in the example that appears in Figure 3-19. In spite of an initial upward trend in the data when contingent reinforcement is first introduced (B), the decreasing trend in the latter half of the phase, which is then followed by a similar decline during the second baseline (A), prevents an analysis of the treatment's controlling ef-

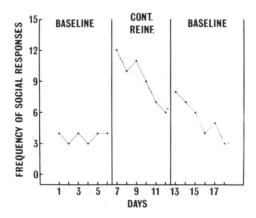

FIGURE 3-18. High-level decreasing treatment phase followed by decreasing baseline. Hypothetical data for frequency of social responses in a schizophrenic patient per 2-hour period of observation.

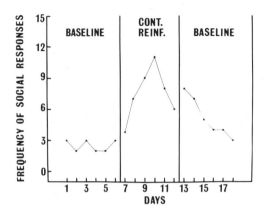

FIGURE 3-19. Increasing-decreasing treatment phase followed by decreasing behavior. Hypothetical data for frequency of social responses in a schizophrenic patient per 2-hour period of observation.

fects. Therefore, the same recommendations made in the case of Figure 3-18 apply here.

Limitations and problems

As mentioned earlier, the applied clinical researcher faces some unique problems when intent on pursuing experimental analysis by withdrawing a particular treatment technique. These problems are heightened in settings where one exerts relatively little control, either with respect to staff cooperation or in terms of other important environmental contingencies (e.g., when dealing with individual problems in the classroom situation, responses of other children throughout the varying stages of experimentation may spuriously affect the results). Although these concerns have been articulated elsewhere in the behavioral literature (Baer et al., 1968; Bijou, Peterson, Harris, Allen, & Johnston, 1969; Hersen, 1982; Kazdin & Bootzin, 1972; Leitenberg, 1973), a brief summary of the issues at stake might be useful at this point.

A frequent criticism leveled at researchers using single-case methodology is that removal of the treatment will lead to the subject's irreversible deterioration (at least in terms of the behavior under study). However, as Leitenberg (1973) pointed out, this is a weak argument with no supporting evidence to be found in the experimental literature. If the technique shows initial beneficial effects and it exerts control over the targeted behavior being examined, then, when reinstated, its controlling effects will be established. To the contrary, Krasner (1971b) reported that recovery of initially low levels of baseline performance often fails to occur in extended applications of the A-B-A design where multiple withdrawals and reinstatements of the treatment technique are instituted (e.g., A-B-A-B-A-B-A-B). Indeed, the possible carryover effects across phases and concomitant environmental events leading to improved conditions contribute to the researcher's difficulties in carrying out scientifically acceptable studies.

A less subtle problem encountered is one of staff resistance. Usually, the researcher working in an applied setting (be it at school, state institution for the retarded, or psychiatric hospital) is consulting with house staff on difficult problems. In efforts to remediate the problem, the experimenter encourages staff to apply treatment strategies that are likely to achieve beneficial results. When staff members are subsequently asked to temporarily withdraw treatment procedures, some may openly rebel. "What teacher, seeing Johnny for the first time quietly seated for most of the day, would like to experience another week or two of bedlam just to satisfy the perverted whim of a psychologist?" (J. M. Johnston, 1972, p. 1035). In other cases the staff member or parent (when establishing parental retraining programs) may be unable to revert to his or her original manner of functioning (i.e., his or her way of previously responding to certain classes of behavior). Indeed, this happened in a study reported by Hawkins, Peterson, Schweid, and Bijou (1966). Leitenberg (1973) argued that "In such cases, where the therapeutic procedure cannot be introduced and withdrawn at will, sequential ABA designs are obviated" (p. 98). Under these circumstances, the use of alternative experimental strategies such as multiple baseline (Hersen, 1982) or alternating-treatment designs (Barlow & Hayes, 1979) obviously are better suited (see chapters 7 and 8).

To summarize, the researcher using the withdrawal design must ensure that (1) there is full staff or parental cooperation on an a priori basis; (2) the withdrawal of treatment will lead to minimal environmental disruptions (i.e., no injury to subject or others in the environment will result) (see R. F. Peterson & Peterson, 1968); (3) the withdrawal period will be relatively brief; (4) outside environmental influences will be minimized throughout baseline, treatment, and withdrawal phases; and (5) final reinstatement of treatment to its logical conclusion will be accomplished as soon as it is technically feasible.

3.6. LENGTH OF PHASES

Although there has been some intermittent discussion in the literature with regard to the length of phases when carrying out single-case experimental research (Barlow & Hersen, 1973; Bijou et al., 1969; Chassan, 1967; J. M. Johnston, 1972; Kazdin, 1982b), a complete examination of the problems faced and the decision to be made by the researcher has yet to appear. Therefore, in this section the major issues involved will be considered including individual and relative length of phases, carryover effects and cyclic variations. In addition, these considerations will be examined as they apply to the study of drugs on behavior.

Individual and relative length

When considering the individual length of phases independently of other factors (e.g., time limitations, ethical considerations, relative length of phases), most experimenters would agree that baseline and experimental conditions should be continued until some semblance of stability in the data is apparent. J. M. Johnston (1972) has examined these issues with regard to the study of punishment. He stated that:

> It is necessary that each phase be sufficiently long to demonstrate stability (lack of trend and a constant range of variability) and to dispel any doubts of the reader that the data shown are sensitive to and representative of what was happening under the described condition (p. 1036).

He notes further:

> That if there is indication of an increasing or decreasing trend in the data or widely variable rates from day to day (even with no trend) then the present condition should be maintained until the instability disappears or is shown to be representative of the current conditions (p. 1036).

The aforementioned recommendations reflect the ideal and apply best when each experimental phase is considered individually and independently of adjacent phases. If one were to fully carry out these recommendations, the possibility exists that widely disparate lengths in phases would result. The strategic difficulties inherent in unequal phases has been noted elsewhere by Barlow and Hersen (1973). Indeed, they cited the advantages of obtaining a relatively equal number of data points for each phase.

Let us illustrate the importance of their suggestions by considering the following hypothetical example, in which the effects of time-out on frequency of hitting other children during a free-play situation are assessed in a 3-year-old child. Examination of Figure 3-20 shows a stable baseline pattern, with a high frequency of hitting behavior exhibited. Data for Days 5–7, when treatment (time-out) is first instigated, show no effects, but on Day 8 a slight decline in frequency appears. If the experimenter were to terminate treatment at this point, it is obvious that few statements about its efficacy could be made. Thus the treatment is continued for an additional 4 days (9–12), and an appreciable decrease in hitting is obtained. However, by extending (doubling) the length of the treatment phase, the experimenter cannot be certain whether additional treatment in itself leads to changes, whether some correlated

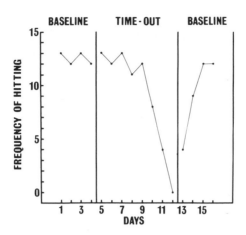

FIGURE 3-20. Extension of the treatment phase in an attempt to show its effects. Hypothetical data in which the effects of time-out on daily frequency of hitting other children (based on a 2-hour free-play situation) in a 3-year-old male child are examined.

variable (e.g., increased teacher attention to incompatible positive behaviors emitted by the child) results in changes, or whether the mere passage of time (maturational changes) accounts for the decelerated trend. Of course, the withdrawal of treatment on Days 13–16 (second baseline) leads to a marked incrased in hitting behavior, thus suggesting the controlling effects of the time-out contingency. However, the careful investigator would reinstate time-out procedures, to dispel any doubts as to its possible controlling effects over the target behavior of hitting. Additionally, once the treatment (time-out) phase has been extended to 8 days, it would be appropriate to maintain equivalence in subsequent baseline and treatment phases by also collecting approximately 8 days of data on each condition. Then, questions as to whether treatment effects are due to maturational or other controllable influences will be satisfactorily answered.

As previously noted, the actual length of phases (as opposed to the ideal length) is often determined by factors aside from design considerations. However, where possible, the relative equivalence of phase lengths is desirable. If exceptions are to be made, either the initial baseline phase should be lengthened to achieve stability in measurement, or the last phase (e.g., second B phase in the A-B-A-B design) should be extended to insure permanence of the treatment effects. In fact, with respect to this latter point, investigators should make an effort to follow their experimental treatments with a full clinical application of the most successful techniques available.

An example of the ideal length of alternating behavior and treatment phases appears in Miller's (1973) analysis of the use of Retention Control

Training (RCT) in a "secondary enuretic" child (see Figure 3-21). Two target behaviors, number of enuretic episodes and mean frequency of daily urination, were selected for study in an A-B-A-B experimental design. During baseline, the child recorded the natural frequency of target behaviors and received counseling from the experimenter on general issues relating to home and school. Following baseline, the first week of RCT involved teaching the child to postpone urination for a 10-minute period after experiencing each urge. Delay of urination was increased to 20 and 30 minutes in the next 2 weeks. During Weeks 7–9 RCT was withdrawn, but was reinstated in Weeks 10–14.

Examination of Figure 3-21 indicates that each of the first three phases consisted of 3 weeks, with data reflecting the controlling effects of RCT on both target behaviors. Reinstatement of RCT in the final phase led to renewed control, and the treatment was extended to 5 weeks to ensure maintenance of gains.

It might be noted that phase and data patterns do not often follow the ideal sequence depicted in the Miller (1973) study. And, as a consequence, experimenters frequently are required to make accommodations for ethical, proce-

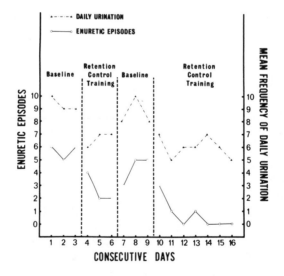

FIGURE 3-21. Number of enuretic episodes per week and mean number of daily urinations per week for Subject 1. (Figure 1, p. 291, from: Miller, P. M. (1973). An experimental analysis of retention control training in the treatment of nocturnal enuresis in two institutionalized adolescents. *Behavior Therapy,* **4**, 288–294. Copyright 1973 by Association for the Advancement of Behavior Therapy. Reproduced by permission.)

dural, or parametric reasons. Moreover, when working in an unexplored area where the issues are of social significance, deviations from some of our proposed rules during the earlier stages of investigation are acceptable. However, once technical procedures and major parametric concerns have been dealt with satisfactorily, a more vigorous pursuit of scientific rigor would be expected. In short, as in any scientific endeavor, as knowledge accrues, the level of experimental sophistication should reflect its concurrent growth.

Carryover effects

A parametric issue that is very much related to the comparative lengths of adjacent baseline and treatment phases is one of overlapping (carryover) effects. Carryover effects in behavioral (as distinct from drug) studies usually appear in the second baseline phase of the A-B-A-B type design and are characterized by the experimenter's inability to retrieve original levels of baseline responding. Not only is the original baseline rate not recoverable in some cases (e.g., Ault, Peterson, & Bijou, 1968; Hawkins et al., 1966), but on occasion (e.g., Zeilberger, Sampen, & Sloane, 1968) the behavior under study undergoes more rapid modification the second time the treatment variable is introduced.

Presence of carryover effects has been attributed to a variety of factors including changes in instructions across experimental conditions (Kazdin, 1973b), the establishment of new conditioned reinforcers (Bijou et al., 1969), the maintenance of new behavior through naturally occurring environmental contingencies (Krasner, 1971b), and the differences in stimulus conditions across phases (Kazdin & Bootzin, 1972). Carryover effects in behavioral research are an obvious clinical advantage, but pose a problem experimentally, as the controlling effects of procedures are then obfuscated.

Proponents of the group comparison approach (e.g., Bandura, 1969) contend that the presence of carryover effects in single-case research is one of its major shortcomings as an experimental strategy. Both in terms of drug evaluation (Chassan, 1967) and with respect to behavioral research (Bijou et al., 1969), short periods of experimentation (application of the treatment variable) were recommended to counteract these difficulties. Examining the problem from the operant framework, Bijou et al. argued that "In studies involving stimuli with reinforcing properties, relatively short experimental periods are advocated, since long ones might allow enough time for the establishment of new conditioned reinforcers" (p. 202). Carryover effects are also an important consideration in alternating treatment designs but are more easily handled through counterbalancing procedures (see chapter 8).

A major difficulty in carrying out meaningful evaluations of drugs on behavior using single-case methodology involves their carryover effects from one phase to the next. This is most problematic when withdrawing active drug treatment (B phase) and returning to the placebo (A₁ phase) condition in the A-A₁-B-A₁-B design. With respect to such effects, Chassan (1967) pointed out that "This, for instance, is thought likely to be the case in the use of monoaminoxidase inhibitors for the treatment of depression" (p. 204). Similarly, when using phenothiazine derivatives, the experimenter must exercise caution inasmuch as residuals of the drugs have been found to remain in body tissues for extended periods of time (as long as 6 months in some cases) following their discontinuance (Ban, 1969).

However, it is possible to examine the short-term effects of phenothiazines on designated target behaviors (Liberman et al., 1973), but it behooves the experimenter to demonstrate, via blood and urine laboratory studies, that controlling effects of the drug are truly being demonstrated. That is to say, correlations (statistical and graphic data patterns) between behavioral changes and drug levels in body tissues should be demonstrated across experimental phases.

Despite the carryover difficulties encountered with the major tranquilizers and antidepressants, the possibility of conducting *extended studies* in long-term facilities should be explored, assuming that high ethical and experimental standards prevail. In addition, study of the *short-term* efficacy of the minor tranquilizers and amphetamines on selected target behaviors is quite feasible.

Cyclic variations

A most neglected issue in experimental single-case research is that of cyclic variations (see chapter 2, sections 2.2 and 2.3, for a more general discussion of variability). Although the importance of cyclic variations was given attention by Sidman (1960) with respect to basic animal research, and J. M. Johnston & Pennypacker (1981) in a more applied context, the virtual absence of serious consideration of this issue in the applied literature is striking. This issue is of paramount concern when using adult female subjects as their own controls in short-term (one month or less) investigations. Despite the fact that the effects of the estrus cycle on behavior are given some consideration by Chassan (1967), he argued that ". . . a 4-week period (with random phasing) would tend to distribute menstrual weeks evenly between treatments" (p. 204). However, he did recognize that "The identification of such weeks in studies involving such patients would provide an added refinement for the statistical analysis of the data" (p. 204).

Whether one is examining drug effects or behavioral interventions, the implications of cyclic variation for single-case methodology are enormous. Indeed, the psychiatric literature is replete with examples of the deleterious effects (leading to increased incidence of psychopathology) of the premenstrual and menstrual phases of the estrus cycle on a wide variety of target behaviors in pathological and nonpathological populations (e.g., Dalton, 1959, 1960a, 1960b, 1961; G. S. Glass, Heninger, Lansky, & Talan, 1971; Mandell & Mandell, 1967; Rees, 1953).

To illustrate, we will consider the following possibility. Let us assume that alternating placebo and active drug conditions are being evaluated (one week each per phase) on the number of physical complaints issued daily by a young hospitalized female. Let us further assume that the first placebo condition coincides with the premenstrual and early part of the subject's menstrual cycle. Instigation of the active drug would then be confounded with cessation of the subject's menstrual phase. Assuming that resulting data suggest a decrease in somatic complaints, it is *entirely possible* that such change is primarily due to correlated factors (e.g., effects of the different portions of the subject's menstrual cycle). Of course, completion of the last two phases (A and B) of this A-B-A-B design might result in no change in data patterns across phases. However, interpretation of data would be complicated unless the experimenter were aware of the role played by cyclic variation (i.e., the subject's menstrual cycle).

The use of extended measurement phases under these circumstances in addition to direct and systematic replications (see chapter 10) across subjects is absolutely necessary in order to derive meaningful conclusions from the data.

3.7. EVALUATION OF IRREVERSIBLE PROCEDURES

There are certain kinds of procedures (e.g., surgical lesions, therapeutic instructions) that obviously cannot be withdrawn once they have been applied. Thus, in assessment of these procedures in single-case research, the use of reversal and withdrawal designs is generally precluded. The problem of irreversibility of behavior has attracted some attention and is viewed as a major limitation of single-case design by some (e.g., Bandura, 1969). The notion here is that some therapeutic procedures produce results in "learning" that will not reverse when the procedure is withdrawn. Thus, one is unable to isolate that procedure as effective. In response to this, some have advocated withdrawing the procedure early in the treatment phase to effect a reversal. This strategy is based on the hypothesis that behavioral improvements may begin as a result of the therapeutic technique but are maintained at a later point by factors in the environment that the investigators cannot remove (see Kazdin, 1973; Leitenberg, 1973, also see chapter 5). The most extreme cases of irreversibility may involve a study of the effects of surgical lesions on behavior, or psychosurgery. Here the effect is clearly irreversible. This problem is easily solved, however, by turning to a multiple baseline design. In fact, the multiple baseline strategy is ideally suited for studying such variables, in that withdrawals of treatment are not required to show the controlling effects of particular techniques (Baer et al., 1968; Barlow & Hersen, 1973; Hersen, 1982; Kazdin, 1982b). A complete discussion of issues related to the varieties of multiple baseline designs currently being employed by applied researchers appears in chapter 7.

In this section, however, the limited use and evaluation of therapeutic instructions in withdrawal designs will be examined and illustrated. Let us consider the problems involved in "withdrawing" therapeutic instructions. In contrast to a typical reinforcement procedure, which can be introduced, removed, and reintroduced at will, an instructional set, after it has been given, technically cannot be withdrawn. Certainly, it can be stopped (e.g., Eisler, Hersen, & Agras, 1973) or changed (Agras et al., 1969; Barlow, Agras, Leitenberg, Callahan, & Moore, 1972), but it is not possible to remove it in the same sense as one does in the case of reinforcement. Therefore, in light of these issues, when examining the interacting effects of instructions and other therapeutic variables (e.g., social reinforcement), instructions are typically maintained constant across treatment phases while the therapeutic variable is introduced, withdrawn, and reintroduced in sequence (Hersen, Gullick, Matherne, & Harbert, 1972).

Exceptions

There are some exceptions to the above that periodically have appeared in the psychological literature. In two separate studies the short-term effects of instructions (Eisler, Hersen, & Agras, 1973) and the therapeutic value of instructional sets (Barlow et al., 1972) were examined in withdrawal designs. In one of a series of analogue studies, Eisler, Hersen and Agras investigated the effects of focused instructions ("We would like you to pay attention as to how much you are looking at each other") on two nonverbal behaviors (looking and smiling) during the course of 24 minutes of free interaction in three married couples. An A-B-A-B design was used, with A consisting of 6 minutes of interaction videotaped between a husband and wife in a small television studio. The B phase also involved 6 minutes of videotaped interaction, but focused instructions on looking were administered three times at 2-minute intervals over a two-way intercom system by the experimenter from the adjoining control room. During the second A phase, instructions were discontinued, while in the second B they were renewed, thus completing 24 minutes of taped interaction.

Retrospective ratings of looking and smiling for husbands and wives (mean data for the three couples were used, as trends were similar in all cases) appear in Figure 3-22. Looking duration in baseline for both spouses was moderate in frequency. In the next phase, focused instructions resulted in a substantial increase followed by a slightly decreasing trend. When instructions were discontinued in the second baseline, the downward trend was maintained. But reintroduction of instructions in the final phase led to an upward trend in looking. Thus, there was some evidence for the controlling effects of introducing, discontinuing, and reintroducing the instructional set. However, data for a second but "untreated" target behavior—smiling— showed almost no parallel effects.

Barlow et al. (1972) examined the effects of negative and positive instructional sets administered during the course of covert sensitization therapy for homosexual subjects. In a previous study (Barlow, Leitenberg, & Agras, 1969), pairing of the nauseous scene with undesired sexual imagery proved to be the controlling ingredient in covert sensitization. However, as the possibility was raised that therapeutic instructions or positive expectancy of subjects may have contributed to the treatment's overall efficacy, an additional study was conducted (Barlow et al., 1972).

The dependent measure in the study by Barlow and his associates was mean percentage of penile circumference change to selected slides of nude males.

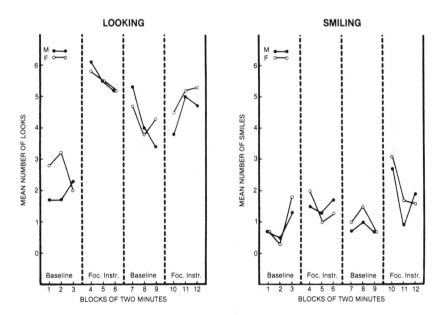

FIGURE 3-22. Mean number of looks and smiles for three couples in 10-second intervals plotted in blocks of 2 minutes for the Focused Instructions Alone Design. (Figure 4, p. 556, from: Eisler, R. M., Hersen, M., & Agras, W. S. (1973). Effects of videotape and instructional feedback on nonverbal marital interactions: An analog study. *Behavior Therapy*, **4**, 551–558. Copyright 1973 by Association for the Advancement of Behavior Therapy. Reproduced by permission.)

Four homosexuals served as subjects in A-BC-A-BD single-case designs. During A (baseline placebo), a positive instructional set was administered, in that subjects were told that descriptions of homosexual scenes along with deep muscle relaxation would lead to improvement. In the BC phase, standard covert sensitization treatment was paired with a negative instructional set (subjects were informed that increased sexual arousal would occur). In the next phase a return to baseline placebo conditions was instituted (A). In the final phase (BD) standard covert sensitization treatment was paired with a positive instructional set (subjects were informed that pairing of the nauseous scene with homosexual imagery, based on a review of their data, would lead to greatest improvement).

Mean data for the four subjects presented in blocks of two sessions appear in Figure 3-23. Baseline data suggest that the positive set failed to effect a decreased trend. In the next phase (BC), a marked improvement was noted as a function of covert sensitization despite the instigation of a negative set. In the third phase (A), some deterioration was apparent although a positive set had been instituted. Finally, in the last phase (BD), covert sensitization coupled with positive expectation of treatment resulted in renewed improvement.

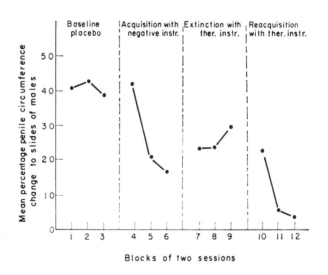

FIGURE 3-23. Mean penile circumference changes to male slides for 4 *Ss*, expressed as a percentage of full erection. In each phase, data from the first, middle, and last pair of sessions are shown. (Figure 1, p. 413, from: Barlow, D. H., Agras, W. S., Leitenberg, H., Callahan, E. J., & Moore, R. C. (1972). The contribution of therapeutic instruction to covert sensitization. *Behaviour Research and Therapy*, **10**, 411–415. Copyright 1972 by Pergamon. Reproduced by permission.)

In summary, data from this study show that covert sensitization treatment is the effective procedure and that therapeutic expectancy is definitely not the primary ingredient leading to success. To the contrary, a positive set paired with a placebo-relaxation condition in baseline did not yield improvement in the target behavior.

Although the design in this study permits conclusions as to the efficacy of positive and negative sets, a more direct method of assessing the problem could have been accomplished in the following design: (1) baseline placebo, (2) acquisition with positive instructions, (3) acquisition with negative instructions, and (4) acquisiton with postive instructions. When labeled alphabetically, it provides an A-BC-BD-BC design. In the event that negative instructions were to exert a negative effect in the BD phase, a reversed trend in the data would appear. On the other hand, should negative instructions have no effect or a negligible effect, then a continued downward linear trend would appear across phases BC, BD and the return to BC.

3.8. ASSESSING RESPONSE MAINTENANCE

In reviewing the theoretical and applied work on single-case strategies, it is clear that most of the attention has been directed to determining the functional relationship between treatment intervention and behavioral change. That is, the emphasis is on response acquisition. (Indeed, this has been the case in behavior therapy in general.) More recently, greater emphasis has been accorded to evaluating and ensuring response maintenance following successful treatment (see Hersen, 1981). Specifically with respect to single-case experimental designs, Rusch and Kazdin (1981) described a methodology for assessing such response maintenance. Techniques outlined are applicable to multiple baseline designs (see chapter 7) but also in some instances to the basic and more complicated withdrawal designs (see chapters 5 and 6).

As noted by Rusch and Kazdin (1981):

> In acquisition studies investigators are interested in demonstrating, unequivocally, that a functional relationship exists between treatment and behavioral change. In maintenance studies, on the other hand, investigators depend on the ability of the subject to discern and respond to changes in the environment when the environment is altered; the latter group relies upon subject's failure to discriminate between those very same stimuli or, possibly, upon the subject's failure to discriminate among functionally similar stimulus [sic] . . . (pp. 131–132)

Rusch and Kazdin referred to three types of response maintenance evaluation strategies: (1) sequential-withdrawal, (2) partial-withdrawal, and (3) partial-sequential withdrawal. In each instance, however, a compound treatment (i.e., one comprised of several elements or strategies) was being evaluated. Let us consider the three response maintenance evaluation strategies in turn.

In *sequential-withdrawal,* one element of treatment is withdrawn subsequent to response acquisition (e.g., reinforcement). In the next phase a second element of the treatment (e.g., feedback) may be withdrawn, and then a third (e.g., prompting). This, then, allows the investigator to determine which, if any, of the treatment elements is required to ensure response maintenance postacquisition. Examples of this strategy appear in Sowers, Rusch, Connis, and Cummings (1980) in a multiple baseline design and in O'Brien, Bugle, and Azrin (1972) in a withdrawal design.

The *partial-withdrawal* strategy requires use of a multiple baseline design. Here a component of treatment from one of the baselines or the entire treatment for one of the baselines is removed (see Russo & Koegel, 1977). This, of course, allows a comparison between untreated and treated baselines following response acquisiton. Thus if removal of a part or all of treatment leads to decremental performance, it would be clear that response maintenance following acquisition requires direct and specific programming. Treatment, then, could be reimplemented or altered altogether. It should be noted, however, that, "The possibility exists that the information obtained from partially withdrawing treatment or withdrawing a component of treatment may not represent the characteristic data pattern for all subjects, behaviors, or situations included in the design" (Rusch & Kazdin, 1981, p. 136).

Finally, in the *partial-sequential withdrawal* strategy, a component of treatment from one of the baselines or the entire treatment for one of the baselines is removed. (To this point, the approach followed *is* identical to the procedures used in the *partial-withdrawal* strategy.) But, this is followed in turn by subsequent removal of treatment in succeeding baselines. Irrespective of whether treatment loss appears across the baselines, Rusch and Kazdin (1981) argued that, "By combining the partial- and sequential-withdrawal design strategies, investigators can predict, with increasing probability, the extent to which they are controlling the treatment environment as the progression of withdrawals is extended to other behaviors, subjects, or settings" (p. 136).

CHAPTER 4

Assessment Strategies

by Donald P. Hartmann

4.1. INTRODUCTION

Assessment strategies that best complement single-case experimental designs are direct, ongoing or repeated, and intraindividual or ideographic rather than interindividual or normative. The search is for the determinants of behavior through examination of the individual's transactions with the social and physical environment. Thus behavior is a *sample*, rather than a sign of the individual's repertoire in the specific assessment setting. This approach, with its various strategies and philosophical underpinnings, has burgeoned of late within the general area of behavioral assessment (Hartmann, Roper, & Bradford, 1979). However, as noted throughout the book, the implementation of these strategies is not in any way limited to behavioral approaches to therapy. The treatment-related functions of assessment are aid in the choice of target behavior(s), selection and refinement of intervention tactics, and evaluation of treatment effectiveness (e.g., Hawkins, 1979; Mash & Terdal, 1981).

The relative emphasis on these treatment-related functions differs depending on whether assessment is serving single-case research or between-group comparison. In the latter case, selection goals—particularly those involving subjects or target behaviors—assume greater importance. In the former case, treatment refinement, or calibration, assumes greater importance. The imple-

Thanks to Lynne Zarbatany for her critical reading of an earlier draft of this chapter and to Andrea Stavros for her typing and editorial assistance.

mentation of treatment-related functions also varies as a function of single-subject versus group design. For example, methods of evaluating treatment effectiveness in single case designs (see chapter 2) place much greater emphasis on repeated measurement (e.g., Bijou, Peterson, & Ault, 1968). Indeed, as described in chapter 3, repeated measurement of the target behavior is a common, critical feature of all single-case experimental designs.

Just as assessment serves diverse functions, it also varies in its focus. Assessment can be used to evaluate overt motor behaviors such as approach responses to feared objects, physiological-emotional reactions such as ectodermal reactions and heart-rate acceleration, or cognitive-verbal responses such as hallucinations and subjective feelings of pain (Nelson & Hayes, 1979).[1] Assessors may be interested in some or all of these components of the *triple response system*, as well as in their covariation (Lang, 1968; also see Cone, 1979). While assessment can accommodate most any potential focus, the most common (and perhaps the most desirable) focus in individual subject research is overt motor behavior.

Because the content focus of assessment may vary widely, a variety of assessment techniques or methods have been developed. These techniques include direct observation, self-reports including self-monitoring, questionnaires, structured interviews, and various types of instrumentation, particularly for the measurement of psychophysiological responding (e.g., Haynes, 1978). Though any technique conceivably could be paired with any content domain, current practices favor certain associations between content and method: motor acts with direct observations, cognitive responses with self-report, and physiological responses with instrumentation.

Just as individual subjects researchers prefer to target motor acts, most also prefer the assessment technique associated with that domain, direct observation. Indeed, direct observation has been referred to as the "hallmark," the "*sine qua non*," and the "greatest contribution" not only of behavioral assessment but of behavior analysis and modification (see Hartmann & Wood, 1982). Though direct observation is indeed overwhelmingly the most popular assessment technique in published work in the area of behavior modification (P. H. Bornstein, Bridgwater, Hickey, & Sweeney, 1980), it is noteworthy that the assessment practices of therapists, even behavior therapists, are considerably more varied (e.g., Wade, Backer, & Hartmann, 1979).

This chapter will address issues of particular importance in using assessment techniques for choosing target behaviors and subsequently tracking them for the purposes of refining and evaluating treatment using repeated measurement strategies. In keeping with their importance in applied behavioral research, these issues will be addressed in the context of the assessment of motor behavior using direct observations. Issues featured include defining target behaviors, selecting response dimensions and the conditions of observation, developing observational procedures, reactivity and other observer effects, selecting and training observers, and assessing reliability and validity. Finally, brief mention will be made of other assessment devices used in the assessment of common target behaviors.

4.2. SELECTING TARGET BEHAVIORS

The phases in assessment, particularly behavioral assessment, have been likened to a funnel (e.g., Cone & Hawkins, 1977). At its inception, assessment is concerned with such general and broad issues as "Does this individual have a problem?", and, if so, "What is the nature and extent of the problem?" Interviews, questionnaires, and other self-report measures often provide initial answers to such questions, with direct observations in contrived settings and norm- or criterion-referenced tests pinpointing the behavioral components requiring remediation and indicating the degree of disturbance (Hawkins, 1979). However, the utility of assessment devices for these purposes has not been established (e.g., Mash & Terdal, 1981). In fact, there is some evidence that the use of behavioral assessment techniques by behavioral assessors produces inconsistent target behavior selection (see Evans & Wilson, 1983).[2]

Disagreements in target behavior selection might be limited if behaviors identified as targets for intervention met one or more of the following criteria (Kazdin, 1982b; Mash & Terdal, 1981; Wittlieb, Eifert, Wilson, & Evans, 1979): (1) The behavior is considered important to the client or to people who are close to the client such as spouse or parent; (2) the activity is dangerous to the client or others; (3) the response is socially repugnant; (4) the actions seriously interfere with the client's functioning; (5) the behavior represents a clear departure from normal functioning. Even if an individual's behavior meets one or more of these criteria, the problem's severity or future course may be unknown or the *specific* intervention target may be unclear. This continued ambiguity might be due to the problem's being poorly defined, or to its representing some unknown component of a chain such as long division, a symptom complex such as depression, or a construct such as social skills. A number of empirical methods may help to clarify the problem in such circumstances.

One method involves comparing the individual's behavior to a standard or norm to determine the nature and extent of the problem (e.g., Hartmann et al., 1979). This *social comparison* procedure was used by Minkin et al. (1976) to identify potential targets to improving the conversational skills of predelinquent girls. Normative conversational samples provided by effectively functioning youth were examined to determine their distinguishing features. These features, including asking questions and providing feedback, were then targeted for the predelinquent girls.

In a second method, *subjective evaluation*, ratings of response adequacy or importance are solicited from qualified judges (see Goldfried & D'Zurilla, 1969). For example, Werner et al. (1975) asked police to identify the behaviors of suspected delinquents that were important in police-adolescent interactions. These behaviors, including responding politely and cooperatively, served as target behaviors in a subsequent training program. Subjective evaluation and social-comparison methods are often referred to as *social validation procedures* (Kazdin, 1977; Wolf, 1978). Methodological appraisals of social validation procedures have been provided (Forehand, 1983).

In a third method, a careful empirical-logical analysis is conducted of the problematic behavior to determine which component or components are performed inadequately (Hawkins, 1975). Task analyses have been conducted on diverse behaviors, including dart throwing (Schleien, Weyman, & Kiernan, 1981) and janitorial skills (Cuvo, Leaf, & Borakove, 1978). This approach bears strong similarity to criterion-referencing testing as used to identify academic deficiencies (e.g., Carver, 1974). Other less-common approaches for clarifying problem behaviors, including those based on component analysis and regression techniques, were reviewed by Nelson and Hayes (1981).

If multiple problem behaviors have been targeted following this winnowing and clarifying procedure, a final decision concerns the order of treating target behaviors. While the existing (and scant) data on this issue suggest that the order of treatment of target behaviors may have no effect on outcome (Eyberg & Johnson, 1974), a number of suggestions have been offered for choosing the first behavior to be treated (Mash & Terdal, 1981; Nelson & Hayes, 1981). Behaviors recommended for initial treatment include those that are (1) dangerous to the client or others; (2) most irritating to individuals in the client's immediate social environment such as spouse or parent; (3) easiest to modify; (4) most likely to produce generalized positive effects; (5) earliest in a chain or prerequisite to other important behaviors; or (6) most difficult to modify. Of course this decision, as well as many others faced by therapists, may have to be based on more mundane considerations, such as skill level of the therapist or demands of the referral source.

4.3. TRACKING THE TARGET BEHAVIOR USING REPEATED MEASURES

The stem of the assessment funnel represents the baseline, treatment, and follow-up phases of an intervention study. Measurement during these phases requires a more narrow focus on the target behavior for purposes of refining, and in some cases, extensively modifying, the intervention and subsequently evaluating its impact.[3] Assessment during these phases typically employs direct observation of the target behavior(s) in either contrived or natural settings (e.g., M. B. Kelly, 1977). A first step in developing or utilizing an existing observational or other assessment procedure is to operationally define the target behavior and select the response dimension or property best suited for the purpose of the study.

Defining the target behavior

After pilot observations have roughly mapped the target behavior by providing a narrative record of the how, what, when, and where of responding (e.g., Hawkins, 1982), the investigator will be ready to develop an operational definition for the behavior. In defining responses, one can either emphasize topography or function (e.g., J. M. Johnston & Pennypacker, 1980). Topographically based definitions emphasize the movements comprising the response, whereas functionally based definitions emphasize the consequences of the behavior (Hutt & Hutt, 1970; Rosenblum, 1978). Thumb-sucking might be defined topographically as "the child having his thumb or any other finger touching or between his lips or fully inserted into his mouth between his teeth" (Gelfand & Hartmann, 1984). On the other hand, aggression might be defined functionally as "an act whose goal response is injury to an organism" (Dollard, Dobb, Miller, Mowrer, & Sears, 1939, p. 11). According to Hawkins (1982), functional units provide more valuable information than do topographical units, but they also tend to entail more assumptions on the part of the instrument developer and more inferences on the part of the observer.

Whether the topographical or functional approach is followed, the definition should provide meaningful and replicable data. *Meaningful*, as used here, is similar in meaning to the term *convergent validity* (e.g., Campbell & Fiske, 1959). The definition of the target behavior should agree or converge with the common uses of the label given the target behavior, and with the definition used by the referral source and in related behavior change studies (e.g., Gelfand & Hartmann, 1984).[4] Replicable refers to the extent to which similar results would be obtained if the measurement were obtained either in another laboratory or by two independent observers in the same laboratory (interobserver agreement).

Interobserver disagreements and other definitional problems can be remedied by making definitions *objective*, *clear*, and *complete* (Hawkins & Dobes, 1977). Objective definitions refer only to observable characteristics of the target behavior; they avoid references to intent, internal states, and other private events. Clear definitions are unambiguous, easily understood, and readily paraphrased. A complete definition includes the boundaries of the behavior, so that an observer can discriminate it from other, related behaviors. Complete definitions include the following components (Hawkins, 1982): a descriptive name; a general definition, as in a dictionary; an elaboration that describes the critical parts of the behavior; typical examples of the

TABLE 4-1. Sample Definition of Peer Interaction

Target Behavior:	Peer interaction.
Definition:	Peer interaction refers to a social relationship between agemates such that they mutually influence each other (Chaplin, 1975).
Elaboration:	Peer interaction is scored when the child is (a) within three feet of a peer and either (b) engaged in conversation or physical activity with the peer or (c) jointly using a toy or other play object.
Example:	"Gimme a cookie" directed at a tablemate. Hitting another child. Shouting to a friend across the playground. Sharing a jar of paint.
Questionable Instances:	Waiting for a turn in a group play activity (scored). Not interacting while standing in line (not scored). Two children independently but concurrently talking to a teacher (not scored).

Note. From Gelfand, D. M. & Hartmann, D. P. Child behavior: Analysis and therapy (2nd ed.). Elmsford, NY: Pergamon Press. Copyright 1984. Reproduced by permission.

behavior; and questionable instances—borderline or difficult examples of both occurrences and nonoccurrences of the behavior. An illustrative definition of peer interaction meeting these requirements is given in Table 4-1.

Selecting observation settings

The settings used for conducting behavioral investigations have been limited only by the creativity of investigators and the location of subjects. Because the occurrences of many behaviors are dependent upon specific environmental stimuli, behavior rates may well vary across settings containing different stimuli (e.g., Kazdin, 1979). Thus, for example, drinking assessed in a laboratory bar may not represent the rate of the behavior observed in more natural contexts (Nathan, Titler, Lowenstein, Solomon, & Rossi, 1970), and cooperative behavior modified in the home may not generalize to the school setting (R. G. Wahler, 1969b). Even within the home, desirable and undesirable child behaviors may vary with temporal and climatic variables (Russell & Bernal, 1977). Thus unless the purpose of an investigation is limited to modifying a behavior in a narrowly defined treatment context, observations need to be extended beyond the setting in which treatment occurs. Observations conducted in multiple settings are required (1) if generalization of treatment effects is to be demonstrated; (2) if a representative portrayal of the target behavior is to be obtained; and (3) if important contextual variables that control responding and that may be used to generate effective interactions are to be identified (e.g., Gelfand & Hartmann, 1984; Hutt & Hutt, 1970). Given the infrequency with which settings are typically sampled (P. H. Bornstein et al., 1980), these issues either have not captured the interests of behavior change researchers, or the cost of conducting observations in multiple settings has exceeded available resources.

While most investigators would prefer to observe behavior as it naturally occurs (e.g., Kazdin, 1982b), a number of factors may require that observations be conducted elsewhere. The reasons for employing contrived or analogue settings include convenience to observers and clients; the need for standardization or measurement sensitivity; or the fact that the target behavior naturally occurs at a low rate, and observations in natural settings would involve excessive dress. All of these factors may have determined R. T. Jones, Kazdin and Haney's (1981b) choice of a contrived setting to assess the effectiveness of a program to improve children's skill in escaping from home emergency fires.

The correspondence between behavior observed in contrived observational settings and in naturalistic settings varies as a function of (1) similarities in their physical characteristics, (2) the persons present, and (3) the control exerted by the observation process (Nay, 1979). Even if assessments are conducted in naturalistic settings, the observations may produce variations in the cues that are normally present in these settings. For example, setting cues may change when structure is imposed on observation settings. Structuring may range from presumably minor restrictions in the movement and activities of family members during home observations to the use of highly contrived situations, as in some assessments of fears and social skills. Haynes (1978), McFall (1977), and Nay (1977, 1979) provided examples of representative studies that employed various levels and types of structuring in observation settings; they also discussed the potential advantages and limitations of structuring relative to cost, measurement sensitivity, and generalizability.

Cues in observation settings may also be affected by the type of observers used and their relationship to the persons observed. Observers can vary in their level of participation with the observed. At the one extreme are nonparticipant (independent) observers whose only role is to gather data. At the other extreme are self-observations conducted by the subject or client. Intermediate levels of participant-observation are represented by significant others, such as parents, peers, siblings, teachers, aides, and nurses, who are normally present in the setting where observations take place (e.g., Bickman, 1976). The major advantages of participant-observers is that they may be present at times that might otherwise be inconvenient for independent observers, and their presence may be less obtrusive. On the other hand, they may be less dependable, more subject to biases, and more difficult to train and evaluate than are independent observers (Nay, 1979).

When observation settings vary from natural life settings either because of the presence of possibly obtrusive external observers or the imposition of structure, the ecological validity of the observations is open to question (e.g.,

Barker & Wright, 1955; Rogers-Warren & Warren, 1977). Methods of limiting these threats to ecological validity are discussed in the section on observer effects.

Though selection of observation settings is an important issue, investigators must also determine how best to sample behaviors within these settings. Sampling of behavior is influenced by how observations are scheduled. Behavior cannot be continuously observed and recorded except by participant-observers and when the targets are low-frequency events (see, for example, the Clinical Frequency Recording System employed by Paul & Lentz, 1977), or when self-observation procedures are employed (see Nelson, 1977). Otherwise, the times in which observations are conducted must be sampled, and decisions must be made about the number of observation sessions to be scheduled and the basis for scheduling. More samples are required when behavior rates are low, variable, and changing (either increasing or decreasing); when events controlling the target behaviors vary substantially; and when observers are asked to employ complex coding procedures (Haynes, 1978).

Once a choice has been made about how frequently to schedule sessions, a session duration must be chosen. In general, briefer sessions are necessary to limit observer fatigue when a complex coding system is used, when coded behaviors occur at high rates, and when more than one subject must be observed simultaneously. Ultimately, however, session duration, as well as the number of observation sessions, should be chosen to minimize costs and to maximize the representativeness, sensitivity, and reliability of data and the output of information per unit of time. For an extended discussion of these issues as they apply to scheduling, see Arrington (1943). If observations are to be conducted on more than one subject, decisions must be made concerning the length of time and the order in which each subject will be observed. Sequential methods, in which subjects are observed for brief periods in a previously randomized, rotating order, are superior to fewer but longer observations or to haphazard sampling (e.g., Thomson, Holmberg, & Baer, 1974).

Selecting a response dimension

Behaviors vary in frequency, duration, and quality. The choice of response dimension(s) ordinarily is based on the nature of the response, the availability of suitable measurement devices, and the purpose of the study (e.g., Bakeman, 1978; Sackett, 1978).

Response frequency is assessed when the target behavior occurs in discrete units that are equal in other important respects, such as duration. Frequency measures have been taken (1) of a variety of freely occurring responses such as conversations initiated and headbangs; (2) with discrete-trial or discrete-category responses such as pitches hit, or instructions complied with; and (3) when individuals are themselves the measurement units, such as the number of individuals who litter, overeat, commit murder, or are in their seats at the end of recess (Kazdin, 1982b). Behaviors such as crying, for which individual incidents vary in temporal or in other important respects or which may be difficult to classify into discrete events, are better evaluated using another response dimension such as duration.

When response occurrences are easily discriminated, and occur at moderate to low rates, frequencies can be tallied conveniently by moving an object, such as a paper clip, from one pocket to another; by placing a check mark on a sheet of paper; or by depressing the knob on a wrist counter. When responses occur at *very* low rates, even a busy participant can record a wide range of behavior for a large number of individuals (e.g., Wood, Callahan, Alevizos, & Teigen, 1979). More complex observational settings require the use of a complicated recording apparatus or of multiple observers; sampling of behaviors, individual or both; or making repeated passes through either video or audio recordings of the target behaviors (e.g., Holm, 1978; Simpson, 1979).

Response duration, or one of its derivatives such as percentage of time spent in an activity, is assessed when a temporal characteristic of a response is targeted such as the length of time required to perform the response, the response latency, or the interresponse time (Cone & Foster, 1982). While duration is less commonly observed than is frequency (e.g., M. B. Kelly, 1977), duration has been measured for a variety of target responses including the length of time that a claustrophobic patient sat in a small room (Leitenberg et al., 1968) and latency to comply with classroom instructions (Fjellstedt & Sulzer-Azaroff, 1973).

Duration measures require the availability of a suitable timing device and a target response with clearly discernible onsets and offsets. In single-variable studies, the general availability and convenience of digital wristwatches with real time and stopwatch functions may enable even a participant observer to serve as the primary source of data. In the case of multiple-target behaviors, a complex timing device such as a multiple-channel event recorder such as a Datamyte is required.

Response quality is typically assessed when target behaviors vary either in (1) intensity or amplitude, such as noise level and penile erection; (2) accuracy, such as descriptions of place and time used to test general orientation; or (3) acceptability, such as the appropriateness of assertion and the intelligibility of speech (Cone & Foster, 1982). These qualitative dimensions may be evaluated on continuous or discrete scales, and the discrete scales can themselves be dichotomous or multi-categorical. For example, assessment of the amount of food spilled by a child could be made by weighing the child and the food on his or her plate before and after each meal (quantitative, continuous), by counting the number of spots on the tablecloth (quantitative, discrete), or by determining for each meal whether or not spilling had occurred (dichotomous, discrete). The selection of a particular measurement scale is determined by the discriminatory capabilities of observers, the precision of information required by the study, the cost factors, and the availability of suitable rating devices (e.g., Gelfand & Hartmann, 1984).

To avoid the problems of bias associated with qualitative ratings, particularly of global ratings (e.g. Shuller & McNamara, 1976), scale values should be anchored or identified in terms of critical incidents or graded behavioral examples. For example, the anchor associated with a value of five on a seven-point scale for rating spelling accuracy might be "two errors, including substitutions, omissions, letter reversals, and excessive letters." P. C. Smith and Kendall (1963) described how to develop behavioral rating scales with empirically formulated anchors, and additional suggestions are given by Cronbach (1970, chapter 17). Examples of how complex qualitative judgments can be made reliably can be found in Goetz and Baer (1973) and in Hopkins, Schutte, and Garton (1971). Because all qualitative scales can be conceived of as either frequency or duration measures, they must conform to the requirements previously described for measurement of these response dimensions.

Selecting observation procedures

Altmann's (1974) description of observation procedures (traditionally called sampling procedures) contained at least five techniques of general use for applied behavioral researchers. Selection of one of these procedures will be determined in part by which response characteristics are recorded, and in turn will determine how the behavioral stream is segregated or divided.

Real-time observations involve recording both event frequency and duration on the basis of their occurrence in the noninterrupted, natural time flow (Sanson-Fisher, Poole, Small, & Fleming, 1979). Data from real-time recording are powerful, rigorous, and flexible, but these advantages may come at the cost of expensive recording devices (e.g., Hartmann & Wood, 1982). The real-time method and *event recording*—the technique discussed next—are the only two procedures commonly employed to obtain unbiased estimates of response frequency, to determine rate of responses, and to calculate conditional probabilities (e.g., Bakeman, 1978).

Event recording, sometimes called *frequency recordings, the tally method*, or *trial scoring* when applied to discrete trial behavior, is used when frequency is the response dimension of interest. With event recording, initiations of the target behavior are scored for each occurrence in an observation session or during brief intervals within a session (H. F. Wright, 1960). Event recording has the overwhelming advantage of simplicity. Its disadvantages include (1) the fragmentary picture it gives of the stream of behavior; (2) the difficulty of identifying sources of disagreements between observers, unless the observations are locked into real time; (3) the unreliability of observations when response onset or offset are difficult to discriminate; and (4) the tendency of observers to nod off when coded events occur infrequently (Nay, 1979; Reid, 1978; Sulzer-Azaroff & Mayer, 197). Despite these disadvantages, event recording is a commonly used method in behavior change research (M. B. Kelly, (1977).

Duration recording is used when one of the previously discussed temporal aspects of responding is targeted. According to M.B. Kelly (1977), duration recording is the least used of the common recording techniques, perhaps in part because of the belief that frequency is a more basic response characteristic (e.g., Bijou et al., 1969), and perhaps in part because of the apparent ease of estimating duration by either of the two methods described next.

Scan sampling, also referred to as *instantaneous time sampling*, *momentary time sampling*, and *discontinuous probe time sampling*, is particularly useful with behaviors for which duration (percentage of time occurrence) is a more meaningful dimension than is frequency. With scan sampling, the observer periodically scans the subject or client and notes whether or not the behavior is occurring at the instant of the observation. The brief observation periods that give this technique its name can be signaled by the beep of a digital watch, an oven timer, or an audiotape played through an earplug, or either a fixed or random schedule. Impressive applications of scan sampling with chronic mental patients were described by Paul and his associates (Paul & Lentz, 1977; Power, 1979).

The final procedure, *interval recording*, is also referred to as *time sampling*, *one-zero recording*, and the Hansen system. It is at the same time one of the most popular recording methods (M. B. Kelly, 1977) and one of the most troublesome (e.g., Altman, 1974; Kraemer, 1979). With this technique, an observation session is divided into brief observe-record intervals, and each interval is scored if the target behavior occurs either throughout the interval, or, more commonly, during any part of the interval (Powell, Martindale, & Kulp, 1975). The observation and recording intervals can be signaled efficiently and unobtrusively by means of an earpiece speaker used in conjunction with a portable cassette audio recorder. The observers listen to an audiotape on which is recorded the *number* of each observation and recording interval, separated by the actual length of these intervals. If data sheets are similarly numbered, the likelihood of observers getting lost is substantially reduced in comparison to the use of other common signaling devices.

While interval recording procedures have been recommended for their ability to measure both response frequency and response duration, recent research indicates that this method may provide seriously distorted estimates of both of these response characteristics (see Hartmann & Wood, 1982). As a measure of frequency, the rate of interval-recorded data will vary depending upon the duration of the observation interval. With long intervals, more than one occurrence of a response may be observed, yet only one response would be scored. With short intervals, a single response may extend beyond an interval and thus would be scored in more than one interval. As a measure of response duration, interval-recorded data also present problems. For example, duration will be overestimated whenever responses are scored, yet occur for only a portion of any observation interval. The interval method will only provide a good estimate of duration when observation intervals are very short in comparison with the mean duration of the target behavior. Under these conditions the interval method becomes procedurally similar to scan sampling.

Despite these and other limitations (see Sackett, 1978; Sanson-Fisher et al., 1979), interval recording continues to enjoy the favor of applied behavioral researchers (Hawkins, 1982). This popularity is due, no doubt, to the technique's ease of application to multiple-behavior coding systems, particularly when some of the behaviors included in the system cannot readily be divided into discrete units, and its convenience for detecting sources of interobserver unreliability (Cone & Foster, 1982). Nonetheless, if accurate estimates of frequency and duration are required, investigators would be well advised to consider alternatives to interval recording. If real-time sampling is not required or is prohibitively expensive, adequate measures of response duration and frequency can result from combining the scan and event recording techniques. However, data produced by combining these two methods do not have the same range of applications as data obtained by the real-time procedure.

More detailed guidelines for selecting an observation procedure were given in Gelfand and Hartmann (1975), in Nay (1979), and in Sulzer-Azaroff and Mayer (1977). Table 4-2 summarizes the most important of these guidelines. Additional suggestions for dealing with special recording problems, such as those involved in observing more than one subject, are available in Bijou et al. (1968), in Boer (1968), and in Paul (1979).

Observer effects

Observer effects represent a conglomerate of systematic or directional errors in behavior observations that may result from using human observers. The most widely recognized and potentially hazardous of these effects include reactivity, bias, drift, and cheating (e.g., Johnson & Bolstad, 1973; Kent & Foster, 1977; Wildman & Erickson, 1977).

Reactivity refers to the fact that subjects may respond atypically as a result of being aware that their behavior is being observed (Weick, 1968). The factors that contribute to reactivity (e.g., Arrington, 1939; Kazdin, 1982a)

TABLE 4-2. Factors to Consider in Selecting an Appropriate Recording Technique

METHOD	ADVANTAGES AND DISADVANTAGES
Real-Time Recording	*Advantages:* —Provides unbiased estimates of frequency and duration. —Data capable of complex analyses such as conditional probability analysis. —Data susceptible to sophisticated reliability analysis. *Disadvantages:* —Demanding task for observers. —May require costly equipment. —Requires responses to have clearly distinguishable beginnings and ends.
Event or Duration Recording	*Advantages:* —Measures are of a fundamental response characteristic (i.e., frequency or duration). —Can be used by participant-observers (e.g., parents or teachers) with low rate responses. *Disadvantages:* —Requires responses to have clearly distinguishable beginnings and ends. —Unless responses are located in real time (e.g., by dividing a session into brief recording intervals), some forms of reliability assessment may be impossible. —May be difficult with multiple behaviors unless mechanical aids are available.
Momentary Time Samples	*Advantages:* —Response duration of primary interest. —Time-saving and convenient. —Useful with multiple behaviors and/or children. —Applicable to responses without clear beginnings or ends. *Disadvantages:* —Unless samples are taken frequently, continuity of behavior may be lost. —May miss most occurrences of brief, rare responses.
Interval Recording	*Advantages:* —Sensitive to both response frequency and duration. —Applicable to wide range of responses. —Facilitates observer training and reliability assessments. —Applicable to responses without clearly distinguishable beginnings and ends. *Disadvantages:* —Confounds frequency and duration. —May under- or overestimate response frequency and duration.

Note. Adapted from Gelfand, D. M. & Hartmann, D. P. (1984). Child behavior: Analysis and therapy (2nd ed.). Elmsford, NY: Pergamon Press. Copyright 1984. Reproduced by permission.

include the following: (1) Socially desirable or appropriate behaviors may be facilitated while socially undesirable or "private" behaviors may be suppressed when subjects are aware of being observed (e.g., Baum, Forehand, & Zegiob, 1979); (2) the more conspicuous or obvious the assessment procedure, the more likely it is to evoke reactive effects; however, numerous contrary findings have been obtained, and such factors as observer proximity to subjects and instructions that alert subjects to observations do not guarantee reactive responding (see Hartmann & Wood, 1982); (3) observer attributes such as sex, activity level/responsiveness, and age appear to influence reactivity in children, whereas adults are influenced by observers' appearance, tact, and public-relations skills (e.g., Haynes, 1978; also see Johnson & Bolstad, 1973); (4) young children under the age of six and subjects who are open and confident or perhaps merely insensitive may react less to direct observation than subjects who do not share these characteristics; and (5) the rationale for observation may affect the degree to which subjects respond in an atypical manner (see discussion by Weick, 1968). Johnson and Bolstad (1973) recommended providing a thorough rationale for observation procedures in order to reduce subject concerns and potential reactive effects due to the observation process. Other methods for reducing reactivity also may prove useful (Kazdin, 1979; 1982a).

1. Use unobtrusive observational procedures (see Sechrest, 1979; Webb et al., 1981). For example, Hollandsworth, Glazeski, and Dressel (1978) evaluated the effects of training on the social-communicative behavior of an anxious, verbally deficient clerk by observing him unobtrusively at work while he interacted with customers.
2. Reduce the degree of obtrusiveness by hiding observers behind one-way mirrors or making them less conspicuous, that is, by having them avoid eye contact with the observee. Table 4-3 lists suggestions for classroom observers that are intended to decrease their obtrusiveness and hence the reactivity of their observations.
3. Increase reliance on reports from informants who are a natural part of the client's social environment.
4. Obtain assessment data from multiple sources differing in method artifact.
5. Allow subjects to adapt to obervations before formal data collection begins. Unfortunately, the length of time or number of observation sessions required for habituation is unclear, and recommended adaptation periods range as high as six hours for observations conducted in homes (see Haynes, 1978).

Observer bias is a systematic error in assessment usually associated with observers' expectancies and prejudices as well as their information-processing

TABLE 4-3. Suggestions for School Observers

---

1. Obtain the caretaker's permission to observe the child in the classroom or other school environment.
2. Consult the classroom teacher prior to making observations and agree upon an acceptable introduction and explanation for your presence in the classroom. Also arrange for mutually agreeable observation times, location, etc.
3. Insofar as possible, coordinate your entry and exit from the classroom with normal breaks in the daily routine.
4. Be inconspicuous in your personal appearance and conduct.
5. Do not strike up conversations with the children.
6. Sit in an inconspicuous location from which you can see but cannot easily be seen.
7. Disguise your interest in the target child by varying the apparent object of your glances.
8. Do not begin systematic behavioral observations until the children have become accustomed to your presence.
9. Minimize disruptions by taking your observations at the same time each day.
10. Thank the teacher for allowing you to visit the classroom.

Note. Adapted from Gelfand, D. M. & Hartmann, D. P. (1984). Child behavior: Analysis and therapy (2nd ed.). Elmsford, NY: Pergamon Press. Copyright 1984. Reproduced by permission.

limitations. Observers may, for example, impose patterns of regularity and orderliness on otherwise complex and unruly behavioral data (Hollenbeck, 1978; Mash & Makohoniuk, 1975). Other systematic errors are due to observers' expectancies including explicit or implicit hypotheses about the purposes of an investigation, how subjects should behave, or perhaps even what might constitute appropriate data (e.g., Haynes, 1978; Kazdin, 1977; Nay, 1979). Observers may also develop biases on the basis of overt expectations resulting from knowledge of experimental hypotheses, subject characteristics, and prejudices conveyed explicitly or implicitly by the investigator (e.g., O'Leary, Kent, & Kanowitz, 1975).

Methods of controlling biases include using professional observers; using videotape recording with subsequent rating of randomly ordered sessions; maintaining experimental naivete among observers; cautioning observers about the potential lethal effects of bias; employing stringent training criteria; and using precise, low-inference operational definitions (Haynes, 1978; Kazdin, 1977; Redfield & Paul, 1976; Rosenthal, 1976; also see Weick, 1968). If there is any reason to doubt the effectiveness with which observer bias is being controlled, investigators should assess the nature and extent of bias by systematically probing their observers (Hartmann, Roper, & Gelfand, 1977; Johnson & Bolstad, 1973).

Observer drift, or instrument decay (Cook & Campbell, 1979; Johnson & Bolstad, 1973), occurs when observer consistency or accuracy decreases, for example, from the end of training to the beginning of formal data collection (e.g., Taplin & Reid, 1973).[5] Drift occurs when a recording-interpretation bias has gradually evolved over time (Arrington, 1939, 1943) or when response definitions or measurement procedures are informally altered to suit novel changes in the topography of some target behavior (Doke, 1976). Drift can also result from observer satiation or boredom (Weick, 1968). Observer drift can cause inflated estimates of interobserver reliability when these estimates are based on data obtained (1) during training sessions, (2) from overt reliability assessment no matter when scheduled, or (3) from a long-standing, familiar team of observers during the course of a lengthy investigation (see Hartmann & Wood, 1982).

Drift can be limited or its effects reduced by providing continuing training throughout a project, by training and recalibrating all observers at the same time, and by inserting random and covert reliability probes throughout the course of the investigation. Alternatively, investigators can take steps to evaluate the presence of observer drift by having observers periodically rate prescored videotapes (sometimes referred to as *criterion videotapes*), by conducting reliability assessment across rotating members of observation teams, and by using independent reliability assessors (see reviews by Cone & Foster, 1982; Hartmann & Wood, 1982; Haynes, 1978).

Observer cheating has been reported only rarely (e.g., Azrin, Holz, Ulrich, & Goldiamond, 1961). More commonly, observers have been known to calculate inflated reliability coefficients, though these calculation mistakes are not necessarily the result of intentional fabrication (e.g., Rusch, Walker, & Greenwood, 1975). Precautions against observer cheating include random, unannounced reliability spot checks; collection of data forms immediately after an observation session ends; restriction of data analysis and reliability calculations to individuals who did not collect the data; provision of pens rather than pencils to raters (obvious corrections might then be evaluated as an indirect measure of cheating); and reminders to observers about the canons of science and the dire consequences of cheating (Hartmann & Wood, 1982). See the section on staging reliability assessments (p. 124) for further suggestions regarding limiting observer drift and observer cheating.

Selecting and training observers

Unsystematic or random observer errors as well as many of the systematic sources of error in observational data just described may be partially controlled by properly selecting observers and training them well.

Behavioral researchers seem unaware of the substantial amount of research on individual differences in observational skills (see Boice, 1983). In general, observational skills increase with age and are better developed in women than in men. There is also some evidence to suggest that the components of social skills, such as the ability to perceive nonverbally communicated affect, may be related to observer accuracy, and that the perceptual-motor skills of observers may prove directly relevant to training efficiency and to the maintenance of desired levels of observer performance (e.g., Nay, 1979). Additional observer attributes that may be important include morale, intelligence, motivation, and attention to detail (e.g., Boice, 1983; Hartmann & Wood, 1982; Yarrow & Waxler, 1979).

Once potential observers are selected, they require systematic training in order to perform adequately. Recent reviews of the observer-training literature (e.g., Hartmann & Wood, 1982; Reid, 1982) suggest that observers should progress through a sequence of training experiences that includes general orientation, learning the observation manual, conducting analogue observations, *in situ* practice, retraining-recalibration, and debriefing. Training should begin with a suitable rationale and introduction that explains to the observers the need for tunnel vision—for remaining naive regarding the purpose of the study and its experimental hypotheses. They should be warned against attempts to generate their own hypotheses and instructed to avoid private discussions of coding procedures and problems. Observers should also become familiar with the APA's *Ethical Principles in the Conduct of Research with Human Participants* (1973); particular emphasis should be placed upon their learning confidentiality, the canons of science, and observer etiquette.

Next, observer trainees should *memorize* verbatim the operational definitions, scoring procedures, and examples of the observation system as presented in a formal observation training manual (Paul & Lentz, 1977). (Suggestions for constructing observation manuals are given by Nay, 1979, p. 237.) Oral drills, pencil-and-paper tests, and scoring of written descriptions of behavioral vignettes can be employed for training and evaluation at this stage. Investigators should utilize appropriate instructional principles such as successive approximations and ample positive reinforcement in teaching their observer trainees appropriate observation, recording, and interpersonal skills. Having passed the written test, observers should next be trained to criterion accuracy and consistency on a series of analogue assessment samples portrayed via film clips or role playing. Training should begin with exposure to simple or artificially simplified behavioral sequences; later material should present rather complex interactional sequences containing unpredictable and variable patterns of responding. The observers should be *overtrained* on these materials in order to minimize later decrements in performance. Immediately after observers complete each training segment, their protocols should be reviewed, and both correct and incorrect entries should be discussed (Reid, 1982). During this phase, observers should recode training segments until 100% agreement with criterion protocols is achieved (Paul & Lentz, 1977). Discussion of procedural problems and confusions should be encouraged throughout this training phase, and all scoring decisions and clarifications should be posted in an observer log or noted in the observation manual that each observer carries.

Practice in the observation setting follows. Practice observations can serve the dual purpose of desensitizing observers to fears about the setting (i.e., inpatient psychiatric unit) and allowing subjects or clients to habituate to the observation procedures. Training considerations outlined in the previous step are also relevant here. Particular attention should be given to observer motivation. Reid (1982) suggests that observer motivation and morale may be strengthened by providing observers with (1) varied forms of scientific stimulation such as directed readings on topics related to the project, and (2) incentives for obtaining reliable and accurate data.

During the course of the investigation, periodic retraining and recalibration sessions should be conducted with all observers: recalibration could include spot tests on the observation manual, coding of prescored videotapes, and covert reliability assessments. If data quality declines, extra retraining sessions should be held. At the end of the investigation, observers should be interviewed to ascertain any biases or other potential confounds that may have influenced their observations. Observers should be informed about the nature and results of the investigation and should receive acknowledgment in technical reports or publications.

Reliability

Observational instruments require periodic assessments to ensure that they promote correct decisions regarding treatment effectiveness. Such evaluations are particularly critical for relatively untried observational instruments, for those that attempt to obtain scores on multiple-response dimensions, and for those that are applied in uncontrolled, naturalistic settings by unprofessional personnel. Traditionally, these evaluations have fallen under the domain of one of the various theories of reliability (or more recently of generalizability) and its associated methods (Cronbach et al., 1972; Nunnally, 1978).

Any reliability analysis requires a series of decisions. These decisions involve selecting the dimensions of observation that require formal assessment; deciding on the conditions under which reliability data will be gathered; choosing a unit of analysis; selecting a summary reliability statistic; interpreting the values of reliability statistics; modifying, if necessary, the data collection plan; and reporting reliability information.

The first step in assessing data quality is to decide the dimensions (or facets) of the data that are important to the research question. Potentially relevant dimensions can include observers, coding categories, occasions, and settings (e.g., Cone, 1977). With the exception of interobserver reliability,[6] these dimensions have not engaged the systematic attention of researchers using observations (Hartmann & Wood, 1982; Mitchell, 1979). This is unfortunate because sessions or occasions clearly deserve as much attention as observers have already received (Mitchell, 1979) and are particularly important in single-case research. Without observation sessions of adequate number and duration, the resulting data will be unstable. Data that are unstable, either because of variability or because of trends in the changeworthy direction, may produce inconclusive tests of treatment effects (see chapter 9). Because of the pivotal importance of observers and sessions to the use of observational codes, the remainder of this section will refer to these two aspects of observational reliability.

Conditions of observation can affect the performance of both subjects and observers and, hence, estimates of data quality or dependability (e.g., Hartmann & Wood, 1982). For example, observer performance improves, sometimes substantially, under overt, in comparison to disguised, reliability assessment conditions. Because most reliability assessments are conducted under overt conditions, much of our observational data are substantially less adequate than our interobserver reliability analyses suggest. The performance by observers also can deteriorate substantially from training to the later phases of an investigation, and in response to increases in the complexity of the behavior displayed by subjects (e.g., Cone & Foster, 1982). The quality of data recorded by observers can also vary as a function of their expectations and biases and as a result of calculation errors and fabrication, as previously discussed.

To counter the distortions that these conditions can produce, (1) subjects and observers should be given time to acclimate to the observational setting before reliability data are collected; (2) observers should be separated and, if possible, kept unaware of both when reliability assessment sessions are scheduled and the purpose of the study; (3) observers should be reminded of the importance of accurate data and regularly retrained with observational stimuli varying in complexity; (4) reliability assessments should be conducted throughout the investigation, particularly in each part of multiphase behavior-change investigations; and (5) the task of calculating reliability should be undertaken by the investigator, not by the observers (Hartmann, 1982).

Before a reliability analysis can be completed, the investigator must determine the appropriate behavioral units (or the levels of data) on which the analysis will be conducted (Johnson & Bolstad, 1973). A common, molar unit is obtained by combining the scores of either empirically or logically related molecular variables. For example, scores on *tease* can be added to scores on *cry, humiliate,* and the like to generate a total aversive behavior score (R. R. Jones, Reid, & Patterson, 1975). Still other composite units can be based on aggregation of scores over time. For example, students' daily question asking can be combined over a 5-day period to generate weekly question-asking scores.

Because the reliability of composites differs from the reliability of their components (e.g., Hartmann, 1976), investigators should be careful not to make inferences about the reliability of composites based upon the reliability of their components, and vice versa. To ensure that reliability is neither overestimated nor underestimated, reliability calculations should be performed on the level of data or units of behavior that will be subjected to substantive analysis. Thus if weekly behavior rate is the focus of analysis, the reliability of the rate measure should be assessed at the level of data summed over the seven days of a week. However, in some situations, it may be useful to assess reliability at a finer level of data than that at which substantive analyses are conducted. For example, even if data are analyzed at the level of daily session totals, assessment of reliability on individual trial scores can be useful in identifying specific disagreements that indicate the need for more observer training, for revision of the observer code, or for modification of recording procedures (Hartmann, 1977).

Investigators have a surfeit of statistical indexes to use in summarizing their reliability data. Berk (1979) described 22 different summary reliability statistics, and both Fleiss (1975) and House, House, and Campbell (1981) discussed 20 partially overlapping sets of procedures for summarizing the reliability of categorical ratings provided by two judges. Still other summary statistics were described by Frick and Semmel (1978), Tinsley and Weiss (1975), and Wallace and Elder (1980). These statistics differ in their appropriateness for various forms of data, their inclusion of correction for chance agreement, their inclusion of factors that lower their numerical value (contribute to error), their underlying measurement scale, their capacity for summarizing scores for the entire observational system with a single index, and their degree of computational complexity and abstractness (Hartmann, 1982).

Observation data are typically obtained in one or both of two forms: (1) categorical data such as occur-nonoccur, correct-incorrect, or yes-no that might be observed in brief time intervals or scored in response to discrete trials; and (2) quantitative data such as response frequency, rate, or duration. Somewhat different summary statistics have been developed for the two kinds of data.

Table 4-4 includes a two-by-two table for summarizing categorical data and the statistics commonly used or recommended for these data. These statistics all are progeny of raw agreement (referred to as *percent agreement* in its common form), the most common index for summarizing the interobserver consistency of categorical judgments (M. B. Kelly, 1977). Raw agreement has been repeatedly criticized, largely because the value of this statistic may be inflated when the target behavior occurs at extreme rates (e.g., Mitchell, 1979). A variety of techniques have been suggested to remedy this problem. Some procedures differentially weight occurrence and nonoccurrence agree-

TABLE 4-4. Two-by-Two Summary Table of Relative Proportion of Occurrence
of a Behavior as Recorded by Two Observers,
with Selected Statistical Procedures Applicable to These Data

SUMMARY TABLE

O_2

		Occurrence	Nonoccurrence	Total
	Occurrence	.60 = a	.05 = b	.65 = p_1
O_1	Nonoccurrence	.10 = c	.25 = d	.35 = q_1
	Total	.70 = p_2	.30 = q_2	1.00

Raw Agreement = $a + d$ = .85

Occurrence Agreement = $a/(a + b + c)$ = .80

Nonoccurrence Agreement = $d/(b + c + d)$ = .63

Kappa = $(a + d - p_1p_2 - q_1q_2)/(1 - p_1p_2 - q_1q_2)$ = .66

Note. Some of the summary statistics described here commonly employ a percentage scale (for example, raw agreement). For convenience, these statistics are defined in terms of a proportion scale. (Adapted from Hartmann, D. P. (1982). Assessing the dependability of observational data. In D. P. Hartmann, (Ed.), *Using observers to study behavior: New directions for methodology of social and behavioral science.* San Francisco: Jossey-Bass. Copyright 1982 by D. P. Hartmann. Reproduced by permission.)

ments (e.g., Cone & Foster, 1982; Hawkins & Dotson, 1975), whereas other procedures provide formal correction for chance agreements. The most popular of these corrected statistics is Cohen's kappa (J. Cohen, 1960). Kappa has been discussed and illustrated by Hartmann (1977) and Hollenbeck (1978), and a useful technical bibliography on kappa appears in Hubert (1977). Kappa may be used for summarizing observer agreement as well as accuracy (Light, 1971), for determining consistency among many raters (A. J. Conger, 1980), and for evaluating scaled (partial) consistency among observers (J. Cohen, 1968).

Table 4-5 includes qualitative data from a subject—scores from six sessions for two observers—and analyses of these data. The percentage agreement for these data, sometimes called *marginal agreement* (Frick & Semmel, 1978), is the ratio of the smaller value (frequency or duration) to the larger value obtained by two observers, multiplied by 100. This form of percentage agreement also has been criticized for potentially inflating reliability estimates (Hartmann, 1977). Berk (1979) advocated use of generalizability coefficients, as these statistics provide more information and permit more options than do either percentage agreement or simple correlation coefficients (also see Hartmann, 1977; Mitchell, 1979; and Shrout & Fleiss, 1979). Despite these advantages, some researchers argue that generalizability and related correlational approaches should be avoided because their mathematical properties may

TABLE 4-5. Days-by-Observers Data and Analysis of These Data

| | OBSERVERS | | |
Sessions	O_1	O_2	"Percentage Agreement"
1	11	9	82%
2	8	6	75%
3	9	7	78%
4	10	9	90%
5	12	11	92%
6	8	8	100%

ANALYSIS OF VARIANCE SUMMARY

Sources	Mean Squares (MS)
Between Sessions (BS)	5.40
Within Sessions (WS)	1.16
Observers (O)	5.33
$S \times O$	.33

GENERALIZABILITY OR INTERCLASS COEFFICIENTS (ICC)

$$\text{ICC }(1,1) = (MS_{BS} - MS_{WS})/[MS_{BS} + (k-1)MS_{WS}] =$$
$$(5.40 - 1.16)/[5.40 + 5(1.16)] = .38$$

$$\text{ICC }(3,1) = (MS_{BS} - MS_{S \times 0})/[MS_{BS} + (k-1)MS_{S \times 0}] =$$
$$(5.40 - .33)/[5.40 + 5(.33)] = .72$$

Note. Adapted from Hartmann, D.P. (1982). Assessing the dependability of observational data. In D. P. Hartmann (Ed.), Using observers to study behavior: New directions for methodology of social and behavioral science. San Francisco: Jossey-Bass. Copyright 1982 by D. P. Hartmann. Reproduced by permission.)

inhibit applied behavior analysis from becoming a "people's science" (Baer, 1977a; Hawkins & Fabry, 1979).

Disagreement about procedures for summarizing observer reliability are also related to differing recommendations for "acceptable values" of observer reliability estimates. Given the variety of available statistics—with various statistics based on different metrics and employing different conceptions of error—a common standard for satisfactory reliability seems unlikely. Nevertheless, recommendations have ranged from .70 to .90 for raw agreement, and from .60 to .75 for kappa-like structures (see Hartmann, 1982). While these recommendations will be adequate for many, even most, research purposes, the overriding basis for judging the adequacy of data is whether they provide a powerful means of detecting experimentally produced or naturally occurring response covariation.

Power depends not only on data quality, but also on the magnitude of covariation to be detected, the number of available investigative units (for example, sessions), and the experimental design. Thus, data quality must be evaluated in the context of these factors (Hartmann & Gardner, 1979). If the performance of observers is adquate, but the target behavior varies substantially across occasions, the researcher may modify the observational setting by removing distracting stimuli or by adding a brief habituation period to each observational session (e.g., Sidman, 1960); increase the length of each observation period until a session duration is discovered that will provide consistent data; or increase the number of sessions required to achieve stable performance. In each case of inconsistency, the option that is selected will depend upon the purpose of the study and on practical considerations, such as the investigator's ability to identify and control undesirable sources of variability and the feasibility of increasing the number of observers or sessions (Hartmann & Gardner, 1981).

If the quality of data is judged unsatisfactory, a number of options are available to the investigator. For example, if consistency across observers is inadequate, the investigator can train observers more extensively, improve observation and recording conditions, clarify definitions, use more than one observer to gather data and analyze the average of the observers' scores, or employ some combination of the options just described (Hartmann, 1982). The option that is selected will again depend upon the purpose of the study and on practical considerations, such as the investigator's ability to identify and control undesirable sources of variability and the feasibility of increasing the number of observers (Hartmann & Gardner, 1981).

Recommendations for reporting reliability information have ranged from the suggestion that investigators embellish their primary data displays with disagreement ranges and chance agreement levels (Birkimer & Brown, 1979) to advocacy of what appear to be cumbersome tests of statistical significance (Yelton, Wildman, & Erickson, 1977). The recommendations that follow were proposed by Hartmann and Wood (1982): (1) Reliability estimates should be reported on interobserver accuracy, consistency, or both, as well as on session reliability; (2) in the case of interobserver consistency or accuracy assessed with agreement statistics, either a chance-corrected index or the chance level of agreements for the index used should be reported; (3) reliability should be reported for covert reliability assessments scheduled periodically throughout the course of the study, for different subjects (if relevant), and across experimental conditions; and (4) reliability should be reported for each variable that is the focus of substantive analysis.

Validity

Validity, or the extent to which a score meausres what it is intended to measure, has not received much attention in observation research (e.g., Johnson & Bolstad, 1973; O'Leary, 1979). In fact, observations have been considered inherently valid insofar as they are based on direct sampling of behavior and they require minimal inferences on the part of observers (Goldfried & Linehan, 1977). According to Haynes (1978) the assumption of inherent validity in observations involves a serious epistemological error. The data obtained by human observers may not be veridical descriptions of behavior. As previously discussed, accuracy of observations can be attenuated by various sources of unreliability and contaminated by reactivity effects and other sources of measurement bias. The occurrence of such measurement-specific sources of variation provides convincing evidence for the need to validate observation scores. Validation is further indicated when observations are combined to measure some higher-level construct such as *deviant behavior* or when observation scores are used to predict other important behaviors (e.g., Hartmann et al., 1979; Hawkins, 1979). Validation may take the form of content, criterion-related (concurrent and predictive), or construct validity.

Although each of the traditional types of validity is relevant to observation systems (e.g., Hartmann et al., 1979), *content validity* is especially important in the initial development of a behavior coding schema. Content validity is assessed by determining the adequacy with which an observation instrument samples the behavioral domain of interest (Cronbach, 1971). According to Linehan (1980), three requirements must be met to establish content validity. *First*, the universe of interest (i.e., domain of relevant events) must be completely and unambiguously defined. Depending upon the nature and purposes of an observation system, this requirement may apply to the behaviors of the target subject, to antecedent and consequent events provided by other persons, or to settings and temporal factors. *Next*, these relevant factors should be representatively sampled for inclusion in the observation system. *Finally*, the method for evaluating and combining observations to form scores should be specified.

The *criterion-related* validity of assessment scores refers primarily to the degree to which one source of behavioral assessment data can be substituted for by another. Though the literature on the consistency between alternative sources of assessment data is small and inconclusive, there is evidence of poor correspondence between observation data obtained in structured (analogue) settings and in naturalistic settings (e.g., Cone & Foster, 1982; Nay, 1979). Poor correspondence has also been shown when contrasting observation data with less reactive assessment data (Kazdin, 1979). These results suggest that behavioral outcome data might have restricted generalizability and underscore the desirability of criterion-related validity studies when observational and alternative data sources are used to assess treatment outcome.

Construct validity is indexed by the degree to which observations accurately measure some psychological construct. The need for construct validity is most apparent when observation scores are combined to yield a measure of some molar behavior category or construct such as "assertion." G. R. Patterson and his colleagues (e.g., Johnson & Bolstad, 1973; R. R. Jones, Reid, & Patterson, 1975; Weinrott, Jones, & Boler, 1981) have illustrated construct validation procedures with their composite, Total Deviancy. Their investigations have demonstrated, for example, that the Total Deviancy score discriminates between clinical and nonclinical groups of children and is sensitive to the social-learning intervention strategies for which it was initially developed. Despite the impressive work done by Patterson and his associates, as well as by other behavioral investigators (e.g., Paul, 1979), the validation of an instrument is an ongoing process. Observations may have impressive validity for one purpose, such as for evaluating the effectiveness of behavioral interventions (see Nelson & Hayes, 1979), but they may be only moderately valid or even invalid measures for subsequent assessment purposes. The validity of observation data for each assessment function must be independently verified (e.g., Mash & Terdal, 1981).

4.4 OTHER ASSESSMENT TECHNIQUES

Target behaviors may be identified for which direct observations are impractical, impossible, or unethical (e.g., Cone & Foster, 1982). In such cases, one or more alternative assessment techniques are required. These techniques may include products of behavior, self-report measures, or physiological procedures. Measurement of behavioral products, such as number of emptied liquor containers, may be particularly useful when the target behavior is relatively inaccessible to direct observation because of its infrequency, subtlety, or private nature; when either the behavior or its observation causes embarrassment to the client; or when observation by others would otherwise disrupt or seriously distort the form, incidence, or duration of the response. Self-report measures also may be useful in such circumstances, though they are prey to a number of distorting influences. At other times, physiological measures may be required, because either the response is ordinarily inaccessible to unaided human observers or observers cannot provide measures of sufficient precision. It is to these classes of measures that we briefly turn next.

Behavioral products

Many target behaviors have relatively enduring effects on the environment. Measuring these behavioral effects or products allows the investigator to make inferences about the target behaviors associated with the products. This *indirect* approach to assessment has several advantages including convenience, nonreactivity, and economy. Because the products remain accessible for some length of time, they can be accurately and precisely measured at a time, and perhaps a location, convenient to the investigator (Nay, 1979). Furthermore, because behavioral products do not require the immediate presence of an observer, they can be measured unobtrusively (and hence nonreactively) and with relatively little cost.

Behavioral products have been used by a large number of behavioral investigators (Kazdin, 1982c). For example, Stuart (1971) used client weight as a measure of eating, and Hawkins, Axelrod, and Hall (1976) assessed various academic behaviors using task-related behavioral products such as number of solved math problems. Webb, Campbell, Schwartz, and Sechrest (1966) lent some order to the array of possible behavioral products by organizing them into three classes: (1) erosion measures such as shortened fingernails used to index nail biting (McNamara, 1972); (2) trace measures such as clothes-on-the-floor to assess "cabin-cleaning" (Lyman, Richard, & Elder, 1975); (3) and archival records such as number of irregular hospital discharges to indicate discontent with the hospital (P. J. Martin & Lindsey, 1976). Both Sechrest (1979) and Webb et al. (1981) presented impressive catalogs of these indirect measures of behavior.

Behavioral by-products, as well as any other indirect or proxy measures, require validation before they can be used with confidence. Until such validation is undertaken, questions remain regarding how accurately the product measure corresponds to the behavior it presumably indexes (J. M. Johnston & Pennypacker, 1981). For example, weight loss, a common index of eating reduction, also may reflect increased exercise and the use of diuretics or stimulants (Haynes, 1978). The distance of behavioral products from their target behaviors also may be troublesome (Nay, 1979). As a result of working with the product, rather than the behavior itself, information on controlling variables may be lost, and changes produced in the target behavior may not be indicated quickly enough. Furthermore, if behavioral products are consequated, the temporal delay of reinforcement may be too great to strengthen appropriate target responding.

Self-report measures

In the tripartite classification of responses (motor, cognitive, and physiological), self-report measures are associated with the assessment of the cognitive domain—thoughts, beliefs, preferences, and other subjective dimensions—because of the inaccessibility of this domain to more direct assessment approaches. However, self-report techniques also can be used to measure motor and physiological responses that potentially could be assessed objectively (e.g., Barrios, Hartmann, & Shigetomi, 1981). The latter use of self-reports is common when cost is a critical concern or when the client is not part of an "observable social system" (Haynes, 1978).

Like other assessment devices, self-report measures can be used to generate information at any part of the assessment funnel, from initial screening decisions to evaluation of treatment outcome. However, they are most popular as an economical means of getting started during the initial phases of assessment (Nay, 1979). The use of self-report procedures in treatment evaluation traditionally has been frowned on by investigators, in large part because of these reports' susceptibility to various forms of bias and distortion, their lack of specificity, and their mediocre correspondence with objective measures (e.g., Bellack & Hersen, 1977). However, more recent *behavioral* self-report procedures have gained in acceptance for the evaluation of behavioral intervention, particularly in pre-post group treatment investigations (e.g., Haynes, 1978) and when used to assess client satisfaction (e.g., Bornstein & Rychtarik, 1983; McMahon & Forehand, 1983).

Self-report measures come in a variety of forms including paper-and-pencil self-rating inventories, surveys and questionnaires, checklists, and self-monitoring procedures. Discussion of these measures will largely be limited to paper-and-pencil questionnaries and self-monitoring techniques, as they have been most widely utilized by behavioral assessors (e.g., Swan & McDonald, 1978).

Numerous *pencil-and-paper self-report questionnaires* are available on which clients are asked to indicate, in response to a series of items (e.g., situations or behaviors) their likelihood of engaging in a response (McFall & Lillisand, 1971), their degree of emotional arousal (e.g., Geer, 1965), or the frequency with which they engage in particular behaviors (e.g., Lewinsohn & Libet, 1972). These inventories or questionnaires provide assessment data on a broad range of target responses including assertive and other forms of social behavior, fears, appetitive or ingestive behaviors such as smoking and drinking, psychophysical responses such as pain, depression, and marital interactions, to name but a few. In fact, if a behavior has been studied by two investigators, the chances are very good that at least two different self-report questionnaires are available for assessing the behavior.[7] For extensive surveys of existing behavioral questionnaires, see Haynes (1978), Haynes and Wilson (1979), and recent reviews of specific content domains published in monographs devoted to behavioral assessment (e.g., Barlow, 1981; Hersen & Bellack, 1981; Mash & Terdal, 1981) and in behavioral assessment journals.

Because self-report inventories vary so substantially in quality and are potentially prey to a variety of distortions, promising inventories should be checked against the following evaluative criteria before a final selection is made (Bellack & Hersen, 1977; Haynes, 1978; Haynes & Wilson, 1979).[8]

1. Can the inventory be administered repeatedly to clients? If the inventory's form or content precludes repeated application, or if the scores change systematically with repeated administration, the self-report procedure is not suitable for tracking the target response in an individual-subject

investigation. However, even if the inventory does not meet this criterion, it may be suitable as an aid to selecting subjects, target behaviors, or treatments (e.g., Hawkins, 1979).

2. Does the questionnaire provide the required degree of specific information regarding the target behavior? Many traditional self-report techniques were based on trait assumptions of temporal, situational, and behavioral (item) homogeneity or consistency that have proven to be incorrect (e.g., Mischel, 1968). Although the increased response and situational specificity of behavioral self-report measures improve their correspondence with objective measures (e.g., Lick, Sushinsky, & Malow, 1977), the term *behavior* in an instrument's title does not guarantee the requisite degree of specificity.

3. Is the inventory sensitive enough to detect changes in performance as a result of treatment? Although most questionnaires evaluated for sensitivity have passed this validity hurdle, not all have done so successfully (e.g., Wolfe & Fodor, 1977).

4. Does the questionnaire guard against the biases common to the self-report genre? Self-report measures are susceptible to a variety of test-related and subject-related distortions. As regards test-related biases, the wording of items may be so ambiguous that idiosyncratic interpretations by respondents are common (e.g., Cronbach, 1970). Furthermore, items may request information that is beyond subjects' discrimination, storage, or recall capabilities, or they may be arranged so as to effect scores (response bias). Scores may also be effected by clients' attempts at impression management. Clients may, for example, endorse socially valued responses (social desirability), agree with strongly worded alternatives (acquiescence), endorse responses that they expect to be positively regarded by the investigator (demand effects), or engage in outright faking or lying. Biases due to impression management are particularly troublesome in the assessment of subjective experiences, as independent verification of the accuracy of responding may be difficult or impossible. Unfortunately, few questionnaires include scales designed to detect biased responding or guard against its occurrence (Evans, 1983).

5. Finally, does the inventory meet expected reliability and validity requirements and possess appropriate norms for the population of interest in the present investigation? Self-report questionnaires may be adequate for one group, but not for another, so an instrument's technical information must be examined with care.

Self-monitoring, the second popular type of self-report among behavioral clinicians, is similar to direct observation, but with one major exception: The client is the observer. Data from self-monitoring have been used for target behavior and treatment selection, as well as for treatment evaluation. However, in the latter case, objective assessments typically play a more important role, except when the target is itself a subjective response.

Self-monitoring has proven particularly useful for assessing rare and sensitive behaviors and responses that are only accessible to the client such as pain due to migraine headaches (Feuerstein & Adams, 1977) and obsessive ruminations (Emmelkamp & Kwee, 1977). Other responses assessed via self-monitoring include appetitive urges, hallucinations, hurt and depressed feelings, sexual behaviors, and waking time (for insomniacs). An array of behaviors more susceptible to direct observations also has been monitored by the client, including weight gain or loss, caloric intake, nail biting, exercise, academic behaviors, alcohol consumption, and whining. Haynes (1978), Haynes and Wilson (1979), Nay (1979), and Nelson (1977) surveyed applications of target behaviors and recording procedures used in self-monitoring.

Self-monitoring procedures share a number of method-related problems. Foremost among these is reactivity (Haynes & Wilson, 1979; Nelson, 1977). Reactivity effects vary as a function of the social desirability of the behavior recorded, with the frequency of positively valued responses likely to increase and negatively valued acts likely to decrease during the course of self-monitoring. The obtrusiveness, the timing, and the frequency of self-monitoring also may influence the level of subject reactivity. Indeed, because of these reactive effects, self-monitoring has been included in a number of treatment packages as an *intervention* technique (e.g., Nay, 1979).

A second, and perhaps more serious, problem is the variable accuracy of self-monitoring (e.g., Haynes & Wilson, 1979; Nelson, 1977). Inaccurate self-monitoring can be improved by many of the same stratagems used to improve the accuracy of direct observation: arrange recording procedures that are convenient, habitual, and generally nonaversive; provide prior training in self-monitoring; and encourage and dispense contingencies for accuracy. Self-monitoring accuracy also can be enhanced by means of various social-influence procedures such as a public commitment to self-monitor (P. H. Bornstein, Hamilton, Carmody, Rychtarik, & Veraldi, 1977). Despite the fact that accuracy can be increased through use of these manipulations, there are numerous factors adversely affecting the validity of self-monitoring; hence this approach should be used with caution when it is the only method available for monitoring the progress or outcome of treatment (Haynes, 1978).

Psychophysiological measures

Psychophysiological measures involve the surface recording of physiological events, most of which are controlled by the autonomic nervous system (Haynes, 1978). The assessment of psychophysiological responses has become increasingly important to behavioral clinicians as a result of the (perhaps premature) popularity of biofeedback training (Bradley & Prokop, 1982) and of the application of behavioral intervention techniques to a variety of physiological responses that can be assessed only imprecisely with self-report measures.

Because of the expense of psychophysiological assessments, their use has been limited largely to the intermediate and lower levels of the behavioral assessment funnel. Their objectivity and precision have made them particularly useful in identifying psychophysiological and psychophysiologically mediated problem behaviors and their etiologies. For example, strain gauges have been used to assess the sexual preferences of males based on their responsiveness to erotic stimuli (e.g., see Freund & Blanchard, 1981), and muscular reactivity (EMG) and temperature measures have been used to distinguish muscular tension from vascular headaches (e.g., see Blanchard, 1981). Other problems assessed with psychophysiological techniques include insomnia, ulcers, hypertension, pain, asthma, inadequate circulation (Raynaud's disease), a variety of sexual dysfunctions (e.g., Haynes, 1978; Haynes & Wilson, 1979) and a variety of anxiety disorders (Mavissakalian & Barlow, 1981c; Taylor & Agras, 1981; Vermilyea, Boice, & Barlow, in press).

Perhaps even more common is the role performed by psychophysiological assessments in monitoring the effects of interventions intended to modify physiological responding. For example, heart rate and blood pressure often have been included in the evaluation of tension reduction techniques like relaxation training (e.g., see Nietzel & Bernstein, 1981), and brain wave patterns (EEG) have been considered *the* criterion for assessing experimental interventions to improve the sleep of insomniacs (e.g., Coates & Thoresen, 1981).

The most common physiological responses recorded by behavioral investigators include muscular activity (EMG), heart rate, and ectodermal responding such as GSR (Haynes & Wilson, 1979). However, other responses such as pupil size, temperature, respiration rate, blood pressure and flow, and EEG also are recorded by behavioral investigators (e.g., Haynes, 1978). EMG recording is used to assess muscle tension, in large part because of the widely held belief that muscle tension mediates anxiety and that muscular relaxation training decreases levels of autonomic arousal. Recordings of muscle tension are particularly common in the assessment of tension headaches and of fears and anxiety (see, for example, Blanchard, 1981; Nietzel & Bernstein, 1981). The popularity of recording heart rate stems from the ease with which this response can be measured and analyzed, and from the apparent relationship of heart rate to stress and anxiety. Despite the utility of this recording to behavioral assessors (see Haynes & Wilson, 1979), caution is required because heart rate is also related to the individual's ". . . evaluation of the situation, his prior experience, and his previously established reaction pattern" (Nay, 1979, p. 262).

The final common physiological measure is of ectodermal activity (EDR)—

usually skin conductance or its reciprocal, skin resistance. EDRs have been viewed as a measure of activation or autonomic arousal; thus, they often are used to monitor changes in response to fear stimuli as a result of behavioral interventions (e.g., Barlow, Leitenberg, Agras, & Wincze, 1969). However, the use of ectodermal responding as a measure of arousal also must be done cautiously, as scores vary depending on the EDR response component measured (conductance, fluctuations, latency, and wave form), the time-sampling parameters utilized, and the specific measurement site and procedures used (e.g., Edelberg, 1972; Venables & Christie, 1973).

Sophisticated uses of physiological measures have been made primarily by laboratory investigators rather than practicing clinicians, due to the expense of the equipment, the inconvenience associated with its use, and the need for extensive knowledge of physiology and electronics (Nietzel & Bernstein, 1981).[9] Equipment for measuring psychophysiological responses includes (1) a sensing device, such as electrodes or some form of transducer for detecting relevant input, (2) a central processor that may include amplifiers for strengthening the incoming signal and filters for removing "noise;" and (3) an output for displaying the electronic signals, such as a pen-tracing or a digitized printout. Because malfunctioning of these components may result in missing data (a particularly serious problem in individual subject investigations), special precautions should be followed in conducting physiological assessments. For example, laboratory assistants should be thoroughly familiar with the equipment, including its maintenance and calibration, and would be well advised to practice with nonclinical subjects before actually monitoring physiological responding during experimental interventions (Hersen & Barlow, 1976).

In conducting any physiological measurement, investigators should be aware of the range of variables that may invalidate their records (e.g., Haynes & Wilson, 1979; Ray & Raczynski, 1981). Aspects of the physical environment, including temperature, lighting, humidity, ambient noise, and unshielded electrical sources, may affect the client's or subject's responding. Control of these variables is necessary, and subjects should be habituated or adapted to the laboratory setting before recording occurs. Similarly, recording techniques, such as the preparation of the recording site, nature of the conductive medium, and type, location, and attachment of electrodes or transducers also can affect the resulting physiological record. Investigators should consult standard references in this area (e.g., Greenfield & Sternbach, 1972; Stern, Ray, & Davis, 1980; Venables & Martin, 1967) in order to avoid problems due to unstandardized recording procedures. Procedural variables also can interact with measurement procedures to determine the nature of clients' responses. Thus aspects of the procedure such as the presence and characteristics of the examiner should be held constant throughout an investigation.

Not surprisingly, the characteristics of the response assessed will determine the nature of the resulting record. For example, some responses display substantial habituation or adaptation effects; that is, the same stimulus evokes lowered levels of responding following repeated stimulation, both within and across sessions (cf. Barlow, Leitenburg, & Agras, 1969; Montague & Coles, 1966). Responsivity to stimulation also will vary inversely with the prestimulus level of that response. According to this "law of initial values," a change in heart rate from 120 to 125 is different from, and probably greater than, a change from 70 to 75. Thus some form of data transformation may be necessary to equate response changes at various ranges of the response dimension (e.g., Ray & Raczynski, 1981). Individuals also may show *response specificity*, or a particular pattern of responding across related stimuli (e.g., Lacey, 1959). Because individuals vary in the response system that is most reactive, investigators should assess their clients' reactivity before selecting a measure that will be sensitive to the changes resulting from treatment. Some physiological systems also may be responsive to circadian rhythms, and to diurnal as well as layer cyclic effects (Haynes & Wilson, 1979); again, familarity with standard technique references is critical to the judicious selection of measurement procedures.

NOTES

1. The by-products, or traces (e.g., Webb, Campbell, Schwartz, Sechrest, & Grove, 1981), of behaviors such as pounds gained and cigarettes smoked also are considered grist for the assessment mill.

2. The inconsistency in target behavior selection is due in part to variations in individual assessors' notions of what is socially important (Baer et al., 1968), their personal values regarding the relative desirability of alternative behaviors, their conceptions of deviancy, and their familiarity with the immediate and long-term consequences of various forms of problem behavior. The operation of these factors can be seen in the recent controversies centering on modifying feminine sex-role behaviors among boys and annoying, but only mildly disruptive, classroom behaviors (e.g., Winett & Winkler, 1972; Winkler, 1977).

3. Not infrequently, additional behaviors will be monitored during one or more of the aforementioned phases. For example, measurements may be regularly or periodically obtained on the independent, or treatment, variable to ensure that it is manipulated in the intended manner. L. Peterson, Homer, and Wonderlich (1982) argued that the infrequent use of independent variable checks seriously threatens the reliability and validity of applied behavior studies. Along with J. M. Johnston and Pennypacker (1980), they suggested a variety of methods of assessing the integrity of independent variable manipulations. Similar recommendations are given in related treatment literatures (e.g., Hartmann, Roper, & Gelfand, 1977; Paul & Lentz, 1977).

   At other times the investigator may choose to measure environmental events such as the opportunities to perform the target response (Hawkins, 1982). For example, when the target is "instruction following," assessing the client's performance may require measurement of the occurrence of each instruction or request.

Without such an assessment, it may be impossible to distinguish changes in compliance by the client from changes in requesting by the client's environment. More complicated sets of environmental events also may be monitored regularly when patterns of responding rather than single events are targeted, as illustrated in the work by Patterson (1982) and by Gottman (1979).

Other client behaviors also may be monitored, including behaviors that might be expected to reflect collateral effects of treatment—either beneficial generalized effects or undesirable side effects (Drabman, Hammer, & Rosenbaum, 1979; Kazdin, 1982c; Stokes & Baer, 1977).

4. A very important, but often overlooked, practical advantage of defining target behaviors consistently with the definitions employed in earlier studies is that the observational systems used in these studies may be readily adapted to current needs. See Haynes (1978, pp. 119–120) and Haynes and Wilson (1979, pp. 49–52) for a sample listing of observational systems; Simon and Boyer (1974) for an anthology; and Barlow (1981), Ciminero, Calhoun, and Adams (1977), Hersen and Bellack (1981); and Mash and Terdal (1981) for surveys of topic-area reviews.

5. When observers perform consistently, yet inaccurately, the phenomenon is labeled *consensual observer drift* (Johnson & Bolstad, 1973).

6. *Reliability* sometimes refers to consistency between standard scores from observers (or settings or occasions), whereas *agreement* refers to consistency between their raw scores (Tinsley & Weiss, 1975). A related term, *observer accuracy,* refers to comparisons between an observer and an established criterion. Various investigators have argued that observer accuracy assessments should be preferred to interobserver reliability or agreement assessments (e.g., Cone, 1982). Possible accuracy criteria include audio- or video-recorded behaviors orchestrated by a predetermined script, mechanically generated responses, and mechanical measurements of behavior (Boykin & Nelson, 1981). However, the development of criterion ratings is infeasible in many situations. Even when it is feasible, agreement with criterion ratings can provide unrepresentative estimates of accuracy if observers can discriminate between accuracy assessments and more typical observations. In such a case, users of observational systems are left with interobserver reliability as an indirect measure of accuracy.

7. Self-report measures have proliferated at such a rapid rate that at least one well-known behavioral assessor suggested that journal editors limit these devices by not considering for publication those studies employing new instruments that are not demonstrably superior to existing ones (see comments by blue-ribbon panelists in Hartmann, 1983).

8. Criteria for selecting or constructing measures of consumer satisfaction with treatment, an increasingly popular complement to objective assessment of treatment outcome, were described in a *Behavior Therapy* miniseries (Forehand, 1983).

9. Though physiological measurement typically occurs in an environmentally controlled context (a laboratory), advances in telemetry have permitted *in situ* recordings of various physiological responses (Rugh & Schwitzgebel, 1977; Vermillyea et al., in press).

CHAPTER 5

Basic A-B-A Withdrawal Designs

5.1. INTRODUCTION

In this chapter we will examine the prototype of experimental single-case research—the A-B-A design—and its many variants. The primary objective is to inform and familiarize the reader as to the advantages and limitations of each design strategy while illustrating from the clinical, child, and behavior modification literatures. The development of the A-B-A design will be traced, beginning with its roots in the clinical case study and in the application of "quasi-experimental designs" (Campbell & Stanley, 1966). Procedural issues discussed at length in chapter 3 will also be evaluated here for each of the specific design options as they apply. Both "ideal" and "problematic" examples, selected from the applied research area, will be used for illustrative purposes.

Since the publication of the first edition of this book (Hersen & Barlow, 1976) the literature has become replete with examples of A-B-A designs. However, there has been very little change with respect to basic procedural issues. Therefore, we have retained most of the original design illustrations but have added some more recent examples from the applied behavioral literature.

Limitations of the case study approach

For many years, descriptions of uncontrolled case histories have predominated in the psychoanalytic, psychotherapeutic, and psychiatric literatures (see chapter 1). Despite the development of applied behavioral methodology (presumably based on sound theoretical underpinnings) in the late 1950s and early to mid-1960s, the case study approach was still the primary method for demonstrating the efficacy of innovative treatment techniques (cf. Ashem, 1963; Barlow, 1980; Barlow et al., 1983; Lazarus, 1963; Ullmann & Krasner, 1965; Wolpe, 1958, 1976).

Although there can be no doubt that the case history method yields interesting (albeit uncontrolled) data, that it is a rich source for clinical speculation, and that ingenious technical developments derive from its application, the multitude of uncontrolled factors present in each study do not permit sound cause-and-effect conclusions. Even when the case study method is applied at its best (e.g., Lazarus, 1973), the absence of experimental control and the lack of precise measures for target behaviors under evaluation remain mitigating factors. Of course, proponents of the case study method (e.g., Lazarus & Davison, 1971) are well aware of its inherent limitations as an evaluative tool, but they show how it can be used to advantage to generate hypotheses that later may be subjected to more rigorous experimental scrutiny. Among their advantages, the case study method can be used to (1) foster clinical innovation, (2) cast doubt on theoretic assumptions, (3) permit study of rare phenomena (e.g., Gilles de la Tourette's Syndrome), (4) develop new technical skills, (5) buttress theoretical views, (6) result in refinement of techniques, and (7) provide clinical data to be used as a departure point for subsequent controlled investigations.

With respect to the last point, Lazarus and Davison (1971) referred to the use of "objectified single case studies." Included are the A-B-A experimental designs that allow for an analysis of the controlling effects of variables, thus permitting scientifically valid conclusions. However, in the more typical case study approach, a subjective description of treatment interventions and resulting behavioral changes is made by the therapist. Most frequently, several techniques are administered simultaneously, precluding an analysis of the relative merits of each procedure. Moreover, evidence for improvement is usually based on the therapist's "global" clinical impressions. Not only is there the strong possibility of bias in these evaluations, but controls for the treatment's placebo value are unavailable. Finally, the effects of time (maturational factors) are confounded with application of the treatment(s), and the specific contribution of each of the factors is obviously not distinguished.

More recently, Kazdin (1981) has pointed out how ". . . the scientific yield from case reports might be improved in clinical practice where methodological alternatives are unavailable" (p. 183). In ascending order of rigor, three types are described: (1) cases with preassessment and postassessment, (2) cases with repeated assessment and marked changes, and (3) multiple cases with continuous assessment and stability information (e.g., no change in a patient's condition over extended periods of time despite prior therapeutic efforts). However, notwithstanding improvements inherent in the aforementioned case approaches, threats to internal validity are still present to one degree or another.

A very modest improvement over the uncontrolled case study method elsewhere (Browning & Stover, 1971) has been labeled the "B Design." In this "design," baseline measurement is omitted, but the investigator monitors one of a number of target measures throughout the course of treatment. One might also categorize this procedure as the simplest of the time series analyses (see G. V. Glass, Willson, & Gottman, 1973). Although this strategy obviously yields a more objective appraisal of the patient's progress, the confounds that typify the case study method apply equally here. In that sense the B Design is essentially an uncontrolled case study with objective measures taken repeatedly. This, of course, is the same as Kazdin's (1981) description of cases with repeated assessment and marked changes.

5.2. A-B DESIGN

The A-B design, although the simplest of the experimental strategies, corrects for some of the deficiencies of the case study method and those of the B Design. In this design the target behavior is clearly specified, and repeated measurement is taken throughout the A and B phases of experimentation. As in all single-case experimental research, the A phase involves a series of baseline observations of the natural frequency of the target behavior(s) under study. In the B phase the treatment variable is introduced, and changes in the dependent measure are noted. Thus, *with some major reservations*, changes in the dependent variable are attributed to the effects of treatment (Barlow & Hersen, 1973; Campbell, 1969; Campbell & Stanley, 1966; Cook & Campbell, 1979; Hersen, 1982; Kazdin, 1982b; Kratochwill, 1978b).

Let us now examine some of the important reservations. In their evaluation of the A-B strategy, Wolf and Risley (1971) argued that "The analysis provided no information about what the natural course of the behavior would have been had we not intervened with our treatment condition" (pp. 314–315). That is to say, it is very possible that changes in the B phase might have occurred regardless of the introduction of treatment or that changes in B might have resulted as a function of correlation with some fortuitous (but uncontrolled) event. When considered in this light, the A-B strategy does not permit a full experimental analysis of the controlling effects of the treatment inasmuch as its correlative properties are quite apparent. Indeed, Campbell and Stanley (1966) referred to this strategy as a "quasi-experimental design."

Risley and Wolf (1972) presented an interesting discussion of the limitations of the A-B design with respect to predicting, or "forecasting," the B phase on the basis of data obtained in A. Two hypothetical examples of the A-B design were depicted, with both showing a mean increase in the amount of behavior in B over A. However, in the first example, a steady and stable trend in baseline is followed by an abrupt increase in B, which is then maintained. In the second case, the upward trend in A is continued in B. Therefore, despite the equivalence of means and variances in the two cases, the importance of the trend in evaluating the data is underscored. Some tentative conclusions can be reached on the basis of the first example, but in the second example the continued linear trend in A permits no conclusions as to the controlling effects of the B treatment variable.

In further analyzing the difficulties inherent in the A-B strategy, Risley and Wolf (1972) contended that:

> The weakness in this design is that the data in the experimental condition is compared with a forecast from the prior baseline data. The accuracy of an assessment of the role of the experimental procedure in producing the change rests upon the accuracy of that forecast. A strong statement of causality therefore requires that the forecast be supported. This support is accomplished by elaborating the A-B design. (p. 5)

Such elaboration is found in the A-B-A design discussed and illustrated in section 5.3 of this chapter.

Despite these aforementioned limitations, it is shown how in some settings (where control-group analysis or repeated introduction and withdrawals of treatment variables are not feasible) the A-B design can be of some utility (Campbell & Stanley, 1966; Cook & Campbell, 1979. For example, the use of the A-B strategy in the private-practice setting has previously been recommended in section 3.2 of chapter 3 (see also Barlow et al., 1983).

Campbell (1969) presented a comprehensive analysis of the use of the A-B strategy in field experiments where more traditional forms of experimentation are not at all possible (e.g., the effects of modifying traffic laws on the documented frequency of accidents). However one uses the quasi-experimental design, Campbell cautioned the investigator as to the numerous threats to internal validity (history, maturation, instability, testing, instrumentation, regression artifacts, selection, experimental mortality, and selection-maturation interaction) and external validity (interaction effects of testing, interaction of selection and experimental treatment, reactive effects of experimental arrangements, multiple-treatment interference, irrelevant responsiveness of measures, and irrelevant replicability of treatments) that may be encountered. The interested reader is referred to Campbell's (1969) excellent article for a full discussion of the issues involved in large-scale retrospective or prospective field studies.

In summary, it should be apparent that the use of a quasi-experimental design such as the A-B strategy results in rather weak conclusions. This design is subject to the influence of a host of confounding variables and is best applied as a last-resort measure when circumstances do not allow for more extensive experimentation. Examples of such cases will now be illustrated.

A-B with single target measure and follow-up

Epstein and Hersen (1974) used an A-B design with a follow-up procedure to assess the effects of reinforcement on frequency of gagging in a 26-year-old psychiatric inpatient. The patient's symptomatology had persisted for approximately 2 years despite repeated attempts at medical intervention. During baseline (A phase), the patient was instructed to record time and frequency of each gagging episode on an index card, collected by the experimenter the following morning at ward rounds. Treatment (B phase) consisted of presenting the patient with $2.00 in canteen books (exchangeable at the hospital store for goods) for a decrease (N − 1) from the previous daily frequency. In addition, zero rates of gagging were similarly reinforced. In order to facilitate maintenance of gains after treatment, no instructions were given as to how the patient might control his gagging. Thus emphasis was placed on self-management of the disorder. At the conclusion of his hospital stay, the patient was requested to continue recording data at home for a period of 12 weeks. In this case, treatment conditions were not withdrawn during the patient's hospitalization because of clinical considerations.

Results of this study are plotted in Figure 5-1. Baseline frequency of gagging fluctuated between 8 and 17 episodes per day but stabilized to some extent in the last 4 days. Institution of reinforcement procedures in the B phase resulted in a decline to zero within 6 days. However, on Day 15, frequency of gagging rose again to seven daily episodes. At this point, the criterion for obtaining reinforcement was reset to that originally planned for

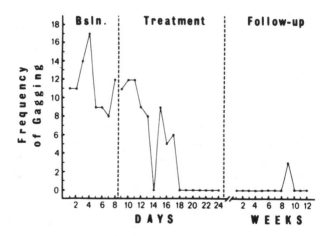

FIGURE 5-1. Frequency of gagging during baseline, treatment, and follow-up. (Figure 1, p. 103, from: Epstein, L. H., & Hersen, M. (1974). Behavioral control of hysterical gagging. *Journal of Clinical Psychology,* **30,** 102–104. Copyright 1974 by American Psychological Association. Reproduced by permission.)

Day 13. Renewed improvement was then noted between Days 15–18, and treatment was continued through Day 24. Thus the B phase was twice as long as baseline, but it was extended for very obvious clinical considerations.

The 12-week follow-up period reveals a zero level of gagging, with the exception of Week 9, when three gagging episodes were recorded. Follow-up data were corroborated by the patient's wife, thus precluding the possibility that treatment only affected the patient's verbal report rather than diminution of actual symptomatology.

Although treatment appeared to be the effective ingredient of change in this study, particularly in light of the longevity of the patient's disorder, it is conceivable that some unidentified variable coincided with the application of reinforcement procedures and actually accounted for observed changes. However, the A-B design does not permit a definitive answer to this question. It might also be noted that the specific use of this design (baseline, treatment, and follow-up) could readily have been carried out in an outpatient facility (clinic or private-practice setting) with a minimum of difficulty and with no deleterious effects to the patient.

Lawson (1983) also used an A-B design with a single target behavior (alcohol consumption) and obtained a follow-up assessment. His case involved a divorced 35-year-old male with a history of problem drinking beginning at age 16. He periodically would experience blackouts as a function of his drinking. But despite the chronicity of his problem, with the exception of a few AA meetings, the subject had not obtained any form of treatment for his alcoholism. Baseline data (based on the subject's self-report) indicated that he consumed an average of 65 drinks per week (see Figure 5-2). This was confirmed by his girlfriend.

Treatment (B phase) began in the third week, and, on the basis of the behavioral analyses performed, three goals were identified: (1) to decrease alcohol consumption, (2) to improve social relationships, and (3) to diminish frequency of anxiety and depression episodes. Thus the comprehensive therapy program involved goal setting with regard to number of drinks consumed, rate-reduction strategies, stimulus-control strategies, development of new social relationships and recreational activities, assertion training, and self-management of depression.

Examination of data in Figure 5-2 indicates that there were substantial improvements in rate of drinking during the course of therapy (to about 10 drinks per week) that appeared to be maintained at the 3-month follow-up (also confirmed by the girlfriend). Indeed, an informal communication received by the therapist 1½ years subsequent to treatment further confirmed that the subject still was drinking in a socially acceptable manner.

Treatment did appear to be responsible for change in Lawson's (1983) alcoholic, particularly given the 19-year history of excessive drinking. This, then, from a design standpoint, fits in nicely with Kazdin's notion of repeated

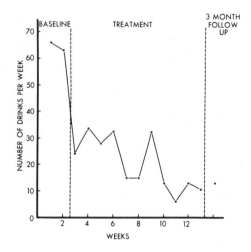

FIGURE 5-2. Weekly self-monitored alcohol consumption during baseline, treatment, and at 3-month follow-up. (Figure 6-1, p. 165, from: Lawson, D. M. Alcoholism. In M. Hersen (Ed.). (1983). Outpatient behavior therapy: A clinical guide. New York: Grune & Stratton. Copyright 1983 by M. Hersen. Reproduced by permission.)

assessment with marked changes and stability information improving the quality of case study. But, in spite of this, the A-B design *does not* allow for a clear demonstration of the controlling effects of the treatment. For that we require an A-B-A or A-B-A-B strategy.

A-B with multiple-target measures

In our next example we will examine the use of an A-B design in which a number of target behaviors were monitored simultaneously (Eisler & Hersen, 1973). The effects of token economy on points earned, behavioral ratings of depression (Williams et al., 1972), and self-ratings of depression (Beck Depressive Inventory—A. T. Beck, Ward, Mendelsohn, Mock, & Erbaugh, 1961) were assessed in a 61-year-old reactively depressed male patient. In this study the treatment variable was not withdrawn due to time limitations. During baseline (A), the patient was able to earn points for a variety of specified target behaviors (designated under general rubrics of *work, personal hygiene,* and *responsibility*), but these earned points were exchangeable for ward privileges and material goods in the hospital canteen. During each phase, the patient filled out a Beck Depressive Inventory (three alternate forms were used to prevent possible response bias) at daily morning "Banking Hours," at which time points previously earned on the token economy were tabulated. In addition, behavioral ratings (talking, smiling, motor activity) of depression (high ratings indicate low depression) were obtained surreptitiously on the average of one per hour between the hours of 8:00 A.M. and 10:00 P.M. during non-work-related activities.

The results of this study appear in Figure 5-3. Inspection of these data indicates that number of points earned in baseline increased slightly but then stabilized. Baseline ratings of depression show stability, with evidence of greater daytime activity. Beck scores ranged from 19–28. Institution of token economy on Day 5 resulted in a marked linear increase in points earned, a substantial increase in day and evening behavioral ratings of depression, and a linear descrease in self-reported Beck Inventory scores.

Thus it appears that token economy effected improvement in this patient's depression as based on both objective and subjective indexes. However, as was previously pointed out, this design does not permit a direct analysis of the controlling effects of the therapeutic variable introduced (token economy), as does our example of an A-B-A design seen in Figure 5-7 (Hersen, Eisler, Alford, & Agras, 1973). Nonetheless, the use of an A-B design in this case proved to be useful for two reasons. *First*, from a clinical standpoint, it was possible to obtain *some* objective estimate of the treatment's success during the patient's abbreviated hospital stay. *Second*, the results of this study prompted the further investigation of the effects of token economic procedures in three additional reactively depressed subjects (Hersen, Eisler, Alford, & Agras, 1973). In that investigation more sophisticated experimental strategies confirmed the controlling effects of token economy in neurotic depression.

A-B with multiple-target measures and follow-up

A more recent and more complicated example of an A-B design with multiple-target measures and follow-up was described by St. Lawrence, Bradlyn, and Kelly (1983). The subject was a 35-year-old male with a 20-year history of homosexual functioning, but whose interpersonal adjustment was unsatisfactory. Treatment, therefore, was directed to enhancing several components of social skill. Five components requiring modification were identified during two baseline assessments: (1) percentage of eye contact, (2) smiles, (3) extraneous movements, (4) appropriate verbal content, and (5) overall social skill. Assessment involved the patient and a male confederate role-playing 16 scenes (8 commendatory; 8 refusal) that were videotaped.

Social skills training was conducted twice a week for nine weeks and consisted of modeling, instructions, behavior rehearsal, cognitive modification, and *in vivo* practice. Training was carried out with half of the commendatory and refusal scenes; the other half served as a measure of generalization. In addition, follow-up sessions were conducted at 1 and 6 months after conclusion of treatment.

The results of this A-B analysis appear in Figure 5-4, with the left half

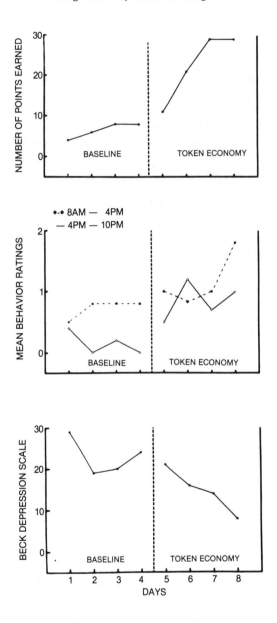

FIGURE 5-3. Number of points earned, mean behavioral ratings, and Beck Depression Scale scores during baseline and token economy in a reactively depressed patient. (Figure 1, from: Eisler, R. M., Hersen, M. (1973). The A-B design: Effects of token economy on behavioral and subjective measures in neurotic depression. Paper presented at the meeting of the American Psychological Association, Montreal, August 29.)

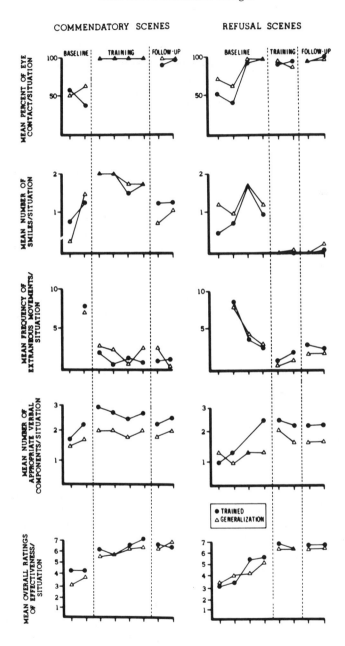

FIGURE 5-4. Mean frequency of targeted behaviors in refused and commendatory role-play situations. (Figure 1, p. 50, from: St. Lawrence, J. S., Bradlyn, A. S., & Kelly, J. A. (1983). Interpersonal adjustment of a homosexual adult: Enhancement via social skills training. *Behavior Modification, 7,* 41–55. Copyright 1983 by Sage Publications. Reproduced by permission.)

portraying commendatory scenes and the right half refusal scenes. In general, improvements during training suggest that the treatment was effective for both categories (commendatory and refusal) and that there was transfer of gains from trained to generalization scenes. Moreover, gains appeared to remain in follow-up, with the exception of smiles (commendatory). However, a closer examination does reveal a number of problems with these data. *First*, for the commendatory scenes there are only one- or two-point baselines. Therefore, complete establishment of baseline trends was not possible. Also, for two of the behaviors (smiles, appropriate verbal content), improvements in training similarly appear to be the continuation of baseline trends. *Second*, this also seemed to be the case with regard to refusal scenes for the following components: eye contact, extraneous movements, appropriate verbal content, and overall social skill. Thus, although the subject was obviously clinically improved, these data do not clearly reflect experimental confirmation of such improvement, given the limited confidence one can ever have with the A-B strategy.

A-B with follow-up and booster treatment

In our next illustration of an A-B design, clinical considerations necessitated a short baseline period and also contraindicated the withdrawal of treatment procedures (Harbert, Barlow, Hersen, & Austin, 1974). However, during the course of extended follow-up assessment, the patient's condition deteriorated and required the reinstatement of treatment in booster sessions. Renewed improvement immediately followed, thus lending additional support for the treatment's efficacy. When examined from a design standpoint, the conditions of the more complete A-B-A-B strategy are approximated in this experimental case study.

More specifically, Harbert et al. (1974) examined the effects of covert sensitization therapy on self-report (card sort technique) and physiological (mean penile circumference changes) indices in a 52-year-old male inpatient who complained of a long history of incestuous episodes with his adolescent daughter. The card sort technique consisted of 10 scenes (typed on cards) depicting the patient and his daughter. Five of these scenes were concerned with normal father-daughter relations; the remaining five involved descriptions of incestuous activity between father and daughter. The patient was asked to rate the 10 scenes, presented in random sequence, on a 0–4 basis, with 0 representing no desire and 4 representing much desire. Thus measures of both deviant and nondeviant aspects of the relationship were obtained throughout all phases of study. In addition, penile circumference changes scored as a percentage of full erection were obtained in response to audiotaped descriptions of incestuous activity and in reaction to slides of the daughter. Three days of self-report data and 4 days of physiological measurements were taken during baseline (A phase).

Covert sensitization treatment (B phase) consisted of approximately 3 weeks of daily sessions in which descriptions of incestuous activity were paired with the nauseous scene as used by Barlow, Leitenberg, and Agras (1969). However, as nausea proved to be a weak aversive stimulus for this patient, a "guilt" scene—in which the patient is discovered engaging in sexual activity with the daughter by his current wife and a respected priest—was substituted during the second week of treatment. The flexibility of the single-case approach is exemplified here inasmuch as a "therapeutic shift of gears" follows from a close monitoring of the data.

Follow-up assessment sessions were conducted after termination of the patient's hospitalization at 2-week, 1-, 2-, 3-, and 6-month intervals. After each follow-up session, brief booster covert sensitization was administered.

The results of this study appear in Figure 5-5 and 5-6. Inspection of Figure 5-5 indicates that mean penile circumference changes to audiotapes in baseline ranged from 18% to 35% (mean = 22·8%). Penile circumference changes to slides ranged from 18% to 75% (mean = 43·5%). Examination of Figure 5-6 shows that nondeviant scores remained at a maximum of 20 for all three baseline probes; deviant scores achieved a level of 17 throughout.

Introduction of standard covert sensitization, followed by use of the guilt imagery resulted in decreased penile responding to audiotapes and slides (see Figure 5-5) and a substantial decrease in the patient's self-reports of deviant

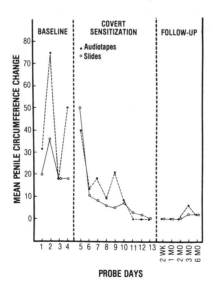

FIGURE 5-5. Mean penile circumference change to audiotapes and slides during baseline, covert sensitization, and follow-up. (Figure 1, p. 83, from: Harbert, T. L., Barlow, D. H., Hersen, M., & Austin, J. B. (1974). Measurement and modification of incestuous behavior: A case study, *Psychological Reports,* **34**, 79–86. Copyright 1974 by Psychological Reports. Reproduced by permission.)

interests in his daughter (see Figure 5-6). Nondeviant interests, however, remained at a high level.

Follow-up data in Figure 5-5 reveal that penile circumference changes remained at zero during the first three probes but increased slightly at the 3-month assessment. Similarly, Figure 5-6 data show a considerable increase in deviant interests at the 3-month follow-up. This coincides with the patient's reports of marital disharmony. In addition, nondeviant interests diminished during follow-up (at that point the patient was angry at his daughter for rejecting his positive efforts at being a father).

As there appeared to be some deterioration at the 3-month follow-up, an additional course of outpatient covert sensitization therapy was carried out in three weekly sessions. The final assessment period at 6 months appears to reflect the effects of additional treatment in that (1) penile responding was negligible, and (2) deviant interests had returned to a zero level.

5.3. A-B-A DESIGN

The A-B-A design is the simplest of the experimental analysis strategies in which the treatment variable is introduced and then withdrawn. For this reason, this strategy as well as those that follow, are most often referred to as *withdrawal designs*. Whereas the A-B design permits only tentative conclusions as to a treatment's influence, the A-B-A design allows for an analysis of the controlling effects of its introduction and subsequent removal. If after

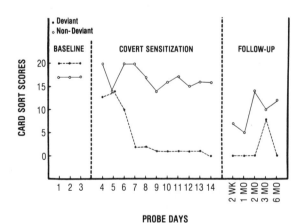

FIGURE 5-6. Card sort scores on probe days during baseline, covert sensitization, and follow-up. (Figure 2, p. 84, from: Harbert, T. L., Barlow, D. H., Hersen, M., & Austin, J. B. (1974). Measurement and modification of incestuous behavior: A case study. *Psychological Reports,* **34,** 79–86. Copyright 1974 by Psychological Reports. Reproduced by permission.)

baseline measurement (A) the application of a treatment (B) leads to improvement and coversely results in deterioration after it is withdrawn (A), one can conclude with a high degree of certainty that the treatment variable is the agent responsible for observed changes in the target behavior. Unless the natural history of the behavior under study were to follow identical fluctuations in trends, it is *most improbable* that observed changes would be due to any influence (e.g., some correlated or uncontrolled variable) other than the treatment variable that is systematically changed. Also, replication of the A-B-A design in different subjects strengthens conclusions as to power and controlling forces of the treatment (see chapter 10).

Although the A-B-A strategy is acceptable from an experimental standpoint, it has one major undesirable feature when considered from the clinical context. Unfortunately for the patient or subject, this paradigm ends on the A or baseline phase of study, therefore denying him or her the full benefits of experimental treatment. Along these lines, Barlow and Hersen (1973) have argued that:

> On an ethical and moral basis it certainly behooves the experimenter-clinician to continue some form of treatment to its ultimate conclusion subsequent to completion of the research aspects of the case. A further design, known as the A-B-A-B design, meets this criticism as study ends on the B or treatment phase. (p. 321).

However, despite this limitation, the A-B-A design is a useful research tool when time factors (e.g., premature discharge of a patient) or clinical aspects of a case (e.g., necessity of changing the level of medication in addition to reintroducing a treatment variable after the second A phase) interfere with the correct application of the more comprehensive A-B-A-B strategy.

A second problem with the A-B-A strategy concerns the issues of multiple-treatment interference, particularly sequential confounding (Bandura, 1969; Cook & Campbell, 1979). The problem of sequential confounding in an A-B-A design and its variants also somewhat limits generalization to the clinic. As Bandura (1969) and Kazdin (1973b) have noted, the effectiveness of a therapeutic variable in the final phase of an A-B-A design can only be interpreted in the context of the previous phases. Change occurring in this last phase may not be comparable to changes that would have occurred if the treatment had been introduced initially. For instance, in an A-B-BC-B design, when A is baseline and B and C are two therapeutic variables, the effects of the BC phase may be more or less powerful than if they had been introduced initially. This point has been demonstrated in studies by O'Leary and his associates (O'Leary & Becker, 1967; O'Leary, Becker, Evans, & Saudargas, 1969), who noted that the simultaneous introduction of two variables produced greater change than the sequential introduction of the same two variables.

Similarly, the second introduction of variable A in a withdrawal A-B-A design may affect behavior differently than the first introduction. (Generally our experience is that behavior improves more rapidly with a second introduction of the therapeutic variable.) In any case, the reintroduction of therapeutic phases is a feature of A-B-A designs that differs from the typical applied clinical situation, when the variable is introduced only once. Thus, appropriate cautions must be exercised in generalizing results from phases occurring late in an experiment to the clinical situation.

In dealing with this problem, the clinical researcher should keep in mind that the purpose of subsequent phases in an A-B-A design is to confirm the effects of the independent variable (internal validity) rather than to generalize to the clinical situation. The results that are most generalizable, of course, are data from the first introduction of the treatment. When two or more variables are introduced in sequence, the purpose again is to test the separate effects of each variable. Subsequently, order effects and effects of combining the variable can be tested in systematic replication series, as was the case with the O'leary, Becker, Evans, and Saudergas (1969) study.

Two examples of the A-B-A design, one selected from the clinical literature and one from the child development area, will be used for illustration. Attention will be focused on some of the procedural issues outlined in chapter 3.

A-B-A from clinical literature

In pursuing their study of the effects of token economy on neurotic depression, Hersen and his colleagues (Hersen, Eisler, Alford, & Agras, 1973) used A-B-A strategies with three reactively depressed subjects. The results for one of these subjects (52-year-old, white, married farmer who became depressed after the sale of his farm) appear in Figure 5-7. As in the Eisler and Hersen (1973) study, described in detail in section 5.2 of this chapter, points earned in baseline (A) had no exchange value, but during the token reinforcement phase (B) they were exchangeable for privileges and material goods. Unlike the Eisler and Hersen study, however, token reinforcement procedures were withdrawn, and a return to baseline conditions (A) took place during Days 9–12. The effects of introducing and removing token economy were examined on two target behaviors—points earned and behavioral ratings (higher ratings indicate lowered depression).

A careful examination of baseline data reveals a slightly decreased trend in behavioral ratings, thus indicating some very minor deterioration in the patient's condition. As was noted in section 3.3 of chapter 3, the deteriorating baseline is considered to be an acceptable trend. However, there appeared to be a concomitant but slight increase in points earned during baseline. It will be recalled that an improved trend in baseline is not the most desirable trend.

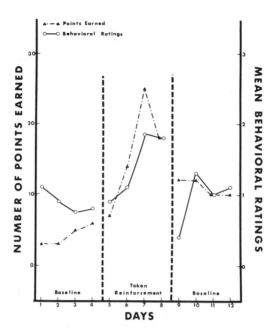

FIGURE 5-7. Number of points earned and mean behavioral ratings for Subject 1. (Figure 1, p. 394, from: Hersen, M., Eisler, R. M., Alford, G. S., & Agras, W. S. (1973). Effects of token economy on neurotic depression: An experimental analysis, *Behavior Therapy*, 4, 392-397. Copyright 1973 by Association for the Advancement of Behavior Therapy. Reproduced by permission.)

However, as the slope of the curve was not extensive, and in light of the primary focus on behavioral ratings (depression), we proceeded with our change in conditions on Day 5. Had there been unlimited time, baseline conditions would have been maintained until number of points earned daily stabilized to a greater extent.

We might note parenthetically at this point that all of the ideal conditions (procedural rules) outlined in our discussion in chapter 3 are rarely approximated when conducting single-case experimental research. Our experience shows that procedural variations from the ideal are required, as data simply *do not* conform to theoretical expectation. Moreover, experimental finesse is sometimes sacrificed at the expense of time and clinical considerations.

Continued examination of Figure 5-7 indicates that instigation of token economic procedures on Day 5 resulted in a marked linear increase in both points earned and behavioral ratings. The abrupt change in slope of the curves, particularly in points earned, strongly suggests the influence of the token economy variable, despite the slightly upward trend initially seen in baseline. Removal of token economy on Day 9 led to an initially large drop in behavioral ratings, which then stabilized at a somewhat higher level. Points earned also declined but maintained stability throughout the second 4-day baseline period. The obtained decrease in target behaviors in the second baseline phase confirms the controlling effects of token economy over neurotic depression in this paradigm. We might also point out here that an equal number of data points appears in each phase, thus facilitating interpretation of the trends.

These results were replicated in two additional reactively depressed subjects (Hersen, Eisler, Alford, & Agras, 1973), lending further credence to the notion that token economy exerts a controlling influence over the behavior of neurotically depressed individuals.

A-B-A from child literature

Walker and Buckley (1968) used an A-B-A design in their functional analysis of the effects of an individualized educational program for a 9½-year-old boy whose extreme distractibility in a classroom situation interfered with task-oriented performance (see Figure 5.8). During baseline assessment (A), percentage of attending behavior was recorded in 10-minute observation sessions while the subject was engaged in working on programmed learning materials. Following baseline measurement, a reinforcement contingency (B) was instituted whereby the subject earned points (exchangeable for a model of his choice) for maintaining his attention (operationally defined for him) to the learning task. During this phase, a progressively increasing time criterion for attending behaviors over sessions was required (30 to 600 seconds of attending per point). The extinction phase (A) involved a return to original baseline conditions.

Examination of baseline data shows a slightly decreasing trend followed by a slightly increasing trend, but within stable limits (mean = 33%). Institution of reinforcement procedures led to an immediate improvement, which then increased to its asymptote in accordance with the progressively more difficult criterion. Removal of the reinforcement contingency in extinction resulted in a decreased percentage of attending behaviors to approximately baseline levels. After completion of experimental study, the subject was returned to his classroom where a variable interval reinforcement program was used to increase and maintain attending behaviors in that setting.

With respect to experimental design issues, we might point out that Walker and Buckley (1968) used a short baseline period (6 data points) followed by longer B (15 data points) and A phases (14 data points). However, in view of the fact that an immediate and large increase in attention was obtained during reinforcement, the possible confound of time when using disparate lengths of phases (see section 3.6, chapter 3) *does not* apply here. Moreover, the shape

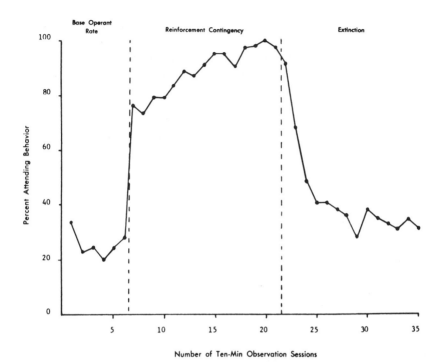

FIGURE 5-8. Percentage of attending behavior in successive time samples during the individual conditioning program. (Figure 2, p. 247, from: Walker, H. M., & Buckley, N. K. (1968). The use of positive reinforcement in conditioning attending behavior. *Journal of Applied Behavior Analysis,* 1, 245–250. Copyright 1968 by Society for the Experimental Analysis of Behavior, Inc. Reproduced by permission.)

of the curve in extinction (A) and the relatively equal lengths of the B and A phases further dispel doubts that the reader might have as to the confound of time.

Secondly, with respect to the decreasing-increasing baseline obtained in the first A phase, although it might be preferable to extend measurement until full stability is achieved (see section 3.3, chapter 3), the range of variability is very constricted here, thus delimiting the importance of the trends.

5.4. A-B-A-B DESIGN

The A-B-A-B strategy, referred to as an *equivalent time-samples design* by Campbell and Stanley (1966), controls for the deficiencies present in the A-B-A design. Specifically, the A-B-A-B design ends on a treatment phase (B), which then can be extended beyond the experimental requirements of study for clinical reasons (e.g., Miller, 1973). In addition, this design strategy provides for *two* occasions (B to A and then A to B) for demonstrating the positive effects of the treatment variable. This, then, strengthens the conclusions that can be derived as to its controlling effects over target behaviors under observation (Barlow & Hersen, 1973).

In the succeeding subsections we will provide four examples of the use of the A-B-A-B strategy. In the first we will present examples from the child literature which illustrate the ideal in procedural considerations. In the second we will examine the problems encountered in interpretation when improvement fortuitously occurrs during the second baseline period. In the third we will illustrate the use of the A-B-A-B design when concurrent behaviors are monitored in addition to targeted behaviors of interest. Finally, in the fourth we will examine the advantages and disadvantages of using the A-B-A-B strategy without the experimenter's knowledge of results throughout the different phases of study.

A-B-A-B from child literature

An excellent example of the A-B-A-B design strategy appears in a study conducted by R. V. Hall et al. (1971). In this study the effects of contingent teacher attention were examined in a 10-year-old retarded boy whose "talking-out" behaviors during special education classes proved to be disruptive, as other children then emulated his actions. Baseline observations of *talk-outs* were recorded by the teacher (reliability checks indicated 84% to 100% agreement) during five daily 15-minute sessions. During these first five sessions, the teacher responded naturally to talk-outs by paying attention to them. However, in the next five sessions, the teacher was instructed to ignore talk-outs but to provide increased attention to the child's productive behaviors. The third series of five sessions involved a return to baseline conditions, and the last series of five sessions consisted of reinstatement of contingent attention.

The results of this study are plotted in Figure 5-9. The presence of equal phases in this study facilitates the analysis of results. Baseline data are stable and range from three to five talk-outs, with three of the five points at a level of four talk-outs per session. Institution of contingent attention resulted in a marked decrease that achieved a zero level in Sessions 9 and 10. Removal of contingent attention led to a linear increase of talk-outs to a high of five. However, reinstatement of contingent attention once again brought talk-outs under experimental control. Thus application and withdrawal of contingent attention clearly demonstrates its controlling effects on talk-out behaviors.

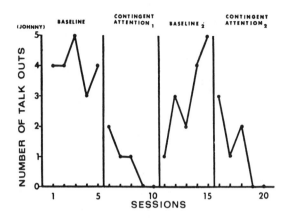

FIGURE 5-9. A record of talking out behavior of an educable mentally retarded student. Baseline₁—before experimental conditions. Contingent Teacher Attention₁—systematic ignoring of talking out and increased teacher attention to appropriate behavior. Baseline₂—reinstatment of teacher attention to talking out behavior. (Figure 2, p. 143, from: Hall, R. V., Fox, R., Willard, D., Goldsmith, L., Emerson, M., Owen, M., Davis, T., & Porcia, E. (1971). The teacher as observer and experimenter in the modification of disputing and talking-out behaviors. *Journal of Applied Behavior Analysis, 4*, 141–149. Copyright 1971 by Society for the Experimental Analysis of Behavior, Inc. Reproduced by permission.)

This is twice-documented, as seen in the decreasing and increasing data trends in the second set of A and B phases.

Let us now consider a more recent example of an A-B-A-B design taken from the child literature. In this experimental analysis, Hendrickson, Strain, Tremblay, and Shores (1982) documented how a normally functioning pre-school child (the peer confederate) was taught to make specific initiations toward three "withdrawn" preschool boys (each four years of age). This peer confederate was a 4-year-old female, with a well-developed repertoire of expressive language and social interaction skills. Prebaseline observation indicated no evidence of physically aggressive behavior. She interacted primarily with adults, and infrequently initiated positive behavior to other children. She did, however, respond positively and consistently when other children initiated play to her. This child was involved in the treatment program as a "model" youngster (p. 327).

During baseline and intervention phases the children were brought to a playroom for two 15-minute sessions. Three behaviors were observed and coded during these sessions: (1) initiations of play organizers (proposes a role or activity in a game), (2) shares (offers or gives toy to another child), and (3) assists (provides help to another child).

Examination of baseline data in Figure 5-10 indicates that the peer confe-

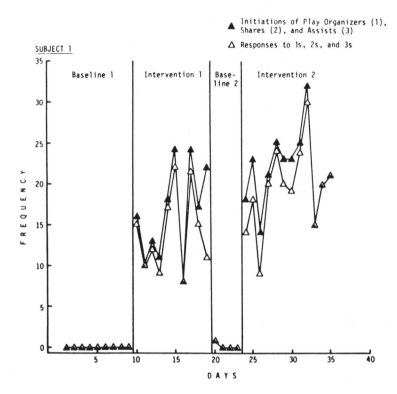

FIGURE 5-10. Experiment 1: Frequency of confederate initiations of play organizers, shares, and assists and subject's positive responses to these approach behaviors. (Figure 1, p. 335, from: Hendrickson, J. M., Strain, P. S., Tremblay, A., & Shore, R. E. (1982). Interactions of behaviorally handicapped children: Functional effects of peer social initiations. *Behavior Modification, 6*, 323–353. Copyright 1982 by Sage Publications. Reproduced by permission.)

derate neither initiated any of the three targeted behaviors nor responded to any initiations of the three withdrawn children. However, during the first intervention phase, when the confederate was prompted, instructed, and reinforced for playing, there was a marked increase in the three categories of behavior. This was noted both in terms of *initiations* and *responses*. When intervention was removed in the second baseline, frequency of such initiating and responding returned to the original baseline level. Finally, in the second intervention phase, high levels of initiating and responding were easily reinstated. Throughout this study, mean interobserver agreement for behaviors targeted was 89% for all subjects.

With respect to design considerations, we have here a very clear demonstration of the efficacy of the intervention on two occasions. As was the case in our prior example (R. V. Hall et al., 1971) baselines (especially the second) were shorter than treatment phases. However, in light of the zero level of baseline responding and the immediate and dramatic improvements as a result of the intervention, the possible confound of time and length of adjacent phases does not apply in this analysis.

A-B-A-B with unexpected improvement in baseline

In our next example we will illustrate the difficulties that arose in interpretation when unexpected improvement took place during the latter half of the second series of baseline (A) measurements. Epstein, Hersen, and Hemphill (1974) used an A-B-A-B design in their assessment of the effects of feedback on frontalis muscle activity in a patient who had suffered from chronic headaches for a 16-year period. EMG recordings were taken for 10 minutes following 10 minutes of adaptation during each of the six baseline (A) sessions. EMG data were obtained while the patient relaxed in a reclining chair in the experimental laboratory. During the six feedback (B) sessions, the patient's favorite music (prerecorded on tape) was automatically turned on whenever EMG activity decreased below a preset criterion level. Responses above that level conversely turned off recordings of music. Instructions to the patient during this phase were to "keep the music on." In the next six sessions baseline (A) conditions were reinstated, while the last six sessions involved a return to feedback (B). Throughout all phases of study, the patient was asked to keep a record of the intensity of headache activity.

Examination of Figure 5-11 indicates that EMG activity during baseline ranged from 28 to 50 seconds (mean = 39·18) per minute that contained integrated responses above the criterion microvolt level. Institution of feed-

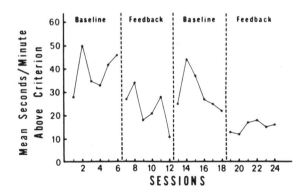

FIGURE 5-11. Mean seconds per minute that contained integrated responses above criterion microvolt level during baseline and feedback phases. (Figure 1, p. 61, from: Epstein, L. H., Hersen, M., & Hemphill, D. P. (1974). Music feedback as a treatment for tension headache: An experimental case study. *Journal of Behavior Therapy and Experimental Psychiatry, 5,* 59–63. Copyright 1974 by Pergamon. Reproduced by permission.)

back procedures resulted in decreased activity (mean = 23·18). Removal of feedback in the second baseline initially resulted in increased activity in Sessions 13–15. However, an *unexplained but decreased trend* was noted in the last half of that phase. This downward trend, to some extent, detracts from the interpretation that music feedback was the responsible agent of change during the first B phase. In addition, the importance of maintaining equal lengths of phases is highlighted here. Had baseline measurement been concluded on Day 15, an unequivocal interpretation (though probably erroneous) would have been made. However, despite the downward trend in baseline, mean data for this phase (30·25) were higher than for the previous feedback phase (23·18).

In the final phase, feedback resulted in a further decline that was generally maintained at low levels (mean = 14·98). Unfortunately, it is not fully clear whether this further decrease might have occurred naturally without the benefits of renewed introduction of feedback. Therefore, despite the presence of statistically significant differences between baseline and feedback phases and confirmation of EMG differences by self-reports of decreased headache intensity during feedback, the downward trend in the second baseline prevents a definitive interpretation of the controlling effects of the feedback procedure.

When the aforementioned data pattern results, it is recommended, where possible, that variables possibly leading to improvement in baseline be examined through additional experimental analyses. However, time limitations and pressing clinical needs of the patient or subject under study usually preclude such additional study. Therefore, the next best strategy involves a replication of the procedure with the same subject—or with additional subjects bearing the same kind of diagnosis (see chapter 10).

A-B-A-B with monitoring of concurrent behaviors

When using the withdrawal strategy, such as the A-B-A-B design, most experimenters have been concerned with the effects of their treatment variable on one behavior—the targeted behavior. However, in a number of reports (Kazdin, 1973a; Kazdin, 1973b; Lovaás & Simmons, 1969; Risley, 1968; Sajwaj, Twardosz, & Burke, 1972; Twardosz & Sajwaj, 1972) the importance of monitoring concurrent (nontargeted) behaviors was documented. This is of particular importance when side effects of treatment are possibly negative (see Sajwaj, Twardosz, & Burke, 1972). Kazdin (1973b) has listed some of the potential advantages in monitoring the multiple effects of treatment on operant paradigms.

One initial advantage is that such assessment would permit the possibility of determining response generalization. If certain response frequencies are increased or decreased, it would be expected that other related operants would be influenced. It would be a desirable addition to determine generalization of beneficial response changes by looking at behavior related to the target response. In addition, changes in the frequency of responses might also correlate with topographical alterations. (p. 527)

We might note here that the examination of collateral effects of treatment should not be restricted to operant paradigms when using experimental single-case designs.

In our following example the investigators (Twardosz & Sajwaj, 1972) used an A-B-A-B design to evaluate the efficacy of their program to increase sitting in a 4-year-old, hyperactive, retarded boy who was enrolled in an experimental preschool class. In addition to assessment of the target behavior of interest (sitting), the effects of treatment procedures on a variety of concurrent behaviors (posturing, walking, use of toys, proximity of children) were monitored. Observations of this child were made during a free-play period (one-half hour) in which class members were at liberty to choose their playmates and toys. During baseline (A), the teacher gave the child instructions (as she did to all others in class) but did not prompt him to sit or praise him when he did. Institution of the sitting program (B) involved prompting the child (placing him in a chair with toys before him on the table), praising him for remaining seated and for evidencing other positive behaviors, and awarding him tokens (exchangeable for candy) for in-seat behavior. In the third phase (A) the sitting program was withdrawn and a return to baseline conditions took place. Finally, in phase four (B) the sitting program was reinstated.

The results of this study appear in Figure 5-12. Examination of the top part of the graph shows that the sitting program, with the exception of the last day in the first treatment phase, effected improvement over baseline conditions on both occasions. Continued examination of the figure reveals that posturing decreased during the sitting program, but walking remained at a consistent rate throughout all phases of study. Similarly, use of toys and proximity to children increased during administrations of the sitting program. In discussing their results, Twardosz and Sajwaj (1972) stated that:

This study . . . points out the desirability of measuring several child behaviors, although a modification procedure might focus on only one. In this way the preschool teacher can assess the efficacy of her program based upon changes in other behaviors as well as the behavior of immediate concern. (p. 77)

However, in the event that nontargeted behaviors remain unmodified or that deterioration occurs in others, additional behavioral techniques can then be applied (Sajwaj, Twardosz, & Burke, 1972). Under these circumstances it

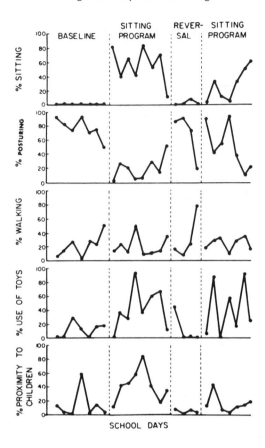

FIGURE 5-12. Percentages of Tim's sitting, posturing, walking, use of toys, and proximity to children during freeplay as a function of the teacher's ignoring him when he did not obey a command to sit down. (Figure 1, p. 75, from: Twardosz, S., & Sajwaj, T. (1972). Multiple effects of a procedure to increase sitting in a hyperactive retarded boy. *Journal of Applied Behavior Analysis, 5,* 73–78. Copyright 1972 by Society for the Experimental Analysis of Behavior, Inc. Reproduced by permission.)

might be preferable to use a multiple baseline strategy (Barlow & Hersen, 1973) in which attention to each behavior can be programed in advance (see chapter 7).

A-B-A-B with no feedback to experimenter

A major advantage of the single-case strategy (cited in section 3.2 of chapter 3) is that the experimenter is in a position to alter therapeutic approaches in accordance with the dictates of the case. Such flexibility is possible because repeated monitoring of target behaviors is taking place.

Thus changes from one phase to the next are accomplished with the experimenter's full knowledge of prior results. Moreover, specific techniques are then applied with the expectation that they will be efficacious. Although these factors are of benefit to the experimental clinician, they present certain difficulties from a purely experimental standpoint. Indeed, critics of the single-case approach have concerned themselves with the possibilities of bias in evaluation and in actual application and withdrawal of specified techniques. One method of preventing such "bias" is to determine lengths of baseline and experimental phases on an a priori basis, while keeping the experimenter uninformed as to trends in the data during their collection. A problem with this approach, however, is that decisions regarding choice of baselines and those concerned with appropriate timing of institution and removal of therapeutic variables are left to change.

The above-discussed strategy was carried out in an A-B-A-B design in which target measures were rated from video tape recordings for all phases on a postexperimental basis. Hersen, Miller, and Eisler (1973) examined the effects of varying conversational topics (nonalcohol and alcohol-related) on duration of looking and duration of speech in four chronic alcoholics and their wives in *ad libitum* interactions videotaped in a television studio. Following 3 minutes of "warm-up" interaction, each couple was instructed to converse for 6 minutes (A phase) about any subject *unrelated* to the husband's drinking problem. Instructions were repeated at 2-minute intervals over a two-way intercom from an adjoining room to ensure maintenance of the topic of conversation. In the next 6 minutes (B phase) the couple was instructed to converse *only* about the husband's drinking problem (instructions were repeated at 2-minute intervals). The last 12 minutes of interaction consisted of identical replications of the A and B phases.

Mean data for the four couples are presented in Figure 5-13. Speech duration data show no trends across experimental phases for either husbands or wives. Similarly, duration of looking for husbands across phases does not vary greatly. However, duration of looking for wives was significantly greater during alcohol- than nonalcohol-related segments of interaction. In the first nonalcohol phase, looking duration ranged from 26 to 43 seconds, with an upward trend in evidence. In the first alcohol phase (B), duration of looking ranged from 57 to 70 seconds, with a continuation of the upward linear trend. Reintroduction of the nonalcohol phase (A) resulted in a decrease of looking (38 to 45 seconds). In the final alcohol segment (B), looking once again increased, ranging from 62 to 70 seconds.

An analysis of these data does not allow for conclusions with respect to the initial A and B phases inasmuch as the upward trend in A continued into B. However, the decreasing trend in the second A phase succeeded by the increasing trend in the second B phase suggests that topic of conversation had a controlling influence on the wives' rates of looking. We might note here that

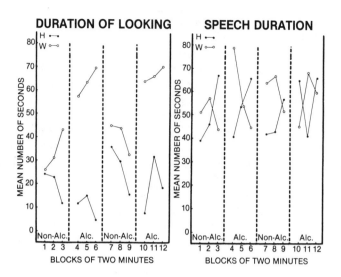

FIGURE 5-13. Looking and speech duration in nonalcohol- and alcohol-related interactions of alcoholics and their wives. Plotted in blocks of 2 minutes. Closed circles—husbands; open circles—wives. (Figure 1, p. 518, from: Hersen, M., Miller, P. M., & Eisler, R. M. (1973). Interactions between alcoholics and their wives: A descriptive analysis of verbal and non-verbal behavior. *Quarterly Journal of Studies on Alcohol,* **34,** 516–520. Copyright 1973 by Journal of Studies on Alcohol, Inc. New Brunswick, N.J. 08903. Reproduced by permission.)

if the experimenters were in position to monitor their results throughout all experimental phases, the initial segment probably would have been extended until the wives' looking duration achieved stability in the form of a plateau. Then the second phase would have been introduced.

5.5. B-A-B DESIGN

The B-A-B design has frequently been used by investigators evaluating effectiveness of their treatment procedures (Agras, Leitenberg, & Barlow, 1968; Ayllon & Azrin, 1965; Leitenbert et al., 1968; Mann & Moss, 1973; Rickard & Saunders, 1971). In this experimental strategy the first phase (B) usually involves the application of a treatment. In the second phase (A) the treatment is withdrawn and in the final phase (B) it is reinstated. Some investigators (e.g., Agras et al., 1968) have introduced an abbreviated baseline session prior to the major B-A-B phases. The B-A-B design is superior to the A-B-A design, described in section 5.3, in that the treatment variable is in effect in the terminal phase of experimentation. However, absence of an initial baseline measurement session precludes an analysis of the effects of treatment over the natural frequency of occurrence of the targeted behaviors under study (i.e., baseline). Therefore, as previously pointed out by Barlow and Hersen (1973), the use of the more complete A-B-A-B design is preferred for assessment of singular therapeutic variables.

We will illustrate the use of the B-A-B strategy with one example selected from the operant literature and a second drawn from the Rogerian framework. In the first, an entire group of subjects underwent introduction, removal, and reintroduction of a treatment procedure in sequence (Ayllon & Azrin, 1965). In the second, a variant of the B-A-B design was imployed by proponents of client-centered therapy (Truax & Carkhuff, 1965) in an attempt to experimentally manipulate levels of therapeutic conditions.

B-A-B with group data

Ayllon and Azrin (1965) used the B-A-B strategy on a group basis in their evaluation of the effects of token economy on the work performance of 44 "backward" schizophrenic subjects. During the first 20 days (B phase) of the experiment, subjects were awarded tokens (exchangeable for a large variety of "backup" reinforcers) for engaging in hospital ward work activities. In the next 20 days (A phase) subjects were given tokens on a noncontingent basis, regardless of their work performance. Each subject received tokens daily, based on the mean daily rate obtained in the initial B phase. In the last 20 days (second B phase) the contingency system was reinstated. We might note at this point that this design could alternately be labeled B-C-B, as the middle phase is not a true measure of the natural frequency of occurrence of the target measure (see section 5.6).

Work performance data (total hours per day) for the three experimental phases appear in Figure 5-14. During the first B phase, total hours per day worked by the entire group averaged about 45 hours. Removal of the contingency in A resulted in a marked linear decrease to a level of one hour per day on Day 36. Reinstitution of the token reinforcement program in B led to an immediate increase in hours worked to a level approximating the first B phase. Thus, Ayllon and Azrin (1965) presented the first experimental demonstration of the controlling effects of token economy over work performance in state hospital psychiatric patients.

It should be pointed out here that when experimental single-case strategies, such as the B-A-B design, are used on a group basis, it behooves the experimenter to show that a majority of those subjects exposed to and then withdrawn from treatment provide supporting evidence for its controlling effects. Individual data presented for selected subjects can be quite useful, particularly if data trends differ. Otherwise, difficulties inherent in the traditional group comparison approach (e.g., averaging out of effects, effects due

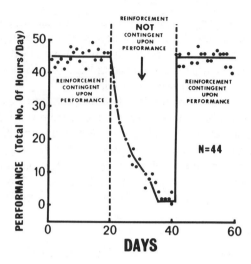

FIGURE 5-14. Total number of hours of on-ward performance by a group of 44 patients, Exp. III (Figure 4, p. 373, redrawn from: Ayllon, T., & Azrin, N. H. (1965). The measurement and reinforcement, of behavior of psychotics. *Journal of the Experimental Analysis of Behavior*, **8**, 357–383. Copyright 1965 by Society for the Experimental Analysis of Behavior, Inc. Reproduced by permission.)

to a small minority while the majority remains unaffected by treatment) will be carried over to the experimental analysis procedure. In this regard, Ayllon and Azrin (1965) showed that 36 of their 44 subjects decreased their performance from contingent to noncontingent reinforcement. Conversely, 36 of 44 subjects increased their performance from noncontingent to contingent reinforcement. Eight subjects were totally unaffected by contingencies and maintained a zero level of performance in all phases.

B-A-B from Rogerian framework

Although the withdrawal design has been used in physiological research for years, and has been associated with the operant paradigm, the experimental strategies that are applied can easily be employed in the investigation of nonoperant (both behavioral and traditional) treatment procedures. In this connection, Truax and Carkhuff (1965) systematically examined the effects of high and low "therapeutic conditions" on the responses of 3 psychiatric patients during the course of initial 1-hour interviews. Each of the interviews consisted of the three 20-minute phases. In the first phase (B) the therapist was instructed to evidence high levels of "accurate empathy" and "unconditional positive warmth" in his interactions with the patient. In the following

A phase the therapist experimentally lowered these conditions, and in the final phase (B) they were reinstated at a high level.

Each of the three interviews was audiotaped. From these audiotapes, five 3-minute segments for each phase were obtained and rerecorded on separate spools. These were then presented to raters (naive as to which phase the tape originated in) in random order. Ratings made on the basis of the Accurate Empathy Scale and the Unconditional Positive Regard Scale confirmed (graphically and statistically) that the therapist followed directions as indicated by the dictates of the experimental design (B-A-B).

The effects of high and low therapeutic conditions were then assessed in terms of depth of the patient's intrapersonal exploration. Once again, 3-minute segments from the A and B phases were presented to "naive" raters in randomized order. These new ratings were made on the basis of the Truax Depth of Interpersonal Exploration Scale (reliability of raters per segment = ·78). Data with respect to depth of intrapersonal exploration are plotted in Figure 5-15. Visual inspection of these data indicates that depth of intrapersonal exploration, despite considerable overlapping in adjacent phases, was *somewhat* lowered during the middle phase (A) for each of the three patients.

Although these data are far from perfect (i.e., overlap between phases), the study does illustrate that the controlling effects of *nonbehavioral therapeutic variables can be investigated systematically using the experimental analysis of behavior model.* Those of nonbehavioral persuasion might be encouraged to assess the effects of their technical operations more frequently in this fashion.

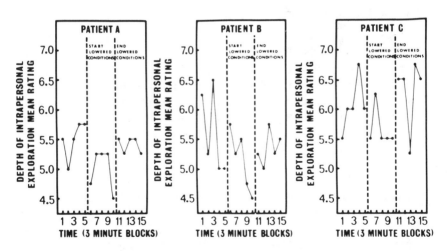

FIGURE 5-15. Depth of intrapersonal exploration. (Figure 4, p. 122, redrawn from: Truax, C.B., & Carkhuff, R. R. (1965). Experimental manipulation of therapeutic conditions, *Journal of Consulting Psychology, 29,* 119–124. Copyright 1965 by the American Psychological Association. Reproduced by permission.)

5.6. A-B-C-B DESIGN

The A-B-C-B design, a variant of the A-B-A-B design, has been used to evaluate the effects of reinforcement procedures. Whereas in the A-B-A-B strategy, baseline and treatment (e.g., contingent reinforcement) are alternated in sequence, in the A-B-C-B strategy only the first two phases of experimentation consist of baseline and contingent reinforcement. In the third phase (C), instead of returning to baseline observation, reinforcement is administered in proportions equal to the preceding B phase but on a totally *noncontingent* basis. This phase controls for the added attention ("attention-placebo") that a subject receives for being in a treatment condition and is analogous to the A_1 phase (placebo) used in drug evaluations (see chapter 6). In the final phase, contingent reinforcement procedures are reinstated. Thus the last three phases of study are identical to those used by Ayllon and Azrin (1965) in the example described in section 5.5 (however, there the study is labeled B-A-B).

In the A-B-C-B design the A and C phases are not comparable, inasmuch as experimental procedures differ. Therefore, the main experimental analysis is derived from the B-C-B portion of study. However, baseline observations are of some value, as the effects of B over A are suggested (here we have the limitations of the A-B analysis). We will illustrate the use of the A-B-C-B design with one example concerned with the control of drinking in a chronic alcoholic.

A-B-C-B with a biochemical target measure

Miller, Hersen, Eisler, and Watts (1974) examined the effects of monetary reinforcement in a 48-year-old "skid row" alcoholic. During all phases of study, a research assistant obtained breathalyzer samples, analyzed biochemically shortly thereafter for blood alcohol concentration, from the subject (psychiatric outpatient) in various locations in his community. To avoid possible bias in measurement, the subject was not informed as to specific times that probe measures were to be taken. In fact, these times were randomized in all phases to control for measurement bias.

During baseline (A phase), eight probe measures were obtained. During contingent reinforcement (B), the subject was awarded $3.00 in canteen booklets (redeemable at the hospital commissary for material goods) whenever a negative blood alcohol sample was obtained. In the noncontingent reinforcement phase (C), reinforcement ($3.00 in centeen booklets) was administered regardless of blood alcohol concentration. In the final phase, contingent reinforcement was reinstituted.

Inspection of Figure 5-16 reveals a variable baseline pattern ranging from a ·00 to ·27 level of blood alcohol. In contingent reinforcement, five of the six

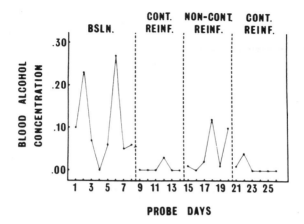

FIGURE 5-16. Biweekly blood-alcohol concentrations for each phase. (Figure 1, p. 262, from: Miller, P. M., Hersen, M., Eisler, R. M., & Watts, J. G. (1974). Contingent reinforcement of lowered blood/alcohol levels in an outpatient chronic alcoholic. *Behaviour Research and Therapy*, **12**, 261–263. Copyright 1974 by Pergamon. Reproduced by permission.)

probe measures attained a ·00 level. During noncontingent reinforcement, blood alcohol concentration measures rose, but to lower levels than in baseline. When contingent reinforcement was reinstated, four of the six probe measures yielded ·00 levels of blood alcohol. Therefore, it appears that monetary reinforcement resulted in decreases in drinking in this chronic alcoholic while the contingency was in effect.

A-B-C-B in a group application and follow-up

A most interesting application of the A-B-C-B design to a group of subjects was reported by Porterfield, Blunden, and Blewitt (1980). Subjects in this experimental analysis were "profoundly mentally handicapped" adults attending a center for the retarded. The behavior targeted for modification was participation in activities during a 1-hour period so designated during the 19 days of the study. Participation was defined by 12 separate activities and involved some of the following: watching television, dancing, responding to a verbal command, talking to another subject, and eating without assistance.

The baseline phase (A) lasted 3 days, with three staff members interacting with subjects in normal fashion. No specific instructions were given at this point. The B phase (room manager) lasted 5 days, with two staff members alternating for half-hour periods. Subjects in this condition were prompted and differentially reinforced for their participation. The C phase (no distraction) lasted 6 days and involved a maximum of two prompts to engage in activity, but subjects were not differentially reinforced. In the fourth phase (B) the room manager condition was reinstated. Then there was a 69-day follow-up period involving the room manager condition in the absence of the experimenter.

Data appear in Figure 5-17 and are presented as the percentage of subjects (i.e., trainees) engaged in activity. It is clear that baseline (A) functioning was poor, ranging from 25.7% to 37.9% participation. Introduction of the room manager (B) condition led to marked increases in participation (72.9% to 90.9%).

However, when the no-distraction (C) condition was introduced, participation decreased to near baseline levels (21.5% to 48.0%). When the room manager condition was reintroduced, in the second B phase, level of participation once again increased to 84.7% to 88.1%. This second application of the room manager condition clearly documented the controlling effects of the contingency. Furthermore, data in follow-up confirmed that participation

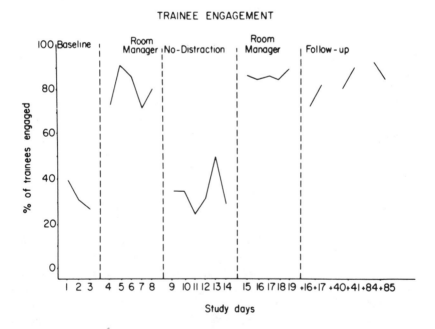

FIGURE 5-17. Percentage of trainees engaged during the activity hour for 19 days and follow-up days. (Figure 1, p. 236 from: Porterfield, J., Blunden, R., & Blewitt, E. (1980). Improving environments for profoundly handicapped adults: Using prompts and social attention to maintain high group engagement. *Behavior Modification, 4,* 225–241. Copyright 1980 by Sage Publications. Reproduced by permission.)

could be maintained (71.5% to 91.1%) in the absence of experimental prompting.

There are two noteworthy features in this particular example of the A-B-C-B design. *First*, even though the A and C phases were technically dissimilar, they certainly were functionally alike. That is, the resulting data pattern was the same as an A-B-A-B design. However, contrary to the A-B-A-B design, where there are two instances of confirmation of the contingency, only the B-C-B portion of the design truly reflected the controlling aspects of the room manager intervention. *Second*, by making the dependent measure the "percentage of trainees engaged," the experimenters obviated the necessity of providing individual data. However, from a single-case perspective, data as to percentage of time active *for each trainee* would be most welcome indeed.

CHAPTER 6

Extensions of the A-B-A Design, Uses in Drug Evaluation and Interaction Design Strategies

6.1. EXTENSIONS AND VARIATIONS OF THE A-B-A WITHDRAWAL DESIGN

The applied behavioral literature is replete with examples of extensions and variations of the more basic A-B-A experimental design. These designs can be broadly classified into five major categories. The first category consists of designs in which the A-B pattern is replicated several times. Advantages here are that (1) repeated control of the treatment variable is demonstrated, and (2) extended study can be conducted until full clinical treatment has been achieved. An example of this type of strategy appears in Mann's (1972) work, where he used an A-B-A-B-A-B design to study the effects of contingency contracting on weight loss in overweight subjects.

In the second category separate therapeutic variables are compared with baseline performance during the course of experimentation (e.g., R. V. Hall et al., 1972; Pendergrass, 1972; Wincze, Leitenberg, & Agras, 1972). Subsumed under this category are the A-B-A-C-A designs discussed in section 3.4 of chapter 3. There it was pointed out that comparison of differential effectiveness of B and C variables is difficult when both variables appear to effect change over baseline levels. However, in the A-B-A-B-A-C-A design the individual controlling effects of B and C variables can be determined. A careful distinction should be made between these kinds of designs and designs where the interactive effects of variables are investigated (e.g., A-B-A-B-BC-B-BC). In the latter design the effects of C above those of B can be assessed experimentally. Once again, in the A-B-A-C-A design the effects of B and C over A can be evaluated. However, interpreting the relative efficacy of B and C is problematic in this strategy.

In the third category specific variations of the treatment procedure are examined during the course of experimentation (e.g., Bailey, Wolf, & Phillips, 1970; Coleman, 1970; Conrin, Pennypacker, Johnston, & Rast, 1982; Hopkins et al., 1971; Kaufman & O'Leary, 1972; McLaughlin & Malaby, 1972; Wheeler & Sulzer, 1970). For example, in some operant paradigms the treatment procedure may be faded out (e.g., Bailey, Wolf, & Phillips, 1970). In other paradigms, differing amounts of reinforcement may be assessed experimentally or in graduated progression (Hopkins et al., 1971) following demonstration of the controlling effects of variables in the A-B-A-B portion of the design. This experimental strategy is occasionally termed a *parametric* one.

In a fourth category, the interaction of additive effects of two or more variables are examined through variations in the basic A-B-A design (e.g., Agras et al., 1974; Bernard, Kratochwill, & Keefauver, 1983; Hersen et al., 1972; Leitenberg et al., 1968; Turner, Hersen, & Alford, 1974). Such analysis is accomplished by examining the effects of both variables alone and in combination, to determine the interaction. This extends beyond analysis of the separate effects of two therapeutic variables over baseline as represented by the A-B-A-C-A type design described in the second category. It also extends a stop beyond merely adding a variation of a therapeutic variable on the end of an A-B-A-B series (e.g., A-B-A-B-BC), since no experimental analysis of the additive effects of BC is performed. Properly run, interaction designs are complex and usually require more than one subject (see section 6.5.).

The fifth category consists of the changing-criterion design (Hartmann & Hall, 1976) and its variant, the periodic-treatments design (cf. Hayes, 1981). Basically, in the changing-criterion design, baseline is followed by treatment until a preset criterion is met. This then becomes the new baseline (A'), and a new criterion is set. Such repetition, of course, continues until eventually the final criterion is reached (see Hersen, 1982).

The following subsections present examples of extensions and variations, with illustrations selected from each of the five major categories.

6.2. A-B-A-B-A-B DESIGN

Mann (1972) repeatedly introduced and withdrew a treatment variable (contingency contracting) during extended study with overweight subjects who had agreed, prior to experimentation, to achieve a designated weight loss within a specified time period. At the beginning of study, each subject entered into a formal contractual arrangement with the experimenter. In each case the subject agreed to surrender a number of his prized possessions (valuables) to the experimenter. During contingency conditions, the subject was able to regain possession of each valuable (one at a time) by evidencing a 2-pound weight loss over his previous low weight. A further 2-pound weight loss over that resulted in the return of still another valuable, and so on. Conversely, a 2-pound weight gain over the previous low weight led to the subject's permanently losing one of the valuables. In addition to these short-term contingency arrangements, 2-week and terminal contingencies (using similar principles) were put into effect during treatment phases. Valuables lost by each subject were subsequently disposed of by the experimenter in equitable fashion (i.e., he did not profit from or retain them). During baseline and "reversal" conditions contractual arrangements were temporarily suspended.

The results of this study for a prototypical subject are plotted in Figure 6-1. Inspection of that figure clearly shows that when contractual arrangements

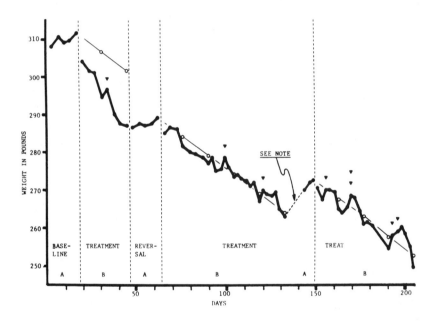

FIGURE 6-1. A record of the weight of Subject 1 during all conditions. Each open circle (connected by the thin solid line) represents a 2-week minimum weight loss requirement. Each solid dot (connected by the thick solid line) represents the subject's weight on each day that he was measured. Each triangle indicates the point at which the subject was penalized by a loss of valuables, either for gaining weight or for not meeting a 2-week minimum weight loss requirement. NOTE: The subject was ordered by his physician to consume at least 2,500 calories per day for 10 days, in preparation for medical tests. (Figure 1a, p. 104, from: Mann, R. A. [1972]. The behavior-therapeutic use of contingency contracting to control an adult behavior problem: Weight control. *Journal of Applied Behavior Analysis*, **5**, 99–109. Copyright 1972 by Society for the Experimental Analysis of Behavior, Inc. Reproduced by permission.)

were in force the subject evidenced a steady linear decrease in weight. By contrast, during baseline conditions, weight loss ceased, as indicated by a plateau and slightly upward trend in the data. In short, the effects of the treatment variable were repeatedly demonstrated in the alternately increasing and decreasing data trends.

6.3. COMPARING SEPARATE THERAPEUTIC VARIABLES, OR TREATMENTS

A-B-A-C-A-C′-A design

Wincze et al. (1972) conducted a series of 10 experimental single-case designs in which the effects of feedback and token reinforcement were examined on the verbal behavior of delusional psychiatric patients. In one of these studies an A-B-A-C-A-C′-A design was used, with B and C representing feedback and token reinforcement phases, respectively. During all phases of study, a delusional patient was questioned daily (15 questions selected randomly from a pool of 105) by his therapist to elicit delusional material. Percentage of responses containing delusional verbalizations was recorded. In addition, percentage of delusional talk on the ward (token economy unit) was monitored by nursing staff on a randomly distributed basis 20 times per day.

During baseline (A), the patient received "free" tokens as no contingencies were placed with respect to delusional verbalizations. During feedback (B), the patient continued to receive tokens noncontingently, but corrective statements in response to delusional verbalizations were offered by the therapist in individual sessions. The third phase (A) consisted of a return to baseline procedures. In Phase 4 (C) a stringent token economy system embracing all aspects of the patient's ward life was instituted. Tokens could be earned by the patient for "talking correctly" (nondelusionally) both in individual sessions and on the ward. Tokens were exchangeable for meals, luxuries, and privileges. Phase 5 (A) once again involved a return to baseline. In the sixth phase (C′) token bonuses were awarded on a predetermined percentage basis for talking correctly (e.g., speaking delusionally less than 10% of the time during designated periods). This condition was incorporated to counteract the tendency of the patient to earn tokens merely for increasing frequency of nondelusional talk while still maintaining a high frequency of delusional verbalizations. In the last phase of experimentation (A), baseline conditions were reinstated for the fourth time.

Results of this experimental analysis for one subject appear in Figure 6-2. Percentage of delusional talk in individual sessions and on the ward did not differ substantially during the first three sessions, thus suggesting the ineffectiveness of the feedback variable. Institution of token economy in Phase 4,

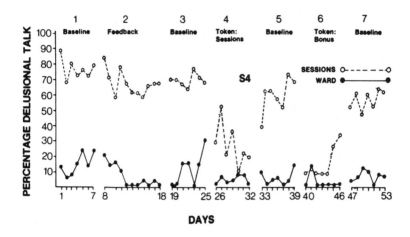

FIGURE 6-2. Percentage of delusional talk of Subject 4 during therapist sessions and on ward for each experimental day. (Figure 4, p. 256, from: Wincze, J. P., Leitenberg, H., & Agras, W. S. [1972]. The effects of token reinforcement and feedback on the delusional verbal behavior of chronic paranoid schizophrenics. *Journal of Applied Behavior Analysis*, **5**, 247–262. Copyright 1972 by Society for the Experimental Analysis of Behavior, Inc. Reproduced by permission.)

however, resulted in a marked decrease of delusional talk in individual sessions. But it failed to effect a change in delusional talk on the ward. Removal of token economy in Phase 5 led to a return to initial levels of delusional talk during individual sessions. Throughout the first five phases, percentage of delusional talk on the ward was consistent, ranging from 0% to 30%. Introduction of the token bonus in Phase 6 again resulted in a drop of delusional verbalizations in individual sessions. Additionally, percentage of delusional talk on the ward decreased to zero. In the last phase (baseline) delusional verbalizations rose both on the ward and in individual sessions.

In this case, feedback (B) proved to be an ineffective therapeutic agent. However, token economy (C) and token bonuses (C'), respectively, controlled percentage of delusional talk in individual sessions and on the ward. Had feedback also effected changes in behavior, the comparative efficacy of feedback and token economy would be difficult to ascertain using this design. Such analysis would require the use of a group comparison design. This is because one variable, token reinforcement, follows the other variable, feedback. Therefore, it is conceivable that tokens were effective only if instituted *after* a feedback phase and would not be effective if introduced initially. Thus a possible confound of order effects exists. Of course, the more usual case is that the first treatment would be effective to an extent that it would not leave much room for improvement in the second treatment. In other words, a "ceiling" effect would prevent a proper comparison between treatments, due to the order of their introduction.

To compare two treatments in this fashion, the investigator would have to administer two treatments with baseline interspersed to two different individuals (and their replications), with the order of treatments counterbalanced. For example, 3 subjects could receive A-B-A-C-A, where B and C were two distinct treatments, and 3 could receive A-C-A-B-A. In fact, Wincze et al. (1972) carried out this necessary counterbalancing with half of their subjects in order to analyze the effects of feedback on token reinforcement.

This design, then, approximates the group crossover design or the counterbalanced within-subject group comparison (e.g., Edwards, 1968), with the exception of the presence of repeated measures and individual analyses of the data. Each design option suffers from possible multiple-treatment interference or carryover effects (see chapter 8 for a discussion of multiple-treatment interference). In group designs, any carryover effects are averaged into group differences and treated statistically as part of the error. In the A-B-A-C-A single-case design, on the other hand, data are usually presented more descriptively, with visual analysis sometimes combined with statistical descriptions (rather than inferences) to estimate the effect of each treatment. Wincze et al. (1972) did an excellent job of this in their series, which is fully described in chapter 10. But analysis depends on comparing individuals experiencing different orders of treatments. Thus the functional analysis cannot be carried out within one individual with all of the experimental control that it affords. Other alternatives to comparing two treatments include a between-groups comparison design or an alternating-treatments design (see chapter 8).

As noted above, this direct replication series will be discussed in greater detail in chapter 10.

6.4. PARAMETRIC VARIATIONS OF THE BASIC THERAPEUTIC PROCEDURES A-B-A-B′-B″-B‴ DESIGN

Our example from the third category of extensions of the A-B-A design is drawn from the child classroom literature. Hopkins et al. (1971) systematically assessed the effects of access to a playroom on the rate and quality of writing in rural elementary schoolchildren. Target measures selected for study were most relevant in that these children came from homes where learning was not a high priority (parents were migrant or seasonal farm workers). Throughout all phases of study, first- and second-grade students were given daily standard written assignments during class periods (class periods were 50 minutes long during the first four phases).

In baseline (A), after each child had completed the assignment, handed it to the teacher, and waited for it to be scored, he or she was expected to return to his or her seat and remain there quietly until *all* others in class had turned in their papers. In the next phase (B) each child was permitted access to an adjoining playroom, containing attractive toys, after his or her paper was scored. The child was allowed to remain there until the 50-minute period was terminated, unless he or she became too noisy; then he or she was required to return to his or her seat. The next two phases (A and B) were identical to the first two. In the last three phases each child was permitted access to the playroom after his or her paper had been scored, but the length of class periods was gradually decreased (45, 40, 35 minutes). A procedural exception to the aforementioned was made in the last phase on Days 47–54 inasmuch as the teacher noted that a concomitant of increased speed was decreased quality (number of errors) in writing. Therefore, during the last 8 days a quality criterion was imposed before the child gained access to the playroom. In some cases the child was required to recopy a portion of writing.

Data for first-grade children are plotted in Figure 6-3. Examination of the bottom half of the figure shows that access to the playroom (50-minute period) increased the rate of letter writing over baseline levels. This was confirmed on two occasions in the A-B-A-B portion of study. When total time of classroom periods systematically decreased, a corresponding increase in rate of writing resulted. However, data for the last three phases are correlative, as an experimental analysis was not performed. For example, a sequential comparison of 50-, 45- and 50-minute periods was not made. Therefore, the controlling effects of time differences were not fully documented.

Examination of the top part of the graph shows considerable fluctuation with respect to mean number of errors per letter. However, this did not appear to represent a systematic increase when class periods were shortened. To the contrary, there was a general decrease in error rate from the first to the last phase of study. Nonetheless, the effects of practice cannot be discounted when total length of the investigation is considered.

A-B-B′-B″-A-B′ design

A more recent example of a study involving variations of the basic therapeutic procedure appears in a study by Conrin et al. (1982), in which differential reinforcement of other behaviors (DRO) was used to treat chronic rumination in mentally retarded individuals. In this study an A-B-B′-B″-A-B′ design was followed. The subject (Bob) was a 19-year-old male (53 in. tall, 56 lbs. at baseline) who was profoundly retarded and who ruminated (emesis of previously chewed food, rechewing food, and reswallowing food). The disorder had begun some 17 years earlier.

Baseline (A) observations took place one hour after the subject had consumed his meal. After each meal Bob was brought to the cottage lounge and observed. Duration of rumination (cheek swelling, chewing, and swallowing)

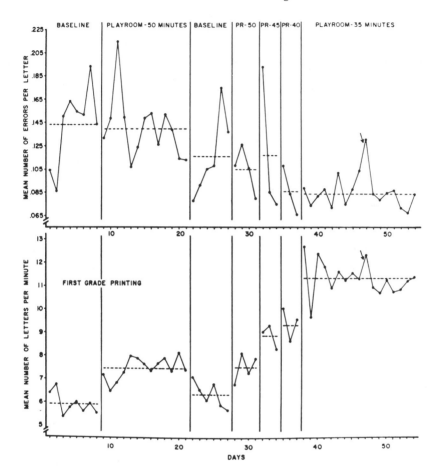

FIGURE 6-3. The mean number of letters printed per minute by first-grade children are shown on the lower coordinates, and the mean proportion of letters scored as errors are on the upper coordinates. Each data point represents the mean averaged over all children for that day. The horizontal dashed lines are the means of the daily means averaged over all days within the experimental conditions noted by the legends at the top of the figure. (Figure 1, p. 81, from: Hopkins, B. L., Schutte, R. C., & Garton, K. L. [1971]. The effects of access to a playroom on the rate and quality of printing and writing of first- and second-grade students. *Journal of Applied Behavior Analysis*, 4, 77–87. Copyright 1971 by Society for the Experimental Analysis of Behavior, Inc. Reproduced by permission.)

was timed. In the second phase (B) a DRO procedure was implemented. This consisted of giving Bob small portions of cookies or bits of peanut butter contingent on *no rumination*. In the B phase reinforcement was provided if no rumination occurred for 15 seconds or more (IRT > 15″). In the next phase

(B') this was increased to 30 seconds (IRT>30"), followed by an IRT>60"
in phase B". Then there was a return to baseline (A) and reintroduction of
IRT>30".

Interrater agreement for behavioral observations ranged from 94% to
100%. Examination of data in Figure 6-4 reveals a high duration of rumination (5 to 22 minutes; mean = 7 minutes) during baseline (A). Introduction
of DRO (IRT>15") resulted in a zero duration after 18 sessions, which was
maintained during the thinning of the reinforcement schedule in B'
(IRT>30") and B" (IRT>60"). A return to baseline conditions (A) resulted
in marked increases in rumination (mean = 10 minutes per session), but was
once again reduced to zero when DRO procedures (IRT>30") were reintroduced in the B' phase.

In summary, this experimental analysis clearly documents the controlling
effects of DRO over duration of rumination. It also shows how it was
possible to thin the reinforcement schedule from IRT>15" to IRT>60" and
still maintain rumination at near zero levels.

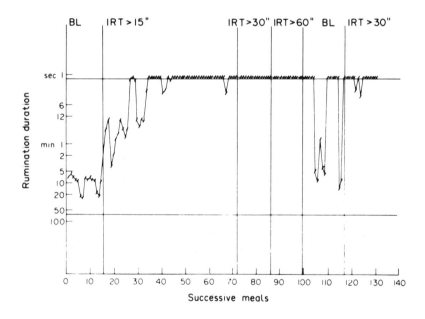

FIGURE 6-4. Duration of ruminations after meals by Bob. (Figure 2, p. 328, from: Conrin, J.,
Pennypacker, H. S., Johnston, J. M., & Rast, J. [1982]. Differential reinforcement of other
behaviors to treat chronic rumination of mental retardates. *Journal of Behavior Therapy and
Experimental Psychiatry,* **13,** 325–329. Copyright 1982 by Pergamon. Reproduced by permission.

6.5. DRUG EVALUATIONS

The group comparison approach generally has predominated in the examination of the effects of drugs on behavior. However, examples in which the subjects have served as their own controls in the experimental evaluation of pharmacological agents are now seen more frequently in the psychological and psychiatric literatures (e.g., Agras, Bellack, & Chassan, 1964; Chassan, 1967; K. V. Davis, Sprague, & Werry, 1969; Grinspoon, Ewalt, & Shader, 1967; Hersen & Breuning, in press; Liberman et al., 1973; Lindsley, 1962; McFarlain & Hersen, 1974; Roxburgh, 1970). Indeed, Liberman et al. (1973) have encouraged researchers to use the within-subject withdrawal design in assessing drug-environment interactions. In support of their position they contend that:

> Useful interactions among the drug-patient-environment system can be obtained using this type of methodology. The approach is reliable and rigorous, efficient and inexpensive to mount, and permits sound conclusions and generalizations to other patients with similar behavioral repertoires when systematic replications are performed . . . (p. 433)

There is no doubt that this approach can be of value in the study of both the major forms of psychopathology and those of more exotic origin (Hersen & Breuning, in press). The single-case experimental strategy is especially well suited to the latter, as control group analysis in the rarer disorders is obviously not feasible.

Specific issues

It should be pointed out that all procedural issues discussed in chapter 3 pertain equally to drug evaluation. In addition, there are a number of considerations specific to this area of research: (1) nomenclature, (2) carryover effects, and (3) single- and double-blind assessments.

With respect to nomenclature, A is designated as the baseline phase, A_1 as the placebo phase, B as the phase evaluating the first active drug, and C as the phase evaluating the second active drug. The A_1 phase is an intermediary phase between A (baseline) and B (active drug condition) in this schema. This phase controls for the subject's expectancy of improvement associated with mere ingestion of the drug rather than for its contributing pharmacological effects.

Some of the above-mentioned considerations have already been examined in section 3.4 of chapter 3 in relation to changing one variable at a time across experimental phases. With regard to this one-variable rule, it becomes apparent, then, that A-B, A-B-A, B-A-B, and A-B-A-B designs in drug research involve the manipulation of two variables (expectancy and condition) at one time across phases. However, under certain circumstances where time limitations and clinical considerations prevail, this type of experimental strategy is justified. Of course, when conditions permit, it is preferable to use strategies in which the systematic progression of variables across phases is carefully followed (see Table 6-1, Designs 4, 6, 7, 9–13). For example, this would be the case in the A_1-B-A_1 design strategy, where only one variable at a time is manipulated from phase to phase. Further discussion of these issues will appear in the following section, in which the different design options available to drug researchers will be outlined.

The problem of carryover effects from one phase to the next has already been discussed in section 3.6 of chapter 3. There some specific recommendations were made with respect to short-term assessments of drugs and the concurrent monitoring of biochemical changes during different phases of study. In this connection, Barlow and Hersen (1973) have noted that "Since continued measurements are in effect, length of phases can be varied from experiment to experiment to determine precisely the latency of drug effects after beginning the dosage and the residual effects after discontinuing the dosage" (p. 324). This may, at times, necessitate the inequality of phase lengths and the suspension of active drug treatment until biochemical measurements (based on blood and urine studies) reach an acceptable level. For example, Roxburgh (1970) examined the effects of a placebo and thiopropazate dihydrochloride on phenothiazine-induced oral dyskinesia in a double-blind crossover in two subjects. In both cases, placebo and active drug treatment were separated by a 1-week interruption during which time no placebo or drug was administered.

A third issue specific to drug evaluation involves the use of single- and double-blind assessments. The double-blind clinical trial is a standard precautionary measure designed to control for possible experimenter bias and patient expectations of improvement under drug conditions when drug and placebo groups are being contrasted. "This is performed by an appropriate method of assigning patients to drugs such that neither the patient nor the investigator observing him knows which medication a patient is receiving at any point along the course of treatment" (Chassan, 1967, pp. 80–81). In these studies, placebos and active drugs are identical in size, shape, markings, and color.

While the double-blind procedure is readily adaptable to group comparison research, it is difficult to engineer for some of the single-case strategies and impossible for others. Moreover, in some cases (see Table 6-1, Designs 1, 2, 4, 5, 8) even the single-blind strategy (where only the subject remains unaware of differences in drug and placebo manipulations) is not applicable. In these designs the changes from baseline observation to either placebo or drug conditions obviously cannot be disguised in any manner.

TABLE 6-1. Single-Case Experimental Drug Strategies

NO.	DESIGN	TYPE	BLIND POSSIBLE
1.	A-A$_1$	Quasi-experimental	None
2.	A-B	Quasi-experimental	None
3.	A$_1$-B	Quasi-experimental	Single or double
4.	A-A$_1$-A	Experimental	None
5.	A-B-A	Experimental	None
6.	A$_1$-B-A$_1$	Experimental	Single or double
7.	A$_1$-A-A$_1$	Experimental	Single or double
8.	B-A-B	Experimental	None
9.	B-A$_1$-B	Experimental	Single or double
10.	A-A$_1$-A-A$_1$	Experimental	Single or double
11.	A-B-A-B	Experimental	None
12.	A$_1$-B-A$_1$-B	Experimental	Single or double
13.	A-A$_1$-B-A$_1$-B	Experimental	Single or double
14.	A-A$_1$-A-A$_1$-B-A$_1$-B	Experimental	Single or double
15.	A$_1$-B-A$_1$-C-A$_1$-C	Experimental	Single or double

Note: A = no drug; A$_1$ = placebo; B = drug 1; C = drug 2.

A major difficulty in obtaining a true double-blind trial in single-case research is related to the experimenter's monitoring of data (i.e., making decisions as to when baseline observation is to be concluded and when various phases are to be introduced and withdrawn) throughout the course of investigation. It is possible to program phase lengths on an a priori basis, but then one of the major advantages of the single-case strategy (i.e., its flexibility) is lost. However, even though the experimenter is fully aware of treatment changes, the spirit of the double-blind trial can be maintained by keeping the observer (often a research assistant or nursing staff member) unaware of drug and placebo changes (Barlow & Hersen, 1973). We might note here additionally that despite the use of the double-blind procedure, the side effects of drugs in some cases (e.g., Parkinsonism following administration of large doses of phenothiazines) and the marked changes in behavior resulting from removal of active drug therapy in other cases often betray to nursing personnel whether a placebo or drug condition is currently in operation. This problem is equally troublesome for the researcher concerned with group comparison designs (see Chassan, 1967; chap. 4).

Different design options

In some of the investigations in which the subject has served as his or her own control, the standard experimental analysis method of study, where the treatment variable is introduced, withdrawn, and reintroduced following initial measurement, has not been followed rigorously. Thus the controlling effects of the drug under evaluation have not been fully documented. For example, K. V. Davis et al. (1969) used the following sequence of drug and no-drug conditions in studying rate of stereotypic and nonstereotypic behavior in severe retardates: (1) methylphenidate, (2) thioridazine, (3) placebo, and (4) no drug. Despite the fact that thioridazine significantly (at the statistical level) decreased the rate of stereotypic responses, failure to reintroduce the drug in a final phase weakens the conclusions to some extent from an experimental analysis standpoint.

A careful survey of the experimental analysis of behavior literature reveals relatively little discussion with regard to procedural and design issues in the assessment of drugs. Therefore, in light of the unique problems faced by the drug researcher and in consideration of the relative newness of this area, we will outline the basic quasi-experimental and experimental analysis design strategies for evaluating singular application of drugs. Specific advantages and disadvantages of each design option will be considered. Where possible, we will illustrate with actual examples selected from the research literature. However, to date, most of these strategies have not yet been implemented.

A number of possible single-case strategies suitable for drug evaluation are presented in Table 6-1. The first three strategies fall into the A-B category and are really quasi-experimental designs, in that the controlling effects of the treatment variable (placebo or active drug) cannot be determined. Indeed, it was noted in section 5.2 of chapter 5 that changes observed in B might possibly result from the action of a correlated but uncontrolled variable (e.g., time, maturational changes, expectancy of improvement). These quasi-experimental designs can best be applied in settings (e.g., consulting room practice) where limited time and facilities preclude more formal experimentation. In the first design the effects of placebo over baseline conditions are *suggested*; in the second the effects of active drug over baseline conditions are *suggested*; in the third the effects of an active drug over placebo are *suggested*.

Examination of Strategies 4–6 indicates that they are basically A-B-A designs in which the controlling effects of the treatment variable can be ascertained. In Design 4 the controlling effects of a placebo manipulation over no treatment can be assessed experimentally. This design has great potential in the study of disorders such as conversion reactions and histrionic personalities, where attentional factors are presumed to play a major role. Also, the use of this type of design in evaluating the therapeutic contribution of placebos in a variety of psychosomatic disorders could be of considerable importance to clinicians. In Design 5, the controlling effects of an active drug are determined over baseline conditions. However, as previously noted, two variables are being manipulated here at one time across phases. Design 6 corrects for this deficiency, as the active drug condition (B) is preceded and followed by placebo (A_1) conditions. In this design the one-variable rule across phases is carefully observed.

An example of an A_1-B-A_1 design appears in a series of single-case drug evaluations reported by Liberman et al. (1973). In one of these studies the effects of fluphenazine on eye contact, verbal self-stimulation (unintelligible or jumbled speech), and motor self-stimulation were examined in a double-blind trial for a 29-year-old regressed schizophrenic who had been continuously hospitalized for 13 years. Double-blind analysis was facilitated by the fact that fluphenazine (10 mg, b.i.d.) or the placebo could be administered twice daily in orange juice without its being detected (breaking of the double-blind code) by the patient or the nursing staff, as the drug cannot be distinguished by either odor or taste. During all phases of study, 18 randomly distributed 1-minute observations of the patient were obtained daily with respect to incidence of verbal and motor self-stimulation. Evidence of eye contact with the patient's therapist was obtained daily in six 10-minute sessions. Each eye contact was reinforced with candy or a puff on a cigarette.

The results of this study are plotted in Figure 6-5. During the first placebo phase (A_1), stable rates were obtained for each of the target behaviors.

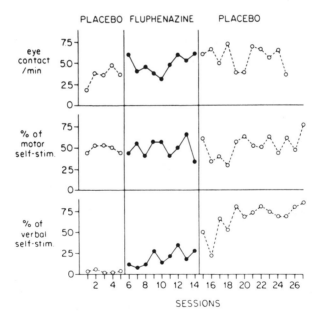

FIGURE 6-5. Interpersonal eye contact, motor, and self-stimulation in a schizophrenic young man during placebo and fluphenazine (20 mg daily) conditions. Each session represents the average of a 2-day block of observations. (Figure 3, p. 437, from: Liberman, R. P., Davis, J., Moon, W., & Moore, J. [1973]. Research design for analyzing drug-environment-behavior interactions. *Journal of Nervous and Mental Disease*, **156**, 432–439. Copyright 1973. Reproduced by permission.)

Introduction of fluphenazine in the second phase (B) resulted in a very slight increase in eye contact, and increased variability in motor self-stimulation, and a linear increase in verbal self-stimulation. Withdrawal of fluphenazine and a return to placebo conditions in the final phase (A_1) failed to yield data trends. On the contrary, eye contact increased slightly while verbal self-stimulation increased dramatically. Motor self-stimulation remained relatively consistent across phases. These data were interpreted by Liberman et al. (1973) as follows: "The failure to gain a reversal suggests a drug-initiated response facilitation which is seen most clearly in the increase of verbal self-stimulation, and less so in rate of eye contact" (p. 437). It was also suggested that residual phenothiazines during the placebo phase may have contributed to the continued increase in eye contact. However, in the absence of concurrent monitoring of biochemical factors (phenothiazine blood and urine levels), this hypothesis cannot be confirmed. In summary, Liberman et al. (1973) were not able to confirm the controlling effects of fluphenazine over any of the target behaviors selected for study in this A_1-B-A_1 design.

Let us now continue our examination of drug designs listed in Table 6-1. Strategies 7–9 can be classified as B-A-B designs, and the same advantages and limitations previously outlined in section 5.5 of chapter 5 apply here. Strategies 10–12 fall into the general category of A-B-A-B designs and are superior to the A-B-A and B-A-B designs for several reasons: (A) The initial observation period involves baseline or baseline-placebo measurement; (2) there are two occasions in which the controlling effects of the placebo or the treatment variables can be demonstrated; and (3) the concluding phase ends on a treatment variable.

Agras (1976) used an A-B-A-B design to assess the effects of chlorpromazine in a 16-year-old, black, brain-damaged, male inpatient who evidenced a wide spectrum of disruptive behaviors on the ward. Included in his repertoire were: temper tantrums, stealing food, eating with his fingers, exposing himself, hallucinations, and begging for money, cigarettes, or food. A specific token economy system was devised for this youth, whereby positive behaviors resulted in his earning tokens, and inappropriate behaviors resulted in his being penalized with fines. Number of tokens earned and number of tokens fined were the two dependent measures selected for study. The results of this investigation appear in Figure 6-6. In the first phase (A) no thorazine was administered. Although improvement in appropriate behaviors was noted, the patient's disruptive behaviors continued to increase markedly, resulting in his being fined many times. This occurred in spite of the addition of a time-out contingency. On Hospital Day 9, thorazine (300 mg per day) was introduced (B phase) in an attempt to control the patient's impulsivity. This dosage was subsequently decreased to 200 mg per day, as he became drowsy. Examination of Figure 6-6 reveals that fines decreased to a zero level whereas tokens earned for appropriate behaviors remained at a stable level. In the

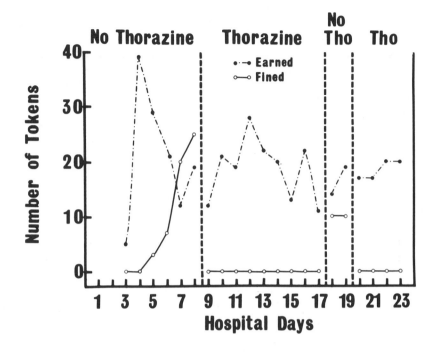

FIGURE 6-6. Behavior of an adolescent as indicated by tokens earned or fined in response to chlorpromazine, which was added to token economy. (Figure 15-3, p. 556, from: Agras, W. S. [1976]. Behavior modification in the general hospital psychiatric unit. In H. Leitenberg [Ed.], *Handbook of behavior modification*. Englewood Cliffs, NJ: Prentice-Hall. Copyright 1976 by H. Leitenberg. Reproduced by permission.)

third phase (A) chlorpromazine was temporarily discontinued, resulting in an increase in fines for disruptive behavior. The no-thorazine condition (A) was only in force for 2 days, as the patient's renewal of disruptive activities caused nursing personnel to demand reinstatement of his medication. When thorazine was reintroduced in the final phase (B), number of tokens fined once again decreased to a zero level. Thus the controlling effects of thorazine over disruptive behavior were demonstrated. But Agras (1976) raised the question as to the possible contribution of the token economy program in controlling this patient's behavior. Unfortunately, time considerations did not permit him to systematically tease out the effects of that variable.

We might also note that in the A-B-A-B drug design, where the single- or double-blind trial is not feasible, staff and patient expectations of success during the drug condition are a possible confound with the drug's pharmacological actions. Designs listed in Table 6-1 that show control for these factors are 12 (A_1-B-A_1-B) and 13 (A-A_1-B-A_1-B). Design 13 is particularly useful in this instance. In the event that administration of the placebo fails to lead to behavioral change (A_1 phase of experimentation) over baseline measurement (A), the investigator is in a position to proceed with assessment of the active drug agent in an experimental analysis whereby the drug is twice introduced and once withdrawn (the B-A_1-B portion of study). If, on the other hand, the placebo exerts an effect over behavior, the investigator may wish to show its controlling effects as in Design 10 (A-A_1-A-A_1), which then can be followed with a sequential assessment of an active pharmacologic agent (Design 14— A-A_1-A-A_1-B-A_1-B). This design, however, does not permit an analysis of the interactive effects of a placebo (A_1) and a drug (B), as this would require the use of an interactive design (see section 6.5).

An example of the A-A_1-B-A_1-B strategy appears in the series of drug evaluations conducted by Liberman et al. (1973). In their study, the effects of a placebo and trifluperazine (stelazine) were examined on social interaction and content of conversation in a 21-year-old, withdrawn, male inpatient whose behavior had progressively deteriorated over a 3-year period. At the time the experiment was begun, the patient was receiving stelazine, 20 mg per day. Two dependent measures were selected for study: (1) willingness to engage in 18 daily, randomly time sampled, one-half minute chats with a member of the nursing staff, and (2) percentage of the chats that contained "sick talk." During the first phase of experimentation (A), the patient's medication was discontinued. In the second phase (A_1) a placebo was introduced, followed by application of stelazine, 60 mg per day, in the next phase (B). Then the A_1 and B phases were repeated. A double-blind trial was conducted, as the patient and nursing staff were not made aware of placebo and drug alternations.

Results of this study with regard to the patient's willingness to partake in brief conversations appear in Figure 6-7. In the no-drug condition (A) a marked linear increase in number of asocial responses was observed. Institution of the placebo in phase two (A_1) first led to a decrease, followed by a renewed increase in asocial responses, suggesting the overall ineffectiveness of the placebo condition. In Phase 3 (B), administration of stelazine (60 mg per day) resulted in a substantial decrease in asocial responses. However, a return to placebo conditions (A_1) again led to an increase in refusals to chat. In the final phase (B), reintroduction of stelazine effected a decrease in refusals. To summarize, in this experimental analysis, the effects of an active pharmacological agent were documented twice, as indicated by the decreasing data trends in the stelazine phases. Data with respect to content of conversation were not presented graphically, but the authors indicated that under stelazine conditions, rational speech increased. However, administration of stelazine did not appear to modify frequency of delusional and hypochondriacal statements in that they remained at a constant level across all phases of study.

Let us now return to and conclude our examination of drug designs in

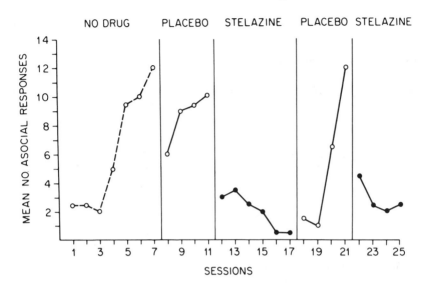

FIGURE 6-7. Average number of refusals to engage in a brief conversation. (Figure 2, p. 435, from: Liberman, R. P., Davis, J., Moon, W., & Moore, J. [1973]. Research design for analyzing drug-environment-behavior interactions. *Journal of Nervous and Mental Disease*, **156**, 432–439. Copyright 1973 Williams & Wilkins. Reproduced by permission.)

Table 6-1. In Design 15 (A_1-B-A_1-C-A_1-C) the controlling effects of two drugs (B and C) over placebo conditions (A_1) can be assessed. However, as in the A-B-A-C-A design, cited in section 6.1, the comparative efficacy of variables B and C are not subject to direct analysis, as a group comparison design would be required.

We should point out here that many extensions of these 15 basic drug designs are possible, including those in which differing levels of the drug are examined. This can be done within the structure of these 15 designs during active drug treatment or in separate experimental analyses where dosages are systematically varied (e.g., low-high-low-high) or where pharmacological agents are evaluated after possible failure of behavioral strategies (or vice versa). However, as in the A-B-A-C-A design cited in section 6.1, the comparative efficacy of variables B and C is subject to a number of restrictions and is, in general, a rather weak method for comparing two treatments.

The following A-B-C-A-D-A-D experimental analysis illustrates how, after two behavioral strategies (flooding, response prevention) failed to yield improvements in ritualistic behavior, a tricyclic (imipramine) led to some behavioral change, but only when administered at a high dosage (Turner, Hersen, Bellack, Andrasik, & Capparell, 1980).

The subject was a 25-year-old woman with a 7-year history of hand-washing and toothbrushing rituals. She had been hospitalized several times, with no treatment proving successful (including ECT). Throughout the seven phases of the study (with the exception of response prevention), mean duration of hand-washing and toothbrushing was recorded. Following a 7-day baseline period (A), flooding (B) was initiated for 8 days, and then response prevention (C) for 7 days. Then there was a 5-day return to baseline (A). Imipramine (C) was subsequently administered in increasing doses (75 mg to 250 mg) over 23 days, followed by withdrawal (A) and then reinstitution (C). In addition, 4 weeks of follow-up data were obtained.

Resulting data in Figure 6-8 are fairly clear-cut. Neither of the two behavioral strategies effected any change in the two behaviors targeted for modification. Similarly, imipramine, until it reached a level of 200 mg per day was ineffective. However, from 200–250 mg per day the drug appeared to reduce the duration of hand-washing and toothbrushing. When imipramine was withdrawn, hand-washing and toothbrushing increased in duration but decreased again when it was reinstated. Improvement was greatest at the higher dosage levels and was maintained during the 4-week follow-up.

From a design perspective, phases 4–7 (A-C-A-C) essentially are the same as Design 11 (A-B-A-B) in Table 6-1. Of course, the problem with the A-B-A-B design is that the intervening A' or placebo phase is bypassed, resulting in two variables being manipulated at once (i.e., ingestion and action of the drug). Therefore, one cannot discount the possible placebo effect in the Turner et al. (1980) analysis, although the long history of the disorder makes this interpretation unlikely.

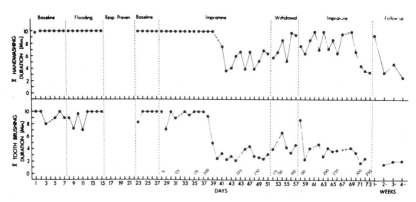

FIGURE 6-8. Mean duration of hand-washing and toothbrushing per day. (Figure 3, p. 654, from: Turner, S. M., Hersen, M., Bellack, A. S., Andrasik, F., & Capparell, H. V. [1980]. Behavioral and pharmacological treatment of obsessive-compulsive disorders. *Journal of Nervous and Mental Disease*, **168**, 651–657. Copyright 1980 The Williams and Wilkins Co., Baltimore. Reproduced by permission.)

6.6. STRATEGIES FOR
STUDYING INTERACTION EFFECTS

Most treatments contain a number of therapeutic components. One task of the clinical researcher is to experimentally analyze these components to determine which are effective and which can be discarded, resulting in a more efficient treatment. Analyzing the separate effects of single therapeutic variables is a necessary way to begin to build therapeutic programs, but it is obvious that these variables may have different effects when interacting with other treatment variables. In advanced stages of the construction of complex treatments it becomes necessary to determine the nature of these interactions. Within the group comparison approach, statistical techniques, such as analysis of variance, are quite valuable in determining the presence of interaction. These techniques are not capable, however, of determining the nature of the interaction or the relative contribution of a given variable to the total effect in an individual.

To evaluate the interaction of two (or more) variables, one must analyze the effects of both variables separately and in combination in one case, followed by replications. However, one must be careful to adhere to the basic rule of not changing more than one variable at a time (see chapter 3, section 3.4).

Before discussing examples of strategies for studying interaction, it will be helpful to examine some examples of designs containing two or more variables that are *not* capable of isolating interactive or additive effects. The first example is one where variations of a treatment are added to the end of a successful A-B-A-B (e.g., A-B-A-B^1-B^2-B^3 described above or an A-B-A-B-BC design in which C is a different therapeutic variable). If the BC variable produced an effect over and above the previous B phase, this would provide a clue that an interaction existed, but the controlling effects of the BC phase would not have been demonstrated. To do this, one would have to return to the B phase and reintroduce the BC phase once again.

A second design, containing two or more variables where analysis of interaction is not possible, occurs if one performs an experimental analysis of one variable against a background of one or more variables already present in the therapeutic situation. For example, O'Leary et al. (1969) measured the disruptive behavior of seven children in a classroom. Three variables (rules, educational structure, and praising appropriate behavior while ignoring disruptive behavior) were introduced sequentially. At this point, we have an A-B-BC-BCD design, where B is rules, C is structure, and D is praise and ignoring. With the exception of one child, these procedures had no effect on disruptive behavior. A fourth treatment—token economy—was then added. In five of six children this was effective, and withdrawal and reinstatement of the token economy confirmed its effectiveness. The last part of the design can be represented as BCD-BCDE-BCD-BCDE, where E is token economy. Although this experiment demonstrated that token economy works in this setting, the role of the first three variables is not clear. It is possible that any one of the variables or all three are necessary for the effectiveness of the token program or at least to enhance its effect. On the other hand, the initial three variables may not contribute to the therapeutic effect. Thus we know that a token program works in this situation, against the background of these three variables, but we cannot ascertain the nature of the interaction, if any, because the token program was not analyzed separately.

A third example, where analysis of interaction is not possible, occurs if one is testing the effects of a composite treatment package. Two examples of this strategy were presented in chapter 3, section 3.4. In one example (see Figure 3-13) the effects of covert sensitization on pedophilic interest were examined (Barlow, Leitenberg, & Agras, 1969). Covert sensitization, where a patient is instructed to imagine both unwanted arousing scenes in conjunction with aversive scenes, contains a number of variables such as therapeutic instruction, muscle relaxation, and instructions to imagine each of the two scenes. In this experiment, the whole package was introduced after baseline, followed by withdrawal and reinstatement of one component—the aversive scene. The design can be represented as A-BC-B-BC, where BC is the treatment package and C is the aversive scene. (Notice that more than one variable was changed during the transition from A-BC. This is in accordance with an exception to the guidelines outlined in chapter 3, section 3.4.)

Figure 3-13 demonstrates that pedophilic interest dropped during the treatment package, rose when the aversive scene was removed, and dropped again after reinstatement of the aversive scene. Once again, these data indicate that the noxious scene is important against the background of the other variables present in covert sensitization. The contribution of each of the other variables and the nature of these interactions with the aversive scene, however, have not been demonstrated (nor was this the purpose of the study). In this case, it would seem that an interaction is present because it is hard to conceive of the aversive scene alone producing these decreases in pedophilic interest. The nature of the interaction, however, awaits further experimental inquiry.

The preceding examples outlined designs where two or more variables are simultaneously present but analysis of interactive or additive effects is not possible. While these designs can hint at interaction and set the stage for further experimentation, a thorough analysis of interaction as noted above requires an experimental analysis of two or more variables, separately and in combination. To illustrate this complex process, two series of experiments will be presented that analyze the same variables—feedback and reinforcement— in two separate populations (phobics and anorexics). One experiment from the first series of phobics was presented in chapter 3, section 3.4, in connection with guidelines for changing one variable at a time.

In that series (Leitenberg et al., 1968) the first subject was a severe knife phobic. The target behavior selected for study was the amount of time (in seconds) that the patient was able to remain in the presence of the phobic object. The design can be represented as B-BC-B-A-B-BC-B, where B represents feedback, C represents praise, and A is baseline. Each session consisted of 10 trials. Feedback consisted of informing the patient after each trial as to the amount of time spent looking at the knife. Praise consisted of verbal reinforcement whenever the patient exceeded a progressively increasing time criterion. The results of the study are reproduced in Figure 6-9. During feedback, a marked upward linear trend in time spent looking at the knife was noted. The addition of praise did not appear to add to the therapeutic effect. Similarly, the removal of praise in the next phase did not subtract from the progress. At this point, it appeared that feedback was responsible for the therapeutic gains. Withdrawal and reinstatement of feedback in the next two

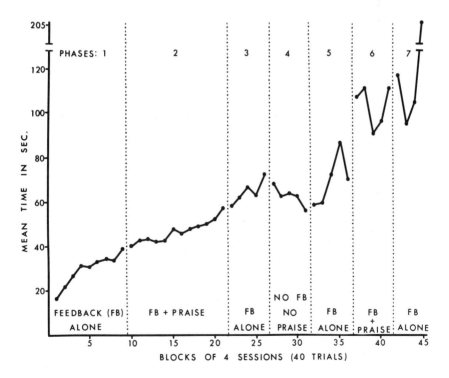

FIGURE 6-9. Time in which a knife was kept exposed by a phobic patient as a function of feedback, feedback plus praise, and no feedback or praise conditions. (Figure 2, p. 136, from: Leitenberg, H., Agras, W. S., Thomson, L. E., & Wright, D. E. [1968]. Feedback in behavior modification: An experimental analysis in two phobic cases. *Journal of Applied Behavior Analysis*, **1**, 131–137. Copyright 1968 by Society for the Experimental Analysis of Behavior, Inc. Reproduced by permission.)

phases confirmed the controlling effects of feedback. Addition and removal of praise in the remaining two phases replicated the beginning of the experiment, in that praise did not demonstrate any additive effect.

This experiment alone does not entirely elucidate the nature of the interaction. At this point, two tentative conclusions are possible. Either praise has no effect on phobic behavior, or praise does have an effect, which was masked or overridden by the powerful feedback effect. In other words, this patient may have been progressing at an optimal rate, allowing no opportunity for a praise effect to appear. In accordance with the general guidelines of analyzing both variables separately as well as in combination, the next experiment reversed the order of the introduction of variables in a second knife phobic patient (Leitenberg, 1973).

Once again, the target behavior was the amount of time the subject was able to remain in the presence of the knife. The design replicated the first experiment, with the exception of the elimination of the last phase. Thus the design can be represented as B-BC-B-A-B-BC. In this experiment, however, B refers to praise or verbal reinforcement and C represents feedback of amount of time looking at the knife, which is just the reverse of the last experiment.

In this subject, little progress was observed during the first verbal reinforcement phase (see Figure 6-10). However, when feedback was added to praise in the second phase, performance increased steadily. Interestingly, this rate of improvement was maintained when feedback was removed. After a sharp gain, performance stabilized when both feedback and praise were removed. Once again, the introduction of praise alone did not produce any further improvement. The addition of feedback to praise for the second time in the experiment resulted in marked improvement in the knife phobic. Direct replication of this experiment with 4 additional subjects, each with a different phobia, produced similar results. That is, praise did not produce improvement when initially introduced, but the addition of feedback resulted in marked improvement. In several cases, however, progress seemed to be maintained in praise after feedback was withdrawn from the package, as in Figure 6-10. In fact, feedback of progress, in its various forms, has come to be a major motivational component within exposure-based programs for phobia (Mavissakalian & Barlow, 1981b).

The overall results of the interaction analysis indicate that feedback is the most active component because marked improvement occurred during both feedback alone and feedback plus praise phases. Praise alone had little or no effect although it was capable of maintaining progress begun in a prior feedback phase in some cases. Similarly, praise did not add to the therapeutic effect when combined with feedback in the first subject. Accordingly, a more efficient treatment package for phobics would emphasize the feedback or knowledge-of-results aspect and deemphasize or possibly eliminate the social reinforcement component. These results have implications for treatments of

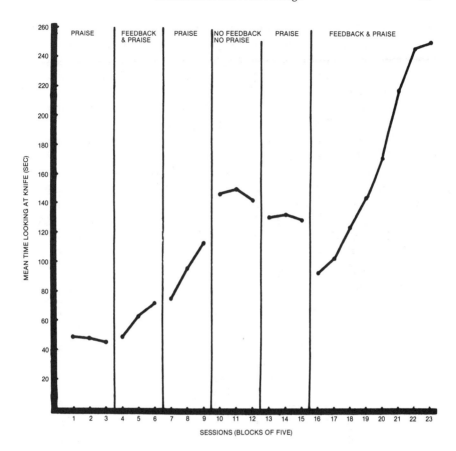

FIGURE 6-10. (Figure 1, from: Leitenberg, H. [1973]. Interaction designs. Paper read at the American Psychological Association, Montreal, August. Reproduced by permission.)

phobics by other procedures such as systematic desensitization, where knowledge of results provided by self-observation of progress through a discrete hierarchy of phobic situations is a major component.

The interaction of reinforcement and feedback was also tested in a series of subjects with anorexia nervosa (Agras et al., 1974). From the perspective of interaction designs, the experiment is interesting because the contribution of a third therapeutic variable, labeled *size of meals*, was also analyzed. To illustrate the interaction design strategy, several experiments from this series will be presented. All patients were hospitalized and presented with 6,000 calories per day, divided into four meals of 1,500 calories each. Two measures of eating behavior—weight and caloric intake—were recorded. Patients were also asked to record number of mouthfuls eaten at each meal. Reinforcement consisted of granting privileges based on increases in weight. If weight gain exceeded a certain criterion, the patient could leave her room, watch television, play table games with the nurses, and so on. Feedback consisted of providing precise information on weight, caloric intake, and number of mouthfuls eaten. Specifically, the patient plotted on a graph the information that was provided by hospital staff.

In one experiment the effect of reinforcement was examined against a background of feedback. The design can be represented as B-BC-BC¹-BC, where B is feedback, C is reinforcement, and C^1 is noncontingent reinforcement. During the first feedback phase (labeled *baseline* on the graph), slight gains in caloric intake and weight were noted (see Figure 6-11). When reinforcement was added to feedback, caloric intake and weight increased sharply. Noncontingent reinforcement produced a drop in caloric intake and a slowing of weight gain, while reintroduction of reinforcement once again produced sharp gains in both measures. These data contain hints of an

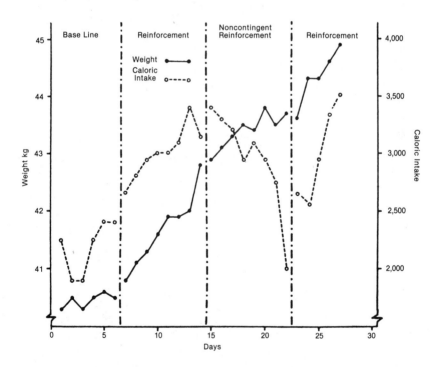

FIGURE 6-11. Data from an experiment examining the effect of positive reinforcement in the absence of negative reinforcement (Patient 3). (Figure 2, p. 281, from: Agras, W. S., Barlow, D. H., Chapin, H. N., Abel, G. G., & Leitenberg, H. [1974]. Behavior modification of anorexia nervosa. *Archives of General Psychiatry*, **30**, 279–286. Copyright 1974 American Medical Association. Reproduced by permission.)

interaction, in that caloric intake and weight rose slightly during the first feedback phase, a finding that replicated two earlier experiments. The addition of reinforcement, however, produced increases over and above those for feedback alone. The drop and subsequent rise of caloric intake and rate of weight gain during the next two phases demonstrated that reinforcement is a controlling variable *when combined with feedback.*

These data only hint at the role of feedback in this study, in that some improvement occurred during the initial phase when feedback alone was in effect. Similarly, we cannot know from this experiment the independent effects of reinforcement because this aspect was not analyzed separately. To accomplish this, two experiments were conducted where feedback was introduced against a background of reinforcement. Only one experiment will be presented, although both sets of data are very similar. The design can be represented as A-B-BC-B-BC, where A is baseline, B is reinforcement, and C is feedback (see Figure 6-12). It should be noted that the patient continued to be presented with 6,000 calories throughout the experiment, a point to which we will return later. During baseline, in which no reinforcement or feedback was present, caloric intake actually declined. The introduction of reinforce-

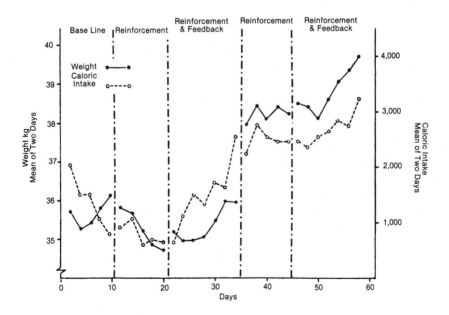

FIGURE 6-12. Data from an experiment examining the effect of feedback on the eating behavior of a patient with anorexia nervosa (Patient 5). (Figure 4, p. 283, from: Agras, W. S., Barlow, D. H., Chapin, H. N., Abel, G. G., & Leitenberg, H. [1974]. Behavior modification of anorexia nervosa. *Archives of General Psychiatry,* **30**, 279–286. Copyright 1974 American Medical Association. Reproduced by permission.)

ment did not result in any increases; in fact, a slight decline continued. Adding feedback to reinforcement, however, produced increases in weight and caloric intake. Withdrawal of feedback stopped this increase, which began once again when feedback was reintroduced in the last phase.

With this experiment (and its replications) it becomes possible to draw conclusions about the nature of what is in this case a complex interaction. When both variables were presented alone, as in the initial phases in the respective experiments, reinforcement produced no increases, but feedback produced some increase. When presented in combination, reinforcement added to the feedback effect and, against a background of feedback, became the controlling variable, in that caloric intake decreased when contingent reinforcement was removed. Feedback, however, also exerted a controlling effect when it was removed and reintroduced against a background of reinforcement. Thus, it seems that feedback can maximize the effectiveness of reinforcement to the point where it is a controlling variable. Feedback alone, however, is capable of producing therapeutic results, which is not the case with reinforcement. Feedback, thus, is the more important of the two variables, although both contribute to treatment outcome.

It was noted earlier that the contribution of a third variable—size of meals—was also examined within the context of this interaction. In keeping with the guidelines of analyzing each variable separately and in combination with other variables, phases were examined when the large amount of 6,000 calories was presented without the presence of either feedback or reinforcement. The baseline phase of Figure 6-12 represents one such instance. In this phase caloric intake declined steadily. Examination of other baseline phases in the replications of this experiment revealed similar results. To complete the interaction analysis, size of meal was varied against a background of both feedback and reinforcement. The design can be represented as ABC-ABC^1-ABC, where A is feedback, B is reinforcement, C is 6,000 calories per day, and C^1 is 3,000 calories per day.

Under this condition, size of meal did have an effect, in that more was eaten when 6,000 calories were served than when 3,000 calories were presented (see Figure 6-13). In terms of treatment, however, even large meals were incapable of producing weight gain in those phases where it was the only therapeutic variable. Thus this variable is not as strong as feedback. The authors concluded this series by summarizing the effects of the three variables alone and in combination across five patients:

> Thus large meals and reinforcement were combined in four experimental phases and weight was lost in each phase. On the other hand, large meals and feedback were combined in eight phases and weight was gained in all but one. Finally, all three variables (large meals, feedback, and reinforcement) were combined in 12 phases and weight was gained in each phase. These findings suggest that informa-

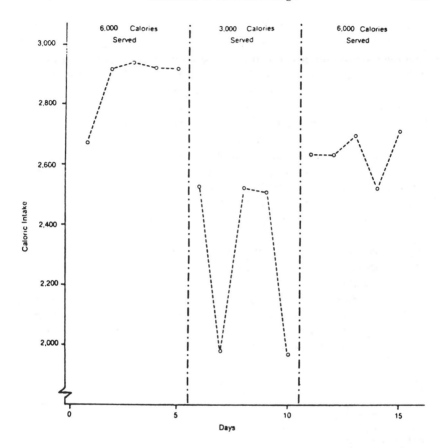

FIGURE 6-13. The effect of varying the size of meals upon the caloric intake of a patient with anorexia nervosa (Patient 5). (Figure 5, p. 285, from: Agras, W. S., Barlow, D. H., Chapin, H. N., Abel, G. G., & Leitenberg, H. [1974]. Behavior modification of anorexia nervosa. *Archives of General Psychiatry*, **30**, 279–286. Copyright 1974 American Medical Association. Reproduced by permission.)

tional feedback is more important in the treatment of anorexia nervosa than positive reinforcement, while serving large meals is least important. However, the combination of all three variables seems most effective. (Agras et al., 1974, p. 285)

As in the phobic series, the juxtaposition of variables within the general framework of analyzing each variable separately and in combination provided information on the interaction of these variables.

Let us now consider two more recent applications of the beginnings of an interaction design strategy in order to illustrate why they are incomplete at this point in time, in contrast with the experiments described above. One example is the evaluation of cognitive strategies (M. E. Bernard et al., 1983) and the other is concerned with the possible combined effects of drugs and behavior therapy (Rapport, Sonis, Fialkov, Matson, & Kazdin, 1983). M. E. Bernard et al. (1983) evaluated the effects of rational-emotive therapy (RET) and self-instructional training (SIT) in an A-B-A-B-BC-B-BC-A design with follow-up. The subject was a 17-year old, overweight female who suffered from trichotillomania (i.e., chronic hair pulling), especially while studying at home. Throughout the study the subject self-monitored time studying and number of hairs pulled out (deposited in an envelope). The dependent variable was the ratio of hairs pulled out per minute of study time.

In baseline (A) the subject simply self-monitored. During the B phase, RET was instituted, followed by a return to baseline (A) and reintroduction of RET (B). In the next phase, (BC), SIT, consisting of problem-solving dialogues, was added to RET. Then, SIT was removed (B) and subsequently reintroduced (BC). In the last phase (A) all treatment was removed, and then follow-up was conducted.

Results of this study appear in Figure 6-14. The first four phases comprise an A-B-A-B analysis and do appear to confirm the controlling effects of RET in reducing hair pulling. However, at this point the subject, albeit improved, still was engaging in the behavior a significant proportion of the time.

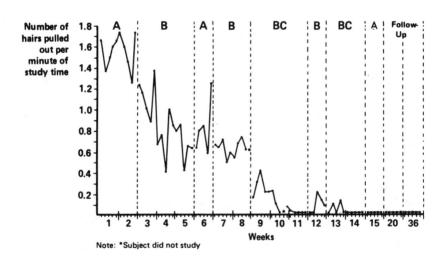

Note: *Subject did not study

FIGURE 6-14. The number of hairs pulled out per minute of study time over baseline treatment and follow-up phases. Missing data (*) reflect times when the subject did not study. (Figure 1, p. 277, from: Bernard, M. E., Kratochwill, T. R., & Keefauver, L. W. [1983]. The effects of rational-emotive therapy and self-instructional training on chronic hair pulling. *Cognitive Therapy and Research*, 7, 273-280. Copyright 1983 Plenum Publishing Corporation. Reproduced by permission.)

Phases 4–7 represent the interaction portion of the design (B-BC-B-BC). In Phase 5, addition of SIT to RET yielded additional improvement to near zero levels. When SIT then was removed in B, a moderate return of hair pulling was noted, which was again decreased to zero levels when SIT was added (BC). These gains subsequently held up in the final A phase and follow-up.

Although these data seem to confirm the therapeutic effect of SIT above and beyond that obtained by RET alone, the reader should be aware of two possible problems. *First*, all data are self-monitored and subject to experimental demand characteristics. *Second*, the BC phases are longer than each B phase; thus, there may be a possible confound with time. That is, a portion of the extra effect brought about by combining RET and SIT simply may be due to increased time of the combined treatment. However, this is unlikely, given the long-standing nature of the disorder.

In addition, a study of the interactional effects is not yet possible because SIT was not analyzed in isolation, but only against a background of RET. Thus it is possible that introducing SIT first would have a somewhat different effect, as would adding RET to SIT rather than the other way around, as in this experiment. While this is a noteworthy beginning, a more thorough evaluation of the interaction of SIT and RET awaits further experimental inquiry. Ideally, this experiment would be directly replicated at least twice, followed by the same experiment with SIT introduced first in three additional subjects. But we do not live in an ideal world, and trichotillomanics are few and far between.

Our final example of an interaction design involves a BC-BC'-B-BC-B-BD design, with two drugs (sodium valproate, carbamazepine) and one behavioral technique (differential reinforcement of other behavior [DRO]) evaluated (Rapport et al., 1983). The subject in this experimental analysis was a 13.7-year-old mentally retarded female who suffered from seizures and exhibited aggressive behavior toward others. She had a long history of hospitalizations and had been tried on a large variety of medications, but with little success. Aggressive behaviors included grabbing, biting, kicking, and hair pulling. Aggression was the primary dependent measure in this study and was recorded by inpatient staff with a high degree of interrater agreement (range = 92%–100%).

The subject received carbamazepine (400 mg, t.i.d.) in each phase of the study. In the first phase (BC) she received sodium valproate (1,200 mg) as well. This was gradually withdrawn in phase 2 (BC') and removed altogether in Phase 3 (B). In Phase 4 (BD) a DRO procedure (edible reinforcements delivered contingently for 15-minute time periods in which no aggression occurred; then increased to 30 and 60 minutes) was added to carbamazepine. DRO was discontinued in Phase 5 (B) and then reinstated in Phase 6 (BD).

Examination of Figure 6-15 shows a high rate of aggressive incidents (mean = 15 per day) in the first phase (BC), which decreased (mean = 3 per day)

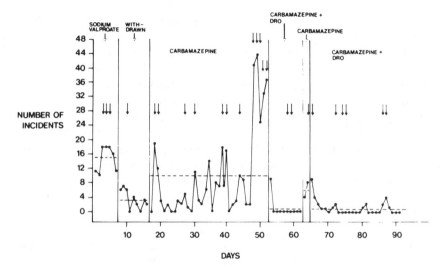

FIGURE 6-15. Data points represent the daily frequency of aggressive behavior during the child's hospital stay. (Arrows indicate days when nocturnal enuresis was observed.) (Figure 1, p. 262, from: Rapport, M. D., Sonis, W. A., Fialkov, M. J., Matson, J. L., & Kazdin, A. E. [1983]. Carbamazepine and behavior therapy for aggressive behavior: Treatment of a mentally retarded, postencephalic adolescent with seizure disorder. *Behavior Modification*, 7, 255–264. Copyright 1983 by Sage Publication. Reproduced by permission.)

when sodium valproate was withdrawn (BC). However, when the patient was totally withdrawn in Phase 3 (B), aggression rose to a mean of 10 a day. Institution of DRO in Phase 4 (BD) led to a dramatic decrease (0), rose to 4–8 when DRO was withdrawn (B) on days 63 and 64, and gradually decreased to zero again when DRO was reintroduced (BD) on days 65–91.

Although there was only a 2-day withdrawal of DRO procedures, this is truly justified given the aggressive nature of the behavior being observed. Indeed, it is quite clear that although the drug, carbamazepine had a minor role in controlling aggression, the addition of DRO was the major controlling force. Moreover, effectiveness of DRO allowed the subject to be discharged to her family, with DRO procedures subsequently implemented at school in order to ensure generalization of treatment gains.

Once again, replication on additional subjects and a subsequent reordering of the experimental strategy so that DRO was analyzed separately and then combined with the drug would be necessary for a more complete study of interactions. Finally, the nature of this experimental strategy deserves some comment, particularly when compared to other strategies attempting to answer the same questions. *First*, in any experiment there are more things interacting with treatment outcome than the two or more treatments or variables under question. Foremost among these are client variables. This, of course, is the reason for direct replication (see chapter 10). If the experimental operations are replicated (in this example the interaction), despite the different experiences clients bring with them to the experiment, then one has increasing confidence in the generality of the interactional finding across subjects.

Second, as pointed out in chapter 5 and discussed more fully in chapter 8, the latter phases of these experiments are subject to multiple-treatment interference. In other words, the effect of a treatment or interaction in the latter phases may depend to some extent on experience in the earlier phases. But if the interaction effect is consistent across subjects, both early and late in the experiment, and across different "orders" of introduction of the interaction, as in the first two examples described in this section (Agras et al., 1974; Leitenberg et al., 1968), then one has greatly increased confidence in both the fact and the generality of the effect. As with A-B-A withdrawal designs, however, the most easily generalizable data from the experiment to applied situations are the early phases before multiple treatments build up. This is because the early phase most closely resembles the applied situation, where the treatment would also be introduced and continued without a prior background of several treatments.

The other popular method of studying interactions is the between-group factorial design. In this case, of course, one group would receive both Treatments A and B, while two other groups would receive just A or just B. (If the factorial were complete, another group would receive no treatment.) Here treatments are not delivered sequentially, but the more usual problems of intersubject variability, inflexibility in altering the design, infrequent measurement, determination of results by statistical inference, and difficulties generalizing to the individual obtain, as discussed in chapter 2. Each approach to studying interactions obviously has its advantages and disadvantages.

6.7. CHANGING CRITERION DESIGN

The changing-criterion design, despite the fact that it has not to date enjoyed widespread application, is a very useful strategy for assessing the shaping of programs to accelerate or decelerate behaviors (e.g., increase interactions in chronic schizophrenics; decrease motor behavior in overactive children). As a specific design strategy, it incorporates A-B design features on a repeated basis. After initial baseline measurement, treatment is carried out until a preset criterion is met, and stability at that level is achieved. Then, a more stringent criterion is set, with treatment applied until this new level is met. If baseline is A and the first criterion is B, when the new criterion is set the former B serves as the new baseline (A^1) with B^1 as the second criterion.

This continues in graduated fashion until the final target (or criterion) is achieved at a stable level. As noted by Hartmann and Hall (1976), "Thus, each phase of the design provides a baseline for the following phase. When the rate of the target behavior changes with each stepwise change in the criterion, therapeutic change is replicated and experimental control is demonstrated" (p. 527).

This design, by its very nature, presupposes ". . . a close correspondence between the criterion and behavior over the course of the intervention phase" (Kazdin, 1982b, p. 160). When such close correspondence fails to materialize, with stability not apparent in each successive phase, unambiguous interpretations of the data are not possible. One solution, of course, is to partially withdraw treatment by returning to a lower criterion, followed by a return to the more stringent one (as in a B-A-B withdrawal design). This adds experimental confidence to the treatment by clearly documenting its controlling effects. Or, on a more extended basis, one can reverse the procedure and experimentally demonstrate successive increases in a targeted behavior following initial demonstration of successive decreases. This is referred to as *bi-directionality*. Finally, Kazdin (1982b) pointed out that some experimenters have dealt with the problem of excessive variability by showing that the mean performance over adjacent subphases reflects the stepwise progression.

None of the aforementioned solutions to variability in the subphases is ideal. Indeed, it behooves researchers using this design to demonstrate close correspondence between the changing criterion and actually observed behavior. Undoubtedly, as this design is employed more frequently, more elegant solutions to this problem will be found.

Hartmann and Hall (1976) presented an excellent illustration of the changing-criterion design in which a smoking-deceleration program was evaluated. Baseline level of smoking is depicted in panel A of Figure 6-16. In the next phase (B treatment), the criterion rate was set at 95% of the baseline rate (i.e., 46 cigarettes a day). An increasing response cost of $1 was established for smoking an additional cigarette (i.e., Number 47) and $2 for Number 48, and on and on. An escalating bonus of $0.10 a cigarette was established if the subject smoked less than the criterion number set. Subsequently, in phases C–G, the criterion for each succeeding phase was established at 94% of the previous one.

Careful examination of Figure 6-16 clearly indicates the success of treatment in reducing cigarette smoking by 2% or more from each preceding phase. Further, from the experimental analysis perspective, there were six replications of the contingencies applied. In each instance, experimental control was documented, with the treatment phase serving as baseline with respect to the decreasing criterion for the next phase, and so on.

Related to the changing criterion design is a strategy that Hayes (1981) has referred to as the *periodic-treatments design*. This design, at our writing, has been used most infrequently and really only has a quasi-experimental basis.

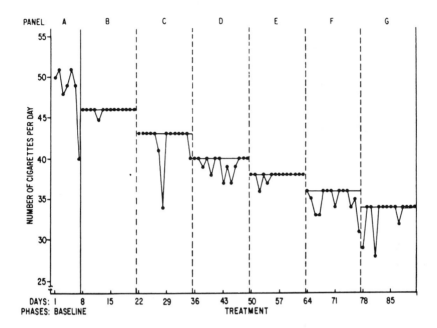

FIGURE 6-16. Data from a smoking-reduction program used to illustrate the stepwise criterion change design. The solid horizontal lines indicate the criterion for each treatment phase. (Figure 2, p. 529, from: Hartmann, D. P., & Hall, R. V. [1976]. The changing criterion design. *Journal of Applied Behavior Analysis*, 9, 527–532. Copyright 1976 by Soc. for the Experimental Analysis of Behavior. Reproduced by permission.)

Indeed, it is best suited for application in the private-practice setting (Barlow et al., 1983).

The logic of the design is quite simple. Frequently, marked improvements in a targeted behavior are seen immediately after a given therapy session. If this is plotted graphically, one can begin to see the relationship between the session (loosely conceptualized as an A phase) and time between sessions (loosely conceptualized as B phases). Thus, if steady improvement occurs, the scalloped display seen in the changing criterion design also will be observed here.

Hypothetical data for this design possibility are presented in Figure 6-17. But, as Hayes (1981) noted:

> These data do not show what about the treatment produced the change (any more than an A-B-A design would). It may be therapist concern or the fact that the client attended a session of any kind. These possibilities would then need to be eliminated. For example, one could manipulate both the periodicity and nature of treatment. If the periodicity of behavior change was shown only when a particular type of treatment was in place, this would provide evidence for a more specific effect. (p. 203)

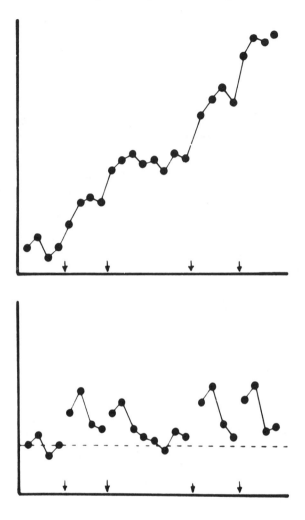

FIGURE 6-17. The periodic treatments effect is shown on hypothetical data. (Data are graphed in raw data form in the top graph.) Arrows on the abscissa indicate treatment sessions. This apparent B-only graph does not reveal the periodicity of improvement and treatment as well as the bottom graph, where each two data points are plotted in terms of the difference from the mean of the two previous data points. Significant improvement occurs only after treatment. Both graphs show an experimental effect; the lower is merely more obvious. (Figure 3, p. 202, from: Hayes, S. C. [1981]. Single case experimental design and empirical clinical practice. [1981]. *Journal of Consulting and Clinical Psychology*, **49**, 193–211. Copyright 1981 by American Psychological Association. Reproduced by permission.)

CHAPTER 7

Multiple Baseline Designs

7.1. INTRODUCTION

The use of sequential withdrawal or reversal designs is inappropriate when treatment variables cannot be withdrawn or reversed due to practical limitations, ethical considerations, or problems in staff cooperation (Baer et al., 1968; Barlow et al., 1977; Barlow & Hersen, 1973; Birnbauer, Peterson, & Solnick, 1974; Hersen, 1982; Kazdin & Kopel, 1975; Van Hasselt & Hersen, 1981). Practical limitations arise when carryover effects appear across adjacent phases of study, particularly in the case of therapeutic instructions (Barlow & Hersen, 1973). A similar problem may occur when drugs with known long-lasting effects are evaluated in single-case withdrawal designs. Despite discontinuation of medication in the withdrawal (placebo) phase, active agents persist psychologically and, with the phenothiazines, traces have been found in body tissues many months later (Goodman & Gilman, 1975). Also, when multiple behaviors within an individual are targeted for change, withdrawal designs may not provide the most elegant strategy for such evaluation.

Ethical considerations are of paramount importance when the treatment variable is effective in reducing self- or other-destructive behaviors in subjects. Here the withdrawal of treatment is obviously unwarranted, even for brief periods of time. Related to the problem of undesirable behavior is the matter of environmental cooperation. Even if the behavior in question does not have immediate destructive effects on the environment, if it is considered to be aversive (i.e., by teachers, parents, or hospital staff) the experimenter will not obtain sufficient cooperation to carry out withdrawal or reversal of treatment procedures. Under these circumstances, it is clear that the applied clinical researcher must pursue the study using different experimental strategies. In still other instances, withdrawal of treatment, despite absence of harm to the subject or others in his or her environment, may be undesirable because of the severity of the disorder. Here the importance of preserving therapeutic gains is given priority, especially when a disorder has a lengthy history and previous efforts at remediation have failed.

Multiple baseline designs and their variants and alternating treatment designs (see chapter 8) have been used by applied clinical researchers with increased frequency when withdrawals and reversals have not been feasible. Indeed, since publication of the first edition of this book in 1976, we find that the pages of our behavioral journals are replete with the innovative use of the multiple baseline strategy, for individuals as well as groups of subjects. A list of some recent, published examples of this design strategy appears in Table 7-1.

In this chapter we will examine in detail the rationale and procedures for multiple baseline designs. Examples of the three principal varieties of multiple baseline strategies will be presented for illustrative purposes. In addition, we will consider the more recent varieties and permutations, including the non-concurrent multiple baseline design across subjects, the multiple-probe technique, and the changing criterion design. Finally, the application of the multiple baseline across subjects in drug evaluations will be discussed.

7.2 MULTIPLE BASELINE DESIGNS

The rationale for the multiple baseline design first appeared in the applied behavioral literature in 1968 (Baer et al.), although a within-subject multiple baseline strategy had been used previously by Marks and Gelder (1967) in their assessment of electrical aversion therapy for a sexual deviate. Baer et al. (1968) point out that:

> In the multiple-baseline technique, a number of responses are identified and measured over time to provide baselines against which changes can be evaluated. With these baselines established, the experimenter then applies an experimental variable to one of the behaviors, produces a change in it, and perhaps notes little or no change in the other baselines. (p. 94)

Subsequently, the experimenter applies the same experimental variable to a second behavior and notes rate changes in that behavior. This procedure is continued in sequence until the experimental variable has been applied to all of the target behaviors under study. In each case the treatment variable is usually not applied until baseline stability has been achieved.

Baseline and subsequent treatment interventions for each targeted behavior can be conceptualized as separate A-B designs, with the A phase further extended for each of the succeeding behaviors until the treatment variable is finally applied. The experimenter is assured that the treatment variable is effective when a change in rate appears after its application while the rate of concurrent (untreated) behaviors remains relatively constant. A basic assumption is that the targeted behaviors are independent from one another. If they should happen to covary, then the controlling effects of the treatment variable are subject to question, and limitations of the A-B analysis fully apply (see chapter 5).

The issue of independence of behaviors within a single subject raises some interesting problems from an experimental standpoint, particularly if the experimenter is involved in a new area of study where no precedents apply. The experimenter is then placed in a position where an a priori assumption of independence cannot be made, thus leaving an empirical test of the proposition. Leitenberg (1973) argued that:

> If general effects on multiple behaviors were observed after treatment had been applied to only one, there would be no way to clearly interpret the results. Such results may reflect a specific therapeutic effect and subsequent response generalization, or they may simply reflect non-specific therapeutic effects having little to do with the specific treatment procedure under investigation. (p. 95)

In some cases, when independence of behaviors is not found, application of the alternating treatment design may be recommended (see chapter 8). In other cases, application of the multiple baseline design across different subjects might yield useful information. Surprisingly, however, in the available published reports the problem of independence has not been insurmountable (Leitenberg, 1973). Although problems of independence of behaviors apparently have been infrequently reported, some of the solutions referred to may not be viable if the experimenter is interested in targeting several behaviors within the same subject for sequential modification.

In attempting to prevent occurrence of the problem in interpretation when "onset of the intervention for one behavior produces general rather than specific changes," Kazdin and Kopel (1975) offered three specific recommendations. *The first*, of course, is to include baselines that topographically are as distinct as possible from one another. But this may be difficult to ascertain on an a priori basis. *The second* is to use four or more baselines rather than two or three. However, there always is the statistical probability that interdependence will be enhanced with a larger number. *The third* (on an *ex post facto* basis) is to withdraw and then reintroduce treatment for the correlated baseline (as in the B-A-B design), thus demonstrating the controlling effects over that targeted response. Even though the multiple baseline strategy was implemented in the first place to avoid treatment withdrawal, as in the A-B-A-B design, the rationale for such temporary (or partial) withdrawal in the multiple baseline design across behaviors seems reasonable when independence of baselines cannot be documented. But, as noted by Hersen (1982), "A problem with the Kazdin and Kopel solution is that in the case of instructions a *true* reversal or withdrawal is not possible. Thus their recommendations apply best to the assessment of such techniques as feedback, reinforcement, and modeling" (p. 191).

The multiple baseline design is considerably weaker than the withdrawal design, as the controlling effects of the treatment on each of the target behaviors are not directly demonstrated (e.g., as in the A-B-A design). As noted earlier, the effects of the treatment variable are inferred from the untreated behaviors. This raises an issue, then, as to how many baselines are needed before the experimenter is able to establish confidence in the controlling effects of his or her treatment. A number of interpretations have appeared in the literature. Baer et al. (1968) initially considered this issue to be an "audience variable" and were reluctant to specify the minimum number of baselines required. Although theoretically only a minimum of two baselines is needed to derive useful information, Barlow and Hersen (1973) argued that ". . . the controlling effects of that technique over at least three target behaviors would appear to be a minimum requirement" (p. 323). Similarly, Wolf and Risley (1971) contended that "While a study involving two baselines can be very suggestive, a set of replications across three or four baselines may be almost completely convincing" (p. 316). At this point, we would recommend a minimum of three to four baselines if practical and experimental considerations permit. As previously noted, Kazdin and Kopel (1975) recommended four or more baselines.

Although demonstration of the controlling effects of a treatment variable is obviously weaker in the multiple baseline design, a major advantage of this strategy is that it fosters the simultaneous measurement of several concurrent target behaviors. This is most important for at least two major reasons. *First*, the monitoring of concurrent behaviors allows for a closer approximation to naturalistic conditions, where a variety of responses are occurring at the same time. *Second*, examination of concurrent behaviors leads to an analysis of covariation among the targeted behaviors. Basic researchers have been concerned with the measurement of concurrent behaviors for some time (Catania, 1968; Herrnstein, 1970; Honig, 1966; G. S. Reynolds, 1968; Sidman, 1960). Applied behavioral researchers also have evidenced a similar interest (Kazdin, 1973b; Sajwaj et al., 1972; Twardosz & Sajwaj, 1972). Kazdin (1973b) underscored the importance of measuring concurrent (untreated) behaviors when assessing the efficacy of reinforcement paradigms in applied settings. He stated that:

> While changes in target behaviors are the *raison d'être* for undertaking treatment or training programs, concomitant changes may take place as well. If so, they should be assessed. It is one thing to assess and evaluate changes in a target

behavior, but quite another to insist on excluding nontarget measures. It may be that investigators are short-changing themselves in evaluating the programs. (p. 527)

As mentioned earlier, there are three basic types of multiple baseline designs. In the first—the multiple baseline design across behaviors—the same treatment variable is applied sequentially to separate (independent) target behaviors in a single subject. A possible variation of this strategy, of course, involves the sequential application of a treatment variable to targeted behaviors for an entire group of subjects (see Cuvo & Riva, 1980). In this connection, R. V. Hall, Cristler, Cranston, and Tucker (1970) note that ". . . these multiple baseline designs apply equally well to the behavior of groups if the behavior of the group members is summed or averaged, and the group is treated as a single organism" (p. 253). However, in this case the experimenter would also be expected to present data for individual subjects, demonstrating that sequential treatment applications to independent behaviors affected most subjects in the same direction.

In the second design—the multiple baseline design across subjects—a particular treatment is applied in sequence across *matched* subjects presumably exposed to "identical" environmental conditions. Thus, as the same treatment variable is applied to succeeding subjects, the baseline for each subject increases in length. In contrast to the multiple baseline design across behaviors (the within-subject multiple baseline design), in the multiple baseline design across subjects a single targeted behavior serves as the primary focus of inquiry. However, there is no experimental contraindication to monitoring concurrent (untreated) behaviors as well. Indeed, it is quite likely that the monitoring of concurrent behaviors will lead to additional findings of merit.

As with the multiple baseline design across behaviors, a possible variation of the multiple baseline design across subjects involves the sequential application of the treatment variable across entire groups of subjects (see Domash et al., 1980). But here, too, it behooves the experimenter to show that a large majority of individual subjects for each group evidenced the same effects of treatment.

We might note that the multiple baseline design across subjects has also been labeled a *time-lagged control* design (Gottman, 1973; Gottman, McFall, & Barnett, 1969). In fact, this strategy was followed by Hilgard (1933) some 50 years ago in a study in which she examined the effects of early and delayed practice on memory and motoric functions in a set of twins (method of co-twin control).

In the third design—the multiple baseline design across settings—a particular treatment is applied sequentially to a single subject or a group of subjects across independent situations. For example, in a classroom situation, one might apply time-out contingencies for unruly behavior in sequence across different classroom periods. The baseline period for each succeeding classroom period, then, increases in length before application of the treatment. As in the across-subjects design, assessment of treatment is usually based on rate changes observed in a selected target behavior. However, once again the monitoring of concurrent behaviors might prove to be of value and should be encouraged where possible.

To recapitulate, in the multiple baseline design across behaviors, a treatment variable is applied sequentially to independent behaviors within the same subject. In the multiple baseline design across subjects, a treatment variable is applied sequentially to the same behavior across different but matched subjects sharing the same environmental conditions. Finally, in the multiple baseline design across settings, a treatment variable is applied se-

TABLE 7-1. Recent Examples of Multiple Baseline Designs

STUDY	DESIGN	SUBJECTS
Alford, Webster, & Sanders (1980)	Across behaviors	Sexual deviate
Allison & Ayllon (1980)	Across subjects Across behaviors	Sports team members
Barmann, Katz, O'Brien, & Beauchamp (1981)	Across subjects	Developmentally disabled enuretics
Bates (1980)	Across behaviors	Retarded adults
Bellack, Hersen, & Turner (1976)	Across behaviors	Schizophrenics
Berler, Gross, & Drabman (1982)	Across behaviors	Learning disabled children
M. R. Bornstein, Bellack, & Hersen (1977)	Across behaviors	Unassertive children
M. R. Bornstein, Bellack, & Hersen (1980)	Across behaviors	Aggressive child inpatients
Breuning, O'Neill, & Ferguson (1980)	Across subjects (groups)	Retarded adults
Bryant & Budd (1982)	Across subjects	Preschoolers
Burgio, Whitman, & Johnson (1980)	Across subjects	Retarded children
Cuvo & Riva (1980)	Across behaviors	Retarded children
Domash et al. (1980)	Across subjects (groups)	Police officers
Dunlap & Koegel (1980)	Across behaviors (groups)	Autistic children
Dyer, Christian, & Luce (1982)	Across subjects	Autistic children
Egel, Richman, & Koegel (1981)	Across subjects	Autistic children
Epstein et al. (1981)	Across subjects	Families of dialectic children
Fairbank & Keane (1982)	Across settings	Vietnam veteran
C. Hall, Sheldon-Wildgen, & Sherman (1980)	Across behaviors (scenes)	Retarded adults
Halle, Baer, & Spradlin (1981)	Across subjects	Developmentally delayed children
Hay, Nelson, & Hay (1980)	Across subjects	Grade-schoolers
Hundert (1982)	Across subjects	Deaf children
R. T. Jones, Kazdin, & Haney (1981a)	Across subjects	Third graders
R. T. Jones, Kazdin, & Haney (1981b)	Across subjects	Third graders

(Continued)

TABLE 7-1. Recent Examples of Multiple Baseline Designs *(Continued)*

STUDY	DESIGN	SUBJECTS
J. A. Kelly, Urey, & Patterson (1980)	Across behaviors	Psychiatric patients
R. E. Kirchner et al. (1980)	Across settings (groups)	High-rate burglary areas
Kistner, Hammer, Wolfe, Rothblum, & Drabman (1982)	Across subjects (groups)	Grade-schoolers
Matson (1981)	Across subjects	Phobic retarded children
Matson (1982)	Across behaviors	Depressed retarded adults
Melin & Götestam (1981)	Across behaviors (groups)	Geriatric patients
Ollendick (1981)	Across settings	Children with nervous tics
Poche, Brouwer, & Swearingen (1981)	Across subjects	Preschoolers
Rosen & Leitenberg (1982)	Across settings (meals)	Anorexia nervosa patient
Russo & Koegel (1977)	Across behaviors	Autistic child
Singh, Dawson, & Gregory (1980)	Across settings	Retarded female
Singh, Manning, & Angell (1982)	Across subjects	Retarded monozygotic twins
Slavin, Wodarski, & Blackburn (1981)	Across subjects (groups)	College dorm residents
Stokes & Kennedy (1980)	Across subjects	Grade-schoolers
Stravynski, Marks, & Yule (1982)	Across behaviors (groups)	Neurotic outpatients
Sulzer-Azaroff & deSantamaria (1980)	Across subjects (groups)	Industrial supervisors
Van Biervliet, Spangler, & Marshall (1981)	Across settings (groups)	Retarded males
Van Hasselt, Hersen, Kazdin, Simon, & Mastantuono (1983)	Across behaviors	Blind adolescents
Whang, Fletcher, & Fawcett (1982)	Across subjects	Counselor trainees
Wong, Gaydos, & Fuqua (1982)	Across behaviors	Mildly retarded pedophile

quentially to the same behavior across different and independent settings in the same subject. Recently published examples of the three basic types of multiple baseline strategies are categorized in Table 7-1 with respect to design type and subject characteristics.

In the following three subsections we will illustrate the use of basic multiple baseline strategies in addition to presenting examples of variations selected from the child, clinical, behavioral medicine, and applied behavioral analysis literatures.

Multiple baseline across behaviors

M. R. Bornstein, Bellack, and Hersen (1977) used a multiple baseline strategy (across behaviors) to assess the effects of social skills training in the role-played performance of an unassertive 8-year-old male third grader (Tom) whose passivity led to derision by peers. Generally, if he experienced conflict with a peer, he cried or reported the incident to his teacher. Three target behaviors were selected for modification as a result of role-played performance in baseline: ratio of eye contact to speech duration, number of words, and number of requests. In addition, independent evaluations of overall assertiveness, based on role-played performance, were obtained. As can be seen in Figure 7-1, baseline responding for targeted behaviors was low and stable. Following baseline evaluation, Tom received 3 weeks of social skills training consisting of three 15–30 minute sessions per week. These were applied sequentially and cumulatively over the 3-week period. Throughout training, six role-played scenes were used to evaluate the effects of treatment. In addition, three scenes (on which the subject received no training) were used to assess generalization from trained to untrained scenes.

The results for training scenes appear in Figure 7-1. Examination of the graph indicates that institution of social skills training for ratio of eye contact to speech duration resulted in marked changes in that behavior, but rates for number of words and number of requests remained constant. When social skills training was applied to number of words itself, the rate for number of requests remained the same. Finally, when social skills training was directly applied to number of requests, marked changes were noted. Thus it is clear that social skills training was effective in increasing the rate of the three target behaviors, but only when treatment was applied directly to each. Independence of the three behaviors and absence of generalization effects from one behavior to the next facilitate interpretation of these data. On the other hand, had nontreated behaviors covaried following application of social skills training, unequivocal conclusions as to the controlling effects of the training could not have been reached without resorting to Kazdin and Kopel's (1975) solution to withdraw and reinstate the treatment.

The reader should also note in Figure 7-1 that, despite the fact that overall assertiveness was not treated directly, independent ratings evinced gradual improvement over the 3-week period, with treatment gains for all behaviors maintained in follow-up.

Examination of data for the *untreated* generalization scenes indicates that similar results were obtained, confirming that transfer of training occurred from treated to untreated items. Indeed, the patterns of data for Figures 7-1 and 7-2 are remarkably alike.

Liberman and Smith (1972) also used a multiple baseline design across behaviors in studying the effects of systematic desensitization in a 28-year-old, multiphobic female who was attending a day treatment center. Four specific phobias were identified (being alone, menstruation, chewing hard foods, dental work), and baseline assessment of the patient's self-report of each was taken for 4 weeks. Subsequently, *in vivo* and standard systematic desensitization (consisting of relaxation training and hierarchical presentation of items in imagination) were administered in sequence to the four areas of

TRAINING SCENES

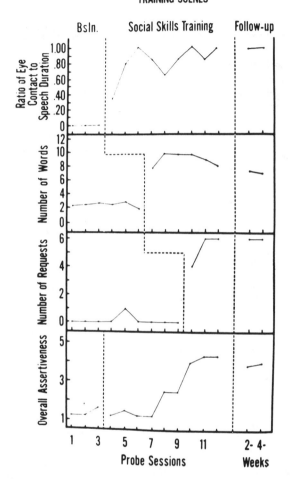

FIGURE 7-1. Probe sessions during baseline, social skills treatment, and follow-up for training scenes for Tom. A multiple baseline analysis of ratio of eye contact while speaking to speech duration, number of words, number of requests, and overall assertiveness. (Figure 3, p. 190, from: Bornstein, M. R., Bellack, A. S., Hersen, M. [1977]. Social-skills training for unassertive children: A multiple-baseline analysis. *Journal of Applied Behavior Analysis*, **10**, 183–195. Copyright 1977 by Society for Experimental Analysis of Behavior. Reproduced by permission.)

phobic concern. Specifically, *in vivo* desensitization was administered in relation to fears of being alone and chewing hard foods, while fears of menstruation and dental work were treated imaginally.

Results of this study, presented in Figure 7-3, indicate that the sequential application of desensitization affected the particular phobia being treated,

GENERALIZATION SCENES

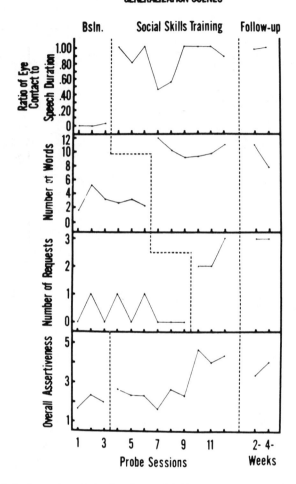

FIGURE 7-2. Probe sessions during baseline, social skills treatment, and follow-up for general-ization scenes for Tom. A multiple baseline analysis of ratio of eye contact while speaking to speech duration, number of words, number of requests and overall assertiveness. (Figure 4, p. 191, from: Bornstein, M. R., Bellack, A. S., & Hersen, M. [1977]. Social-skills training for unassertive children: A multiple-baseline analysis. *Journal of Applied Behavior Analysis*, **10**, 183–195. Copyright 1977 by Society for the Experimental Analysis of Behavior. Reproduced by permission.)

but no evidence of generalization to untreated phobias was noted. Independence of the four target behaviors and rate changes when desensitization was finally applied to each support the conclusion that treatment was effective and that it exerted control over the dependent measures (self-reports of degrees of fear). Although the authors argued that a positive set for improve-

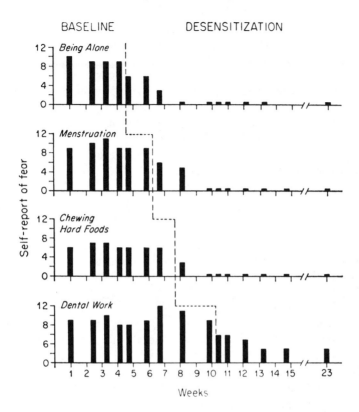

FIGURE 7-3. Multiple baseline evaluation of desensitization in a single case with four phobias. (Figure 1, p. 600, from: Liberman, R. P., & Smith, V. [1972]. A multiple baseline study of systematic desensitization in a patient with multiple phobias. *Behavior Therapy*, **3**, 597–603. Copyright 1972 by Association for the Advancement of Behavior Therapy. Reproduced by permission.)

ment was maintained throughout all phases of study, the possibility that expectancy of improvement and actual treatment effects were confounded cannot be discounted, especially in light of the primary reliance on self-report data. However, casually conducted behavioral observations corroborate self-report data.

Despite the above-mentioned limitations, Liberman and Smith's (1972) investigation is of interest from a number of standpoints. *First*, as most multiple baseline studies emanate from the operant framework, this study lends credence to the notion that nonoperant procedures (e.g., systematic desensitization) can be assessed in this paradigm. *Second*, as the particular dependent measure (ratings of subjective fear on the Target Complaint Scale) is based on the patient's self-report, it would appear that this type of single-case research might easily be carried out in inpatient facilities and even in

consulting room practice (see chapter 3, section 3.2). *Finally*, the treatment
was fully implemented by a mental health paraprofessional who had only one
year's training in psychiatry.

In our next example of a multiple baseline design across behaviors, a
psychological measure (erectile strength as assessed with a penile gauge) was
used to determine efficacy of covert sensitization in the treatment of a 21-
year-old married male, admitted for inpatient treatment of exhibitionism and
obscene phone calling (Alford, Webster, & Sanders, 1980). History of exhibitionism began at age 16, and obscene phone calling had taken place over the
previous year. During baseline assessment:

> Audiotapes of both deviant and nondeviant sexual scenes were used to elicit
> arousal during physiological monitoring sessions. Deviant stimulus material
> included three tapes depicting various obscene phone calls . . . and three tapes of
> exhibitionism. . . . Two nondeviant tapes . . . that depicted normal heterosexual
> behavior were also used. . . . They consisted of verbal descriptions designed to
> closely parallel the patient's own sexual behavior and fantasy. (p. 17)

These included one taped description of intercourse with his wife and another
with different sexual partners.

Covert sensitization sessions were conducted twice daily in the hospital at
various locations. This treatment consisted of imaginally pairing the deviant
sexual approach (i.e., obscene phone calls, exhibitionism) with aversive stimuli such as suffocation, nausea, and arrest. Each session involved 20 pairings
of the deviant scenarios with aversive imagery. Following baseline assessment, covert sensitization was first applied to obscene phone calling and then
to exhibitionism. In addition to therapist-conducted treatment sessions, the
patient was instructed to use covert imagery on his own initiative whenever he
experienced deviant sexual urges.

Data for this multiple baseline analysis are presented in Figure 7-4. During
baseline evaluation, penile tumescence in response to tapes of obscene phone
calling and exhibitionism was quite high. Similarly, tumescence was above
75% in response to nondeviant tapes of sexual activity with females other
than his wife, but only slightly higher than 25% in response to lovemaking
with his wife.

Institution of covert sensitization for obscene phone calling resulted in
marked diminution in penile responsivity to taped descriptions of that behavior, eventually resulting in only a negligible response. However, such treatment also appeared to affect changes in penile response to one of the
exhibitionism tapes (Ex. 1), even though that behavior had not yet been
specifically targeted. (We have here an instance where the baselines are not
independent from one another.) However, when treatment subsequently was
directed to exhibitionism itself, there was marked diminution in penile re-

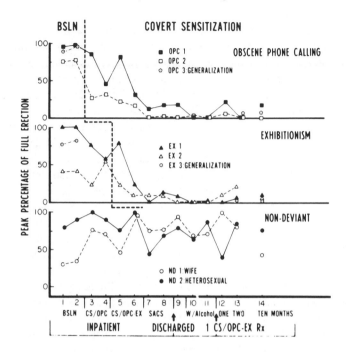

FIGURE 7-4. Percentage of full erection to obscene phone call (OPC) exhibitionistic (EX), and heterosexual stimuli (ND) during baseline, treatment, and follow-up phases. (Figure 1, p. 20, from: Alford, G. S., Webster, J. S., & Sanders, S. H. [1980]. Covert aversion of two interrelated deviant sexual practices: Obscene phone calling and exhibitionism. A single case analysis. *Behavior Therapy*, **11**, 13–25. Copyright 1980 by Association for the Advancement of Behavior Therapy. Reproduced by permission.)

sponse to tapes Ex. 2 and Ex. 3 in addition to continued decreases to tape Ex. 1. During the course of treatment, penile responsivity to nondeviant heterosexual interactions remained high, increasing considerably with respect to lovemaking with the wife.

The reader might note that "the patient was preloaded with 36 oz of beer 90 to 60 minutes prior to Assessments 10 and 11" (Alford et al., 1980, p. 19). This was carried out inasmuch as he had claimed that alcohol had disinhibited deviant sexuality. However, experimental data did not seem to confirm this. One, 2-, and 10-month follow-up assessments indicated that all gains were maintained, with the exception of decreased penile responsivity to taped descriptions of intercourse with the wife. In addition, 10-month collateral information from the patient's wife, parents, and attorney, as well as police, court, and telephone company records revealed no incidents of sexual deviance.

Our illustration reveals a clinically successful intervention evaluated through the multiple baseline strategy. However, because of some correlation between the first two baselines (obscene phone calling and exhibitionism), the experimental control of the treatment over targeted behaviors is somewhat unclear. Retrospectively, a more elegant experimental demonstration might have ensued if the experimenters had temporarily withdrawn treatment from the second baseline and then reinstated it (in B-A-B fashion), in order to show the specific controlling power of the aversive strategy. However, from the clinical standpoint, given the length of the disorder, it is most likely that the aversive intervention was responsible for ultimate change.

The study by Barton, Guess, Garcia, and Baer (1970) illustrates the use of a multiple baseline design in which treatment was applied sequentially to separate targeted behaviors for an entire group of subjects. Sixteen severely and profoundly retarded males served as subjects in an experiment designed to improve their mealtime behaviors through the use of time-out procedures. Several undesirable mealtime behaviors were selected as targets for study during preliminary observations. They included *stealing* (taking food from another resident's tray), *fingers* (eating food with the fingers that should have been eaten with utensils), *messy utensils* (e.g., using a utensil to push food off the dish, spilling food), and *pigging* (eating spilled food from the floor, a tray, etc.; placing mouth directly over food without the use of a utensil). Observations of these behaviors were made 5 days per week during the noon and evening meals by using a time-sampling procedure. Independent observations were also obtained as reliability checks. The treatment—time-out—involved removing the subject (cottage resident) from the dining area for the remainder of a meal or for a designated time period contingent upon his evidencing undesirable mealtime behavior.

The full time-out contingency (removal from the dining area for the entire meal) was initially applied to *stealing* following 6 days of baseline recording. Time-out contingencies for *fingers*, *messy utensils*, and *pigging* were then applied in sequence, each time maintaining the contingency in force for the previously treated behavior. During the application of time-out for *fingers*, the contingency involved time-out from the entire meal for 11 subjects, but only 15 seconds time-out for 5 of the subjects. This differentiation was made in response to nursing staff's concerns that a complete time-out contingency for the five subjects might jeopardize their health. Time-out procedures for *messy utensils* and *pigging* were limited to 15 seconds per infraction for all 16 subjects.

The results of this study are presented in Figure 7-5. Examination of the graph indicates that when time-out was applied to *stealing* and *fingers*, rates for these behaviors decreased. However, application of time-out to *fingers* also resulted in a concurrent increase in the rate for *messy utensils*. But subsequent application of time-out for *messy utensils* effected a decrease in

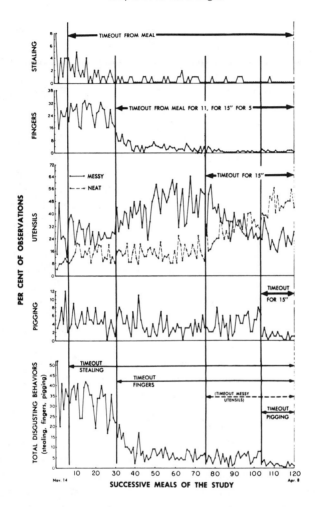

FIGURE 7-5. Concurrent group rates of Stealing, Fingers, Utensils, and Pigging behaviors, and the sum of Stealing, Fingers, and Pigging (Total Disgusting Behaviors) through the baseline and experimental phases of the study. (Figure 1, p. 80, from: Barton, E. S., Guess, D., Garcia, E., & Baer, D. M. [1970]. Improvement of retardates' mealtime behaviors by time-out procedures using multiple baseline techniques. *Journal of Applied Behavior Analysis*, **3**, 77–84. Copyright 1970 by Society for Experimental Analysis of Behavior, Inc. Reproduced by permission.)

rate for that behavior. Finally, application of time-out for *pigging* proved successful in reducing its rate.

Independence of the target behaviors was observed, with the exception of *messy utensils*, which increased in rate when the time-out contingency was applied to *fingers*. Although group data for the 16 subjects were presented, it would have been desirable if the authors had presented data for individual subjects. Unfortunately, the time-sampling procedure used by Barton et al. (1970) precluded obtaining such information. However, this factor should not overshadow the clinical and social significance of this study, in that (1) mealtime behaviors improved significantly; (2) a result of improved mealtime behaviors was a concomitant improvement in staff morale, facilitating more favorable interactions with the subjects; and (3) staff in other cottages were sufficiently impressed with the results of this study to begin to implement similar mealtime programs for their own retarded residents.

A more recent example of a multiple baseline design across behaviors (carried out in group format) was presented by Bates (1980). This study is of particular interest inasmuch as he contrasted the effects of interpersonal skills training (i.e., social skills training) for an experimental group with a control condition that received no treatment. Subjects were moderately and mildly retarded adults (8 in the treatment group, 8 in the control group). Since treatment was carried out sequentially and cumulatively across four behaviors (introductions and small talk, asking for help, differing with others, handling criticism) following initial assessment, a multiple baseline analysis was possible in addition to a controlled group evaluation.

A 16-item role-play test was the dependent measure, with subjects receiving interpersonal skills training for eight of these scenarios. The remaining eight, for which subjects received no training, served as a measure of transfer of training. (But this was only accomplished on a pre-post basis.) Skills training was conducted thrice weekly and consisted of modeling, behavior rehearsal, coaching, feedback, incentives, and homework assignments. After each set of three training sessions an assessment was performed.

Results of this analysis appear in Figure 7-6. As the reader will note, improvements in each of the four targeted behaviors occurred in time-lagged fashion only when treatment was specifically applied to each. Thus there was no evidence of correlated baselines. Data indicate that interpersonal skills training was effective in bringing about behavioral change. Further, results of the group comparison indicated that there were statistically significant differences in favor of the experimental condition.

Although these data are impressive, we would like to identify a few problems. *First*, baseline assessment for introductions and small talk should have been extended to three points, despite the apparent stability. *Second*, a three-point assessment in the treatment phase for handling criticism is warranted considering that there is the beginning of a downward trend in the data. If this trend were to continue, unequivocal statements about the treatment's controlling effects over that behavior could not be made. *Third*, presentation of data for individual subjects in a table would have been useful from the single-subject perspective.

This can be a very useful design, but in co-opting behavior analytic

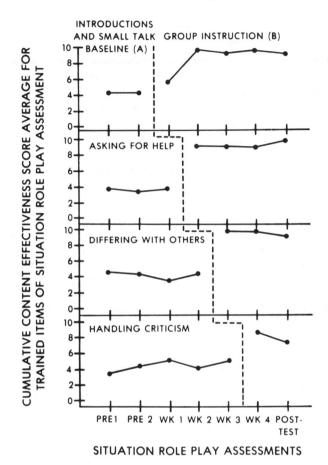

FIGURE 7-6. A multiple baseline analysis of the influence of interpersonal skills training on Exp. 1's cumulative content effectiveness score average across four social skill areas. (Figure 1, p. 244, from: Bates, P. [1980]. The effectiveness of interpersonal skills training on the social skill acquisition of moderately and mildly retarded adults. *Journal of Applied Behavior Analysis*, **13**, 237–248. Copyright 1980 by Society for Experimental Analysis of Behavior. Reproduced by permission.)

procedures, one must be careful to present as much individual data as possible. For example, all of the problems of averaging apply to these data. That is, some subjects could show the very steady changes apparent in the group data across measurement sessions, whereas others might demonstrate very cyclic types of patterns. Presenting data in this way does not allow one the option of examining sources of variability where it might be important. *Finally*, since it is not clear how many individuals changed in clinically significant ways, estimates of the replicability of these procedures across individuals and identification of individual predictors of success and failure are not possible (see chapter 10). Thus, when proceeding in this manner, presentation of as much individual data as possible is strongly recommended.

In an interesting solution to the problem of averaging when a number of subjects are treated simultaneously, Kelly (1980) argued for application of a design referred to as the *Simultaneous Replication Design*. This design is used within a multiple baseline format. The specific example cited involves application of social skills training *in group format* to 6 subjects for three components of social skill on a time-lagged basis. However, although applied on a group basis, behavioral assessment of each subject follows each group session. Thus individual data for each treated subject are available and can be plotted *individually* (see Fig. 10-6). As noted by Kelly (1980):

> The use of this group multiple baseline-simultaneous replication design is particularly useful in applied clinical settings for several reasons. First, it eliminates the need for elaborate and/or untreated control groups to establish group treatment effects and rule out many alternative hypotheses which cannot be adequately controlled by other one group designs. Second, by analyzing the social skills behavior change effects of a *group* treatment procedure, it is possible to demonstrate more compellingly cost- or time-effectiveness than if each subject had been laboriously handled as an *individually treated* case study using single subject procedures. Because subjects all received the same group training but are individually evaluated after each group, it is possible to examine "within subject" response to group treatment with greater specificity than in "between groups" designs. Since data for each subject in the training group is individually measured and graphed, each subject also serves as a simultaneous replication for the training procedure and provides important information on the generality (or specificity) of the treatment. (pp. 206–207)

(See also section 10.2 for a discussion of issues arising from this strategy relevant to replication.)

Although the multiple baseline design is frequently used in clinical research when withdrawal of treatment is considered to be detrimental to the patient, on occasion withdrawal procedures have been instituted following the sequential administration of treatment to target behaviors, particularly when reinforcement techniques are being evaluated (e.g., Russo & Koegel, 1977). If treatment is reintroduced after a withdrawal, a powerful demonstration of its controlling effects can be documented. This type of multiple baseline strategy was used by Russo and Koegel (1977) in their evaluation of behavioral techniques to integrate an autistic child into a normal public school classroom. The subject was a 5-year-old girl who previously had been diagnosed as *autistic*. She evinced limited verbal behavior, failed to respond to the initiatives of others, and, when she did verbalize, her comments reflected pronoun

INTEGRATING AN AUTISTIC CHILD

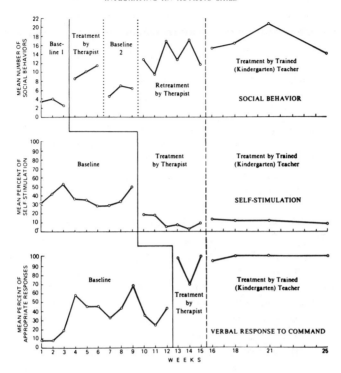

FIGURE 7-7. Social behavior, self-stimulation, and verbal response to command in the normal kindergarten classroom during baseline, treatment by the therapist, and treatment by the trained kindergarten teacher. All three behaviors were measured simultaneously. (Figure 1, p. 585, from: Russo, D. C., & Koegel, R. L. [1977]. A method for integrating an autistic child into a normal public school classroom. *Journal of Applied Behavior Analysis*, 10, 579–590. Copyright 1977 by Society for Experimental Analysis of Behavior. Reproduced by permission.)

reversal. Classroom behavior was characterized by inappropriate actions, tantrums, bizarre mannerisms, and general aloofness.

Three behaviors were targeted for modification by Russo and Koegel (1977) in one of the multiple baseline analyses performed: social behavior, self-stimulation, and verbal response to command. They were all assessed and treated within the context of the child's kindergarten classroom. Examination of Figure 7-7 indicates that rate of social behavior was uniformly low, self-stimulation was quite high, and appropriate responses were low but increasing. Treatment consisted of token reinforcement paired with verbal praise, feedback, and response cost (removal of tokens) for self-stimulation. Tokens were earned contingently upon occurrence of each instance of social behavior and appropriate responses, and they were systematically removed for each occurrence of self-stimulatory behavior. At the end of each training session the child had the opportunity to trade remaining tokens for a menu of backup reinforcers. Three pretraining sessions were carried out to establish the reinforcing value of tokens.

Initial treatment by the therapist for social behaviors resulted in a marked increase in responsivity for that 3-week period. There were no substantial changes in self-stimulatory behavior. However, there was some concurrent increase in rate of appropriate responses, which then decreased somewhat. In Weeks 7–9 the reinforcement contingency for social behaviors was withdrawn, resulting in a marked decrease. However, when reinstated in Weeks 10–15, there once again was a substantial improvement in social responding, thus confirming the controlling effects of reinforcement in A-B-A-B fashion. Concurrent with retreatment of social behavior in Weeks 10–15 was application of the contingency for self-stimulation. This led to marked diminution in such behaviors, with no concurrent changes in the third baseline (appropriate responses). In Weeks 13–16, when treatment was directed specifically to appropriate responses, a marked improvement was observed.

In Weeks 14 and 15 the therapist began training the teacher to apply treatment. From Week 16 through Week 25 the teacher carried out treatment under the supervision of the initial therapist. Over the course of this time period the reinforcement schedule was gradually thinned. Data for Weeks 16–25 indicate that initial improvement was either maintained or enhanced.

In summary, this study illustrates the use of the multiple baseline design across behaviors in a single subject, demonstrating general independence of target behaviors. Sequential application of a reinforcement contingency to individual behaviors showed the controlling effects of the contingency. Additional experimental manipulations (withdrawal and reintroduction of the contingency) for the first baseline (social behavior) further confirmed the controlling effects of the treatment. Finally, data indicate that treatment procedures were effectively taught to the teacher, who was able to maintain the child's improved performance in the last phase of the study.

In our final example of a multiple baseline design across behaviors, the effects of booster treatment subsequent to deterioration during follow-up (after initial success of social skills training) and documented (Van Hasselt, Hersen, Kazdin, Simon, & Mastantuono, 1983). The subject was a blind female child attending a special school for the blind. Baseline assessment of social skills through role playing revealed deficiencies in posture and gaze, a hostile tone of voice, inability to make requests for new behavior, and a general lack of social skills (see Figure 7-8).

The sequential and cumulative application of social skills training resulted in marked improvements in role-played performance, thus documenting the controlling effects of the treatment. However, data for the 4-week posttreat-

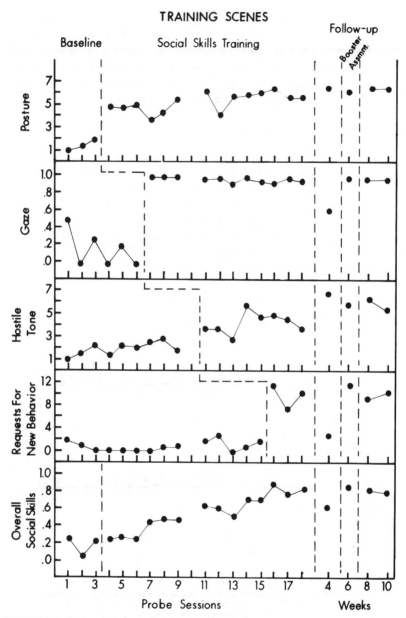

TRAINING SCENES

FIGURE 7-8. Probe sessions during baseline, social skills treatment, follow-up, and booster assessments for training scenes for S1. A multiple baseline analysis of posture, gaze, hostile tone, requests for new behavior, and overall social skill. (Figure 1, p. 201, from: Van Hasselt, V. B., Hersen, M., Kazdin, A. E., Simon, J., & Mastantuono, A. K. [1983]. Social skills training for blind adolescents. *Journal of Visual Impairment and Blindness*, **75**, 199–203. Copyright 1983. Reproduced by permission.)

ment follow-up revealed a decrement for gaze and requests for new behavior. Examination of Figure 7-8 shows that retreatment in booster sessions for those behaviors resulted in a renewed improvement, extending through the 8- and 10-week follow-up assessments. Thus our multiple baseline analysis permitted a clear assessment of which behaviors were maintained after treatment in addition to those requiring booster treatment.

Multiple baseline across subjects

Our first example of the multiple baseline strategy across subjects is taken from the clinical child literature. Barmann, Katz, O'Brien, and Beauchamp (1981) examined the sequential application of overcorrection training for three developmentally disabled children who were diagnosed as irregular enuretics. These children (4-, 7-, and 8-years-old, respectively) had IQs that ranged from 23–41. The first 2 subjects lived at home and the third resided in a home care facility for the developmentally disabled. Subjects 1 and 3 were

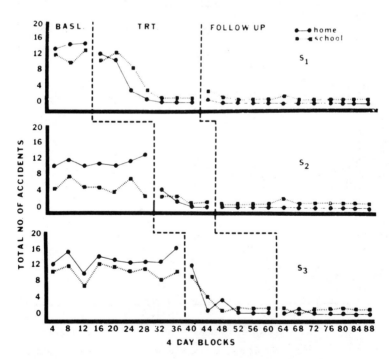

FIGURE 7-9. Total number of accidents at home and school during baseline, treatment, and follow-up conditions. NOTE: Data are collapsed over 4-day periods. (Figure 1, p. 344, from: Barmann, B. C., Katz, R. C., O'Brien, F., & Beauchamp, K. L. [1981]. Treating irregular enuresis in developmentally disabled persons: A study in the use of overcorrection. *Behavior Modification*, 5, 336–346. Copyright 1981 by Sage Publications. Reproduced by permission.)

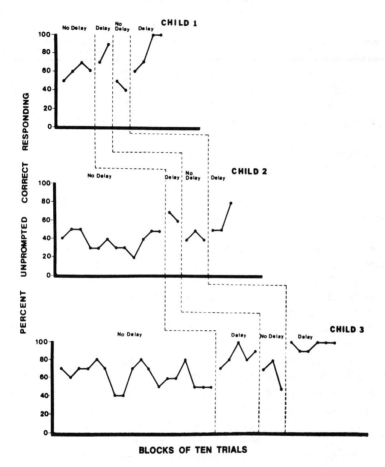

FIGURE 7-10. Results of the multiple baseline analysis with subsequent repeated reversals of the influence of a response-delay requirement of the correct responding of autistic children. (Figure 1, p. 235, from: Dyer, K., Christian, W. P., & Luce, S. C. [1982]. The role of response delay in improving the discrimination performance of autistic children. *Journal of Applied Behavior Analysis*, **15**, 231–240. Copyright 1982 by Society for Experimental Analysis of Behavior. Reproduced by permission.)

enuretic at night at encopretic during the day, in addition to evincing diurnal enuresis. Subject 2 only evidenced diurnal enuresis.

During baseline, hourly pants checks were performed by parents and the teacher, at home and at school respectively. Instances of dry pants were praised at home and at school. Inspection of Figure 7-9 indicates that baseline levels of accidents ranged from 10–15 per child over a 4-day period.

After stable baselines were observed, overcorrection treatment was applied sequentially and cumulatively to the three children. Treatment involved restitution overcorrection when the pants were found to be wet at home. (No treatment was administered at school as this served as a measure of generalization.) Restitutional overcorrection ". . . required the child to (a) obtain a towel, (b) clean up all traces of the accident, (c) go to the bedroom and put on clean pants, and (d) dispose of the wet pants in the diaper pail" (Barmann et al., 1981, p. 341). This was followed by 10 repetitions of positive practice overcorrection in which the child practiced the correct sequence of toileting behavior.

Results of this multiple baseline analysis clearly documented the controlling effects of the treatment, but only when it was directly applied to each child. Indeed, treatment reduced enuretic accidents to near zero levels for each subject and was maintained in a lengthy follow-up evaluation period. Moreover, the effects of treatment generalized from the home to the school setting.

As in the multiple baseline across behaviors, baseline and treatment phases for each subject in this study can be conceptualized as separate A-B designs, with the length of baselines increased for each succeeding subject used in the multiple baseline analysis. The controlling effects of the contingency are inferred from the rate changes in the treated subject, while rates remain unchanged in untreated subjects. When rate changes are sequentially observed in at least 3 subjects, but only after the treatment variable has been directly applied to each, the experimenter gains confidence in the efficacy of the procedure (i.e., overcorrection). Thus we have a direct replication of the basic A-B design in 3 matched subjects exposed to the same environment under "time-lagged" contingency conditions.

Dyer, Christian, and Luce (1982) used an interesting variation of a multiple baseline strategy across subjects in their assessment of response delay to improve the discrimination performance of three autistic children (two 13-year-old girls and one 14-year-old boy). Discrimination tasks for the three children were as follows: Child 1—pointing to a male or female figure; Child 2—describing function of two objects (e.g., a towel and a fork); Child 3—discriminating between right and left. Responses to these tasks were obtained during no-delay and delay conditions, with all experimental sessions conducted in each child's classroom. Treatment (delay) was introduced, withdrawn, and reintroduced, following an initial no-delay condition for each child. This, of course, was conducted sequentially under time-lagged conditions for the three children. Delay consisted of having one child withhold his or her response for 3 to 5 seconds.

Inspection of Figure 7-10 shows that improved performance only occurred when the contingency (i.e., delay) was directly applied to each child, thus documenting the controlling effects of treatment. Data clearly indicate that the three baselines were independent of one another. Moreover, additional confirmation of the controlling effects of delay were noted when introduction

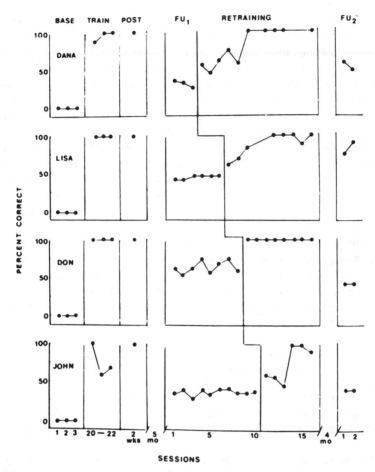

FIGURE 7-11. Percentage of correct emergency escape responses. Baseline—first 3 days of performance from original baseline phase. Training—last 3 days of training from original intervention phase. Post—postcheck assessment 2 weeks after training was terminated. Follow-up—1–5 month follow-up (FU) reassessment when no intervention in effect. Retraining—reinstatement of original training program. Follow-up—2–9 month follow-up (FU) reassessment after original training and 4-month follow-up after retraining. (Figure 1, p. 718, from: Jones, R. T., Kazdin, A. E., & Haney, J. L. [1981]. A follow-up to training emergency skills. *Behavior Therapy*, **12**, 716–722. Copyright 1981 by Association for Advancement of Behavior Therapy. Reproduced by permission.)

of the delay contingency resulted in improved performance, followed by deterioration when withdrawn and renewed improvement when reinstated. Thus, for each child we have an A-B-A-B demonstration, but carried out sequentially and cumulatively across the three. In short, the study by Dyer et al. (1982) is an excellent example of the combined use of the A-B-A-B design in multiple baseline fashion across subjects.

R. T. Jones, Kazdin, and Haney (1981b) used a multiple baseline design across subjects (5 third-grade children) to assess the effects of training (instructions, shaping, modeling, feedback, external, and self-reinforcement) in emergency fire escape skills. The training package in that study proved to be quite effective, as indicated by the increased percentage of correct emergency escape responses accrued by subjects in time-lagged fashion. A portion of these data (first 3 days of performance from original baseline, last 3 days of training from original treatment, and a 2-week follow-up) is presented in the left-hand side of Figure 7-11 for four of these five children. However, a 5-month follow-up (Sessions 1 for Dana, Lisa, Don, and John on the right-hand side of Figure 7-11) indicates some decrement in responding. Therefore, the 5-month reassessment was extended (3 sessions for Dana, 6 for Lisa, 8 for Don, and 10 for John) under time-lagged conditions, in order to evaluate the effects of retraining (R. T. Jones et al., 1981a).

As can be seen in Figure 7-11, such retraining did result in improved performance, but only when treatment was directly applied to each child, thus reconfirming its controlling effects. However, an additional follow-up 4 months after retraining again indicated decrements in performance, particularly for Don and John. R. T. Jones et al. (1981a), on the basis of these results, argue that:

> The present follow-up study has several implications for future research. First, conclusions about the effectiveness of particular procedures need to be tempered unless accompanied by evidence showing maintenance of behavior. The implication of many demonstrations is that an important applied problem has been solved by application of behavioral (or other) procedures. However, durability of behavior change is not an ancillary measure of treatment effects. (p. 721)

Our illustration shows how the multiple baseline strategy allows for (1) an initial demonstration of the controlling effects of a treatment, (2) an assessment at follow-up, (3) a second demonstration of the controlling effects of the treatment, and (4) a second follow-up assessment showing differential responding among subjects.

A three-group application of the multiple baseline strategy across subjects (groups of children with insulin dependent diabetes) was provided by Epstein et al. (1981). The effects of a behavioral treatment program to increase the percentage of negative urine tests were examined in 19 families of such diabetic children. Treatment was directed to decrease intake of simple sugars and saturated fats, decrease stress, increase exercise, and adjust insulin intake. Parents were taught to use praise and token economic techniques to reinforce improvements in the child's self-regulating behavior. When treatment began, 10 of the children (ages 8 to 12) were self-administering their insulin; the remaining 9 were receiving shots from their parents.

The major dependent measure involved a biochemical determination of *any* glucose in the urine. As noted by Epstein et al. (1981), this ". . . suggests that greater than normal glucose concentrations are present in the blood, and the renal threshold has been exceeded" (p. 367). Such testing was carried out on a daily basis during baseline, treatment, and follow-up.

The 19 families were assigned on a random basis to three groups, with treatment begun under time-lagged conditions 2, 4, or 6 weeks after initiation

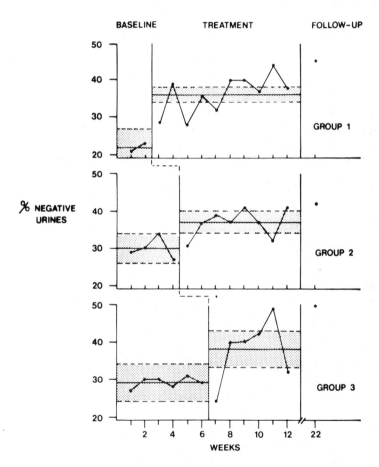

FIGURE 7-12. Percentage of 0% urine concentration tests weekly for children in each group. The mean and standard error of the mean for all the observations in each phase by group are represented by a solid and dotted line, respectively. (Figure 1, p. 371, from: Epstein, L. H., Beck, S., Figueroa, J., Farkas, G., Kazdin, A. E., Daneman, D., & Becker, D. [1981]. The effects of targeting improvements in urine glucose on metabolic control in children with insulin dependent diabetes. *Journal of Applied Behavior Analysis*, **14**, 365–375. Copyright 1981 by Society for Experimental Analysis of Behavior. Reproduced by permission.)

of the 12-week program. Examination of Figure 7-12 indicates that percent-age of negative urines was relatively low for each of the three groups during baseline. Institution of treatment resulted in marked improvements in per-centage of negative urines, indicating the controlling effects of the strategy. Moreover, it appears that these gains were maintained posttreatment, as indicated by the follow-up assessment at 22 weeks.

In summary, Epstein et al. (1981) presented a powerful demonstration of the effects of a behavioral treatment over a biochemical dependent measure (that has serious health implications). From a design standpoint, this study is an excellent illustration of the multiple baseline strategy across small groups of subjects, suggesting how the particular experimental strategy can be used to evaluate treatments in the area of behavioral medicine. However, from the design standpoint, the cautionary note articulated with respect to averaging of data in Bates (1980) certainly applies here.

Sulzer-Azaroff and deSantamaria (1980) also used a multiple baseline strategy across subjects (groups) in their assessment of feedback procedures to prevent and decrease occupational accidents in a small industrial organiza-tion. Six departments were evaluated during baseline for frequency of haz-ards: (1) screen printing, (2) heat sealing, (3) cutting and assembly, (4) credit and ID card manufacturing, (5) packing, and (6) receiving and distributing. Inspection of Figure 7-13 reveals that, in baseline, mean frequency of hazards in Departments 1 and 2 was 30.1 and 28.8, respectively; 13.2 and 14.8 for Departments 4 and 5; and 38.6 and 14.0 for Departments 3 and 6.

The experimental intervention consisted of providing twice-weekly feed-back, specific suggestions for improvement, and positive comments for ac-complishments in the area of safety to supervisors for each of the six departments. This, of course, was carried out in time-lagged fashion 3 weeks after baseline for Departments 1 and 2, 6 weeks after baseline for Depart-ments 4 and 5, and 9 weeks after baseline for Departments 3 and 6.

The effects of the intervention were considerable, resulting in a 60% drop in accidents averaged across departments. The specific controlling effects of the feedback strategy were documented, in that decreased rates occurred in those departments only when the intervention was directly applied. For Department 1, feedback appeared to yield continued improvement, which originally seemed to be occurring during baseline (i.e., downward trend in the data). However, data are more convincing for application of the intervention for Department 2, where such a downward trend was not observed in baseline data.

Data also indicate that the effects of this intervention were maintained during the follow-up phase (2 and 6 weeks and 4 months).

An important feature of the Sulzer-Azaroff and deSantamaria (1980) presentation is that data for each supervisor's department are presented rather than being collapsed across groups. Such data are important, as it is

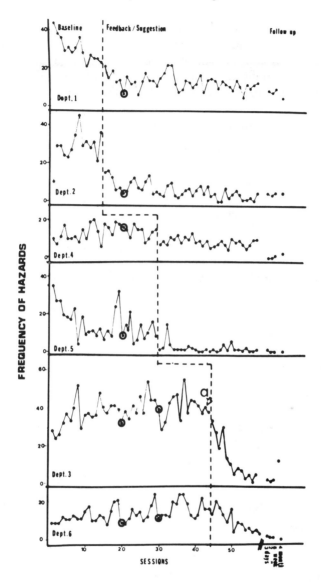

FIGURE 7-13. Frequency of hazards across department as a function of the introduction of the "feedback package." Data for days following unplanned safety meetings are indicated by an open circle. At point "a" there was a change in supervisors. (Figure 1, p. 293, from: Sulzer-Azaroff, B., & deSantamaria, M. C. [1980]. Industrial safety hazard reduction through performance feedback. *Journal of Applied Behavior Analysis*, 13, 287–295. Copyright 1980 by Society for Experimental Analysis of Behavior. Reproduced by permission.)

conceivable (as frequently occurs when a group comparison design is used) that some subjects may be unaffected by the contingency in force. Therefore, once again, we recommend that investigators employing group variations of multiple baseline strategies provide data showing the efficacy of their procedures in a majority of individual subjects in each respective group.

Multiple baseline across settings

Our first example of a multiple baseline strategy across settings involves treatment of eye twitching in an 11-year-old white male (David) whose disorder had been ongoing since age 5 (Ollendick, 1981). Eye twitching began when David entered kindergarten, which was concurrent with his mother's being admitted to a hospital for glaucoma treatments. The child was described as "mommy's boy" and apparently was very dependent on her.

During baseline, David's tics were surreptitiously observed in school by the teacher and at home by his mother. This was accomplished in 20-minute sampling periods. Following a 5-day observation period at school, David was

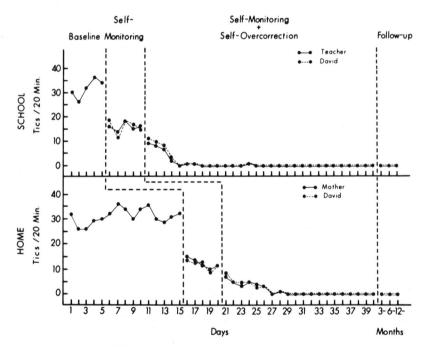

FIGURE 7-14. Effects of self-monitoring and self-administered overcorrection in the school and home: David. (Figure 1, p. 81, from: Ollendick, T. H. [1981]. Self-monitoring and self-administered overcorrection: The modification of nervous tics in children. *Behavior Modification*, **5**, 75–84. Copyright 1981 by Sage Publications. Reproduced by permission.)

taught to self-monitor and record rate of tics. On Day 11 self-overcorrection procedures were added to self-observation. This involved practicing the tensing of muscles that were antagonistic to the tic. Throughout the entire study period, the teacher continued to monitor tic behavior, thus providing a reliability check for David's self-observations.

As can be seen in Figure 7-14, similar self-monitoring and self-overcorrection procedures were carried out by David in the home following 15 days of initial observation by the mother. Here too, mother continued to monitor tic behavior when David began to self-monitor (Day 16) and self-overcorrect (Day 21).

The results of this multiple baseline analysis indicate that self-monitoring resulted in modest improvements followed by marked improvements when overcorrection was added (school). However, there appeared to be no change in tic frequency at home until self-monitoring was specifically applied there (i.e., baselines *are* independent from one another). Also, application of overcorrection in the home led to a continuation of the downward trend to a zero level. Three-, 6- and 12-month follow-ups indicated a complete maintenance of gains.

This study is interesting from a design standpoint for two reasons. *First,* the successive controlling effects of two strategies are nicely documented. *Second,* excellent reliability (teacher and David; mother and David) for the self-monitoring of tics appears for both the school ($r = .88$) and the home ($r = .89$) settings.

Singh, Dawson, and Gregory (1980) employed the withdrawal strategy (A-B-A-B) in an application of the multiple baseline design across settings in a 17½-year-old profoundly retarded female. She suffered from epilepsy (controlled pharmacologically) and had a 6-year history of hyperventilation. Apparently, prior attempts to deal with her symptoms (defined as a single instance of deep, heavy breathing, accompanied by a grunting noise and up-and-down head movements) had failed. Such symptoms were observed in four separate settings (classroom, dining room, bathroom, dayroom) in the residential unit of the state facility in which she lived. Data were recorded in 10-second intervals throughout 30-minute sessions.

Baseline data were obtained for 5 sessions in the classroom, 10 in the dining room, 15 in the bathroom, and 20 in the dayroom. Then, under time-lagged conditions, treatment (B) was introduced. Subsequently it was removed and reintroduced in each setting. (This constitutes the A-B-A-B part of the design). Treatment consisted of the application of response-contingent aromatic ammonia whenever an instance of hyperventilation was observed: ". . . a vial of aromatic ammonia . . . was crushed and held under her nose for more than 3 sec" (Singh et al., 1980, p. 563). Finally, during the 8 weeks of the genralization phase, ward nurses were requested to carry out the punishment procedure on an 8-hour-per-day basis. This is in contrast to original treatment that was carried out for only four 30-minute sessions per day.

Results of this single-case analysis appear in Figure 7-15. Data clearly indicate the controlling effects of the treatment, both in terms of its initial application on a time-lagged basis (baselines were independent) and when it was removed and reintroduced simultaneously in all four settings. Rate of hyperventilation episodes increased dramatically when the punishment contingency was removed in the second baseline and decreased to near zero levels

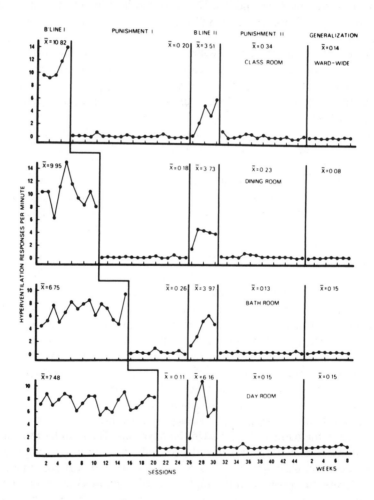

FIGURE 7-15. Number of hyperventilation responses per minute and condition means across experimental phases and settings. (Figure 1, p. 565, from: Singh, N. N., Dawson, J. H., & Gregory, P. R. [1980]. Suppression of chronic hyperventilation using response-contingent dramatic ammonia. *Behavior Therapy*, **11**, 561–566. Copyright 1980 by Association for Advancement of Behavior Therapy. Reproduced by permission.)

when it was reintroduced. Moreover, the positive effects of treatment were prolonged and enhanced as a result of the more extensive punishment approach followed in the generalization phase.

Fairbank and Keane (1982) present an interesting application of the multiple baseline design across settings (i.e., *imaginal* scenes) in a 31-year-old divorced male veteran suffering from a posttraumatic stress disorder following his serving 20 months of combat duty in Vietnam. This subject complained of chronic anxiety, nightmares, and flashback of traumatic events that had occurred during the course of combat. Through careful interviewing, four particularly traumatic scenes were selected as stimulus material for assessment and treatment. During baseline these scenes were presented verbally (with one considerable detail) to the subject in 5- to 10 minute probe evaluations. During presentation of each scene the subject was asked to self-rate the discomfort elicited by the material (0 = lowest, 10 = highest). This is referred to as a *SUDS rating*. The highest of four such SUDS ratings per scene was recorded. Concurrently, heart rate and skin conductance responses to scenes were obtained.

Treatment (i.e., flooding) was applied sequentially and cumulatively to each of the four scenes. Flooding consisted of 60- to 120 minute sessions in which "Stimulus and response cues relevant to the scene were slowly and gradually presented by the therapist, who regularly elicited feedback regarding the next chronological event in the sequence" (Fairbank & Keane, 1982, p. 503). During the course of a session the subject's anxiety level first increased considerably and then dissipated toward the end.

Data in Figure 7-16 clearly confirm the controlling effects of flooding treatment on SUDS ratings. This is indicated by the fact that decreases in SUDS ratings were noted only when treatment was directly applied to each traumatic scene. Moreover, these data are confirmed by concurrent diminution in skin conductance responses during probe sessions following direct application of treatment. Further confirmation of these results was obtained by replicating the procedure with 2 additional posttraumatic stress-disordered patients.

From a design perspective, however, it would have been preferable if the experimenters had obtained more probe measures in Scenes 1 and 2 (i.e., a minimum of three data points for Scene 1) and additional probe measures in treatment for Scenes 3 and 4. This, of course, is in direct reference to the point raised in chapter 3 with regard to obtaining three measurements in order to determine a trend in the data.

A particularly socially relevant example of a multiple baseline design across settings (two high density residential areas) was provided by R. E. Kirchner et al. (1980) (see Figure 7-17). This study also contains A-B-A withdrawal features. In the portion of the study we are to describe, two high-population density areas in Nashville were targeted for study (9.82 and 14.7 square miles;

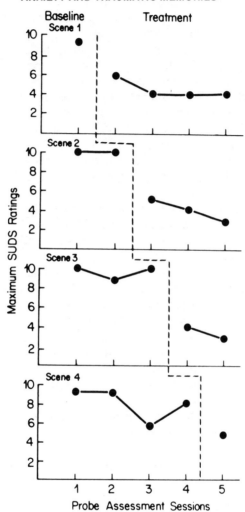

FIGURE 7-16. Maximum SUDS ratings during probe sessions (Subject 2). (Figure 2, p. 505, from: Fairbank, J. A., & Keane, M. [1982]. Flooding for combat-related stress disorders: Assessment of anxiety reduction across traumatic memories. *Behavior Therapy, 13*, 499–510. Copyright 1982 by Association for Advancement of Behavior Therapy. Reproduced by permission.)

populations 49,978 and 65,910). During baseline, the mean number of home burglaries committed per day was computed for each area ($\bar{X}$s = 2.83 and 2.25).

After 17 days of baseline in Area 1 of standard police patrolling, an

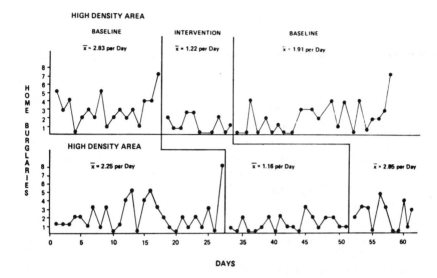

FIGURE 7-17. Number of home burglaries in two high-density areas over baseline and intervention conditions. (Figure 1, p. 145, from: Kirchner, R. E., Schnelle, J. F., Domash, M., Larson, L., Carr, A., & McNees, M. P. [1980]. The applicability of a helicopter patrol procedure to diverse areas: A cost-benefit evaluation. *Journal of Applied Behavior Analysis*, **13**, 143–148. Copyright 1980 by Society for Experimental Analysis of Behavior. Reproduced by permission.)

intervention consisting of close scrutiny with a helicopter patrol was added. This resulted in a decrease in home burglaries to 1.22 per day. However, when the helicopter patrol was discontinued on Day 29, the home burglary rate increased to 1.91 per day. Thus, from the A-B-A aspect of this study, it is clear that the helicopter patrol served to reduce home burglaries in Area 1.

Similarly, on Day 33, when the helicopter patrol was introduced in Area 2, home burglaries dropped from 2.25 to 1.16 per day, but rose to 2.85 per day when it was discontinued on day 52 (control demonstrated in A-B-A fashion for Area 2).

The A-B-A confirmation of the controlling power of the intervention adds substantially to documentation of the time-lagged contingency. That is, for Area 2, change only occurred when the helicopter intervention was directly applied. Baselines were completely independent. R. E. Kirchner et al. (1980) presented yet additional evidence for the efficacy of this intervention. From the cost effectiveness perspective, in baseline, daily burglary costs were $1,376 and $1,094 respectively for the two areas. When the helicopter intervention was instituted, daily burglary costs diminished to $823 and $815. Thus we have a very powerful demonstration of this contingency in a multiple baseline design across settings that incorporates A-B-A withdrawal features.

7.3 VARIATIONS OF MULTIPLE BASELINE DESIGNS

Nonconcurrent multiple baseline design

As noted in section 7.2, in the multiple baseline design across subjects, each individual targeted for treatment is exposed to the same environment. Treatment is delayed for each successive subject in time-lagged fashion because of the increased length of baselines required for each. The functional relationship between treatment and behavior selected for change can be determined only when such treatment is applied to each subject in succession. Thus, since subjects (at least two but usually three or more) are simultaneously available for assessment and treatment, this design is able to control for *history* (cf. Campbell & Stanley, 1963), a possible experimental contaminant.

There are times, however, when one is unable to obtain concurrent observations for several subjects, in that they may be available only in succession (e.g., less frequently seen diagnostic conditions such as hysterical spasmodic torticollis). Following strictures of the multiple baseline strategy across subjects, this design ordinarily would not be considered appropriate under these circumstances. However, more recently Watson and Workman (1981) have proposed an alternative—the nonconcurrent multiple baseline across individuals.

> In this . . . design, the researcher initially determines the length of each of several baseline designs (e.g., 5, 10, 15 days). When a given subject becomes available (e.g., a client referred who has the target behavior of interest, and is amenable to the use of a specific treatment of interest), s(he) is randomly assigned to one of the pre-determined baseline lengths. Baseline observations are then carried out; and assuming the responding has reached acceptable stability criteria, treatment is implemented at the pre-determined point in time. Observations are continued through the treatment phase, as in a simple A-B design. Subjects who fail to display stable responding would be dropped from the formal investigation; however, their eventual reaction to treatment might serve as useful replication data.

The logic of this variation is graphically portrayed in Figure 7-18. Of course, the major problem with this strategy is that the control for history (i.e., the ability to assess subjects concurrently) is greatly diminished (see also Mansell, 1982). Thus we view this approach as less desirable than the standard multiple baseline design across subjects. It should be employed only when the standard approach is not feasible. Moreover, under such circumstances, an increased number of replications (i.e., number of subjects so treated) might enhance the confidence one has in the results. But in the case of rare disorders this may not be possible. In any event, use of this variant is not defensible when it is possible to run all of the subjects concurrently in time-lagged fashion.

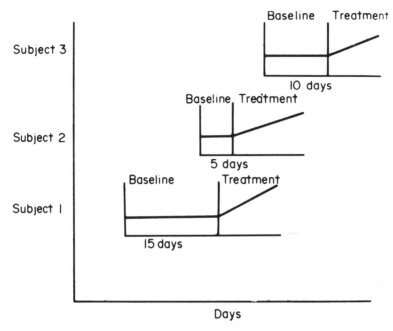

Days

FIGURE 7-18. Hypothetical data obtained through use of a nonconcurrent multiple baseline design. (Figure 1, p. 258, from: Watson, P. J., & Workman, E. A. [1981]. The nonconcurrent multiple baseline across-individuals design: An extension of the traditional multiple baseline design. *Journal of Behavior Therapy and Experimental Psychiatry*, **12**, 257–259. Copyright 1981 by Pergamon. Reproduced by permission.)

Multiple-probe technique

To this point in our descriptions of multiple baseline strategies, baseline measurement has been continuous for all designs, including the nonconcurrent multiple baseline design. However, as noted by Horner and Baer (1978), there are situations in which repeated measurements will result in reactivity (i.e., a change simply as a result of repetition of the assessment). When treatment is subsequently introduced under these circumstances, changes may not be detected or may be masked, due to the inflated or deflated baseline as a function of reactivity. In addition, there are some instances when continuous measurement is not feasible and when (on the basis of prior experimentation) an "*a priori* assumption of stability can be made" (Homer & Baer, 1978, p. 193). This being the case, instead of having 6, 9, and 12 assessments in three successive baselines, these can be more interspersed, resulting in two, three, and four measurement points. An example of this approach is presented in Figure 7-19. Probes (hypothetical) in our example are represented by closed triangles, whereas actual reported data appear as open circles.

In commenting on this graph, Horner and Baer (1978) argued that:

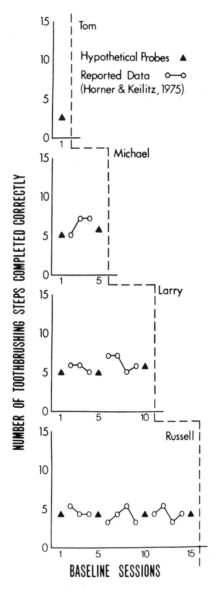

FIGURE 7-19. Number of toothbrushing steps conforming to the definition of a correct response across 4 subjects. (Figure 2, p. 194, from: Horner, R. D., & Baer, D. M. [1978]. Multiple-probe technique: A variation of the multiple baseline. *Journal of Applied Behavior Analysis*, **11**, 189–196. Copyright 1978 by Society for Experimental Analysis of Behavior. Reproduced by permission.)

The multiple-probe technique, with probes every five days, would have provided one, two, three, and five probe sessions to establish baselines across the four subjects. The multiple-probe technique probably could have provided a stable baseline with five or fewer probe sessions for the subject who had 15 days of continuous baseline in the original study. The use of the multiple-probe procedure might have precluded the increase in irrelevant and competing behaviors by this subject because such behavior began to increase after the tenth baseline session. (p. 195)

It should be noted that, over the years, a variety of researchers have applied this variant of baseline assessment in the multiple baseline design (Baer & Guess, 1971; Schumaker & Sherman, 1970; Striefel, Bryan, & Aikins, 1974; Striefel & Wetherby, 1973). In each of these studies the design used was the multiple baseline design across behaviors. But, as in Figure 7-19, it could be across subjects, and it certainly might also be across settings.

If reactivity is the primary reason for using this variant, the probe technique should be continued when treatment is instituted. However, if feasibility is questionable in baseline or if an *a priori* assumption of baseline stability can be made, more frequent measurements during treatment may be desirable.

Kazdin (1982b) recommended use of the probe technique for assessment of behaviors that *were not* targeted for treatment (i.e., evaluation of generalization or transfer of treatment effects, say, in the naturalistic environment). Use of probes here is particularly valuable if reactivity is to be avoided. This was specifically carried out in a multiple baseline design across behaviors evaluating generalization effects of social skill training in three chronic schizophrenics (Bellack, Hersen, & Turner, 1976). In each case, baseline assessment involved evaluation of verbal and nonverbal behaviors from video taped role-play scenarios requiring assertive responding. One set of eight scenarios (Training Scenes) was repeatedly used for assessment during baseline, treatment, and follow-up phases. This also served as the training vehicle (see left side of Figure 7-20). A second set of eight scenarios (Generalization Scenes) also was repeatedly used for assessment during baseline, treatment, and follow-up phases, but the patient *did not* receive training here (see right side of Figure 7-20). However, since the patient was repeatedly exposed to Generalization Scenes, reactivity was considered a good possibility. Therefore, a third set of eight scenarios (Novel Scenes) was used for an additional generalization assessment during baseline, treatment, and follow-up phases on a probe basis (see open circles on the right side of Figure 7-20).

Examination of Figure 7-20 confirms the controlling effects of treatment on individual behaviors in Training Scenes, with the exception of "ratio of words spoken to speech duration." Data also confirm transfer of training from Training to Generalization Scenes, but again with the exception of

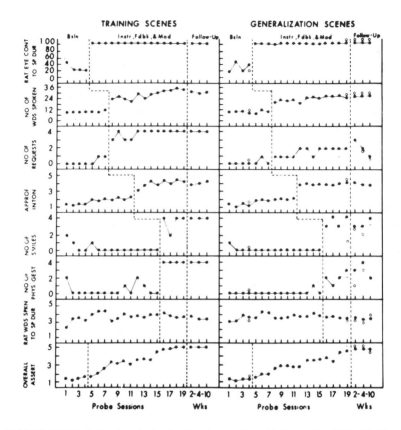

FIGURE 7-20. Probe sessions during baseline, treatment, and follow-up for Subject 3. (Figure 3, p. 396, from: Bellack, A. S., Hersen, M., & Turner, S. M. [1976]. Generalization effects of social skills training in chronic schizophrenics: An experimental analysis. *Behaviour Research and Therapy*, **14**, 391–398. Copyright 1976 by Pergamon. Reproduced by permission.)

"ratio of words spoken to speech duration." Probe data (open circles) suggest that there was further evidence of transfer of training to the Novel Scenes, with the exception of "ratio of words spoken to speech duration." Finally, for the three sets of scenes, data indicate that gradual improvements in overall assertiveness were noted throughout treatment, which appeared to be maintained in follow-up.

As we have seen, the probe technique can be most useful in a number of instances. However, as in the case of the nonconcurrent multiple baseline design, it should not be employed as a substitute for continuous measurement when that is feasible. That is, data accrued from use of probe measures are *suggestive* rather than confirmatory of the controlling effects of a given treatment.

7.4 ISSUES IN DRUG EVALUATIONS

With the exception of the multiple baseline across subjects, the multiple baseline strategies are generally unsuitable for the evaluation of pharmacological agents on behavior. For example, it will be recalled that, in the multiple baseline design across behaviors, the same treatment is applied to independent behaviors within the same individual under time-lagged conditions. Clearly, in the case of drug evaluations this is an impossibility, as no drug is so specific in its action that it can be expected to effect changes in this manner. However, it would be possible to apply *different drugs* under time-lagged conditions to *separate behaviors* following baseline placebo administrations for each. But this kind of design would involve a radical departure from the basic assumptions underlying the multiple baseline strategy across behaviors and would only permit very tentative conclusions based on separate A_1-B designs for each targeted behavior. In addition, the possible interactive effects of drugs might obfuscate specific results. Indeed, the interaction design (see chapter 6) is better suited for evaluation of combined effects of therapeutic strategies.

Similarly, the use of the multiple baseline across different settings in drug evaluations would prove difficult unless the particular drug being applied worked immediately, had extremely short-term effects, and could be rapidly eliminated from body tissues. However, as most drugs used in controlling behavior disorders do not meet these three requirements, this kind of design strategy is not useful in drug research.

Of the three types of multiple baseline strategies currently in use, the multiple baseline across subjects is most readily adaptable to drug evaluations. The application of the multiple baseline design across subjects in drug evaluations could be most useful when withdrawal procedures (return to A_1—baseline placebo) are unwarranted for either ethical or clinical considerations. Using this type of strategy across matched subjects, baseline administration of a placebo (A_1) could be followed by the sequential administration (under time-lagged conditions) of an active drug (B). Thus a series of A_1-B (quasi-experimental) designs would result, with inferences made in accordance with changes observed when the B (drug) condition was applied. Although an approximation of a double-blind procedure is feasible (observer and patient blind to conditions in force), it is more likely that single-blind (patient only) conditions would prevail.

Many other design options are possible in the application of the multiple baseline design across subjects when evaluating pharmacological effects. For example, V. J. Davis, Poling, Wysocki, and Breuning (1981) looked at the effects of decreasing phenytoin drug dosage on the workshop performance of three mentally retarded individuals. Thus one can use the multiple baseline

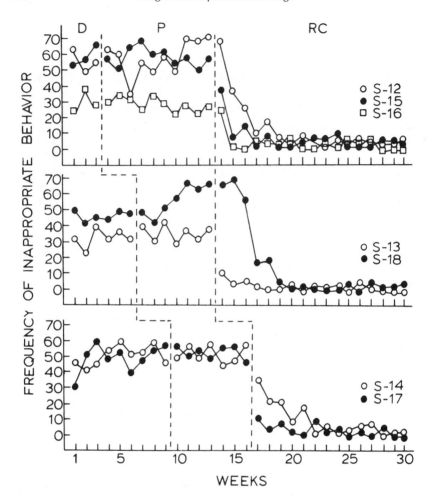

FIGURE 7-21. Frequencies of inappropriate behaviors for Subjects 12–18 plotted as total occurrences per week (summed daily interval totals). During the D condition, the subjects received their drug; during the P condition, the subjects received a placebo, were no longer receiving their drug, and the response cost procedure was not in effect. Drugs were discontinued during the first 3 weeks of the P condition. During the RC condition, the response cost procedure was in effect, and the subjects were not receiving their drug. The dotted vertical lines separate the conditions. (Figure 2, p. 261, from: Breuning, S. E., O'Neill, M. J., & Ferguson, D. G. [1980]. Comparison of psychotropic drug, response cost, and psychotropic drug plus response cost procedures for controlling institutionalized mentally retarded persons. *Applied Research in Mental Retardation*, **1**, 253–268. Copyright 1980. Reproduced by permission.)

design across subjects to examine the effects of drug withdrawal in discrete steps. Another possibility is to evaluate the addition of a behavioral regime to pharmacological maintenance followed by withdrawal of the drug. This results in a B-BC-C design, with drug as B, drug plus behavioral intervention as BC, and the behavioral intervention alone as C (cf. Breuning, O'Neill, & Ferguson, 1980).

Breuning et al. (1980) followed yet a different option of the multiple baseline design across subjects (small groups) in their successive evaluation of drug, placebo, and response cost conditions. This yields a B (drug), A' (placebo), C (response cost) design. Let us consider this study in some detail (see Figure 7-21). Subjects were institutionalized mentally retarded individuals evincing inappropriate behavior. After 3 weeks on active neuroleptic drugs, Subjects 12, 15, and 16 were switched to placebo for 10 weeks. After 6 weeks on active neuroleptic drugs, Subjects 13 and 19 were switched to placebo for 7 weeks. Finally, after 9 weeks on active neuroleptic drugs, Subjects 14 and 17 were switched to placebo for 7 weeks. Examination of drug and placebo data reveals no apparent improvements in inappropriate behavior. However, as might be expected, the switch to placebo for Subject 18 led to an increase in inappropriate behavior, suggesting at least some controlling effects of the drug. When response-cost procedures were instituted in Week 14 for Subjects 12, 13, 15, 16, and 18, and in Week 17 for Subjects 14 and 17, marked improvements in appropriate behavior were observed, beginning almost immediately. Thus this rather complicated experimental analysis confirmed the efficacy of response cost procedures under time-lagged conditions (baseline 3 versus baselines 1 and 2), but only when the contingency was directly applied. However, both neuroleptic drugs and placebo generally seemed to be ineffective.

In this type of drug evaluation it is important to underscore that the prolonged placebo phases are important in that they provide a needed "washout" period for possible carryover effects of drugs. This, of course, would have been much more critical had neuroleptic drugs substantially decreased the behavior targeted for change (i.e., inappropriate behavior).

CHAPTER 8

Alternating Treatments Design

8.1. INTRODUCTION

Few areas of single-case experimental designs have advanced as much as the design strategies to be discussed in this chapter. The strength and underlying logic of these strategies, as well as the fact that some specific questions can only be answered using these approaches, have ensured the rapid development and increasing use of this design, particularly during the last 5 years.

The major question addressed by this design is the relative effectiveness of two (or more) treatments or conditions. The most common experimental approach employed to address this question until now has been the traditional between-group comparison. In this strategy, each of two or more treatments is usually administered to a separate group of subjects, and the outcome of the treatments is compared between groups. Since considerable intersubject variability exists in each group (some subjects change and some do not), inferential statistics are necessary to determine if an effect exists. This leads to problems in generalizing results from the group average to the individual subjects, as discussed in chapter 2. To avoid intersubject variability, an ideal solution would be to divide the subject in two and apply two different treatments simultaneously to each identical half of the same individual. This would eliminate intersubject variability and allow effects, if any, to be directly observed. In fact, this strategy provides one of the most elegant controls for most threats to internal validity or the ability of an experimental design to rule out rival hypotheses in accounting for the difference between the two treatments (Campbell & Stanley, 1966; Cook & Campbell, 1979). Statements about external validity or the generalizability of findings observed in one subject to other similar subjects must be made, of course, through the more usual process of replication and "logical generalization" (Edgington, 1966; see also chapters 2 and 10).

The name that has come to be employed for the experimental design that accomplishes this goal is the *alternating treatments design* (ATD) (Barlow & Hayes, 1979). As the name implies, the basic strategy involved in this design is the rapid alternation of two or more treatments or conditions within a single subject. *Rapid* does not necessarily mean rapid within a fixed period of time; as, for example, every hour or every day. In applied research, rapid might mean that each time the client is seen he or she would receive an alternative treatment. For example, if an experimenter were comparing treatments A and B in a client seen weekly, he or she might apply Treatment A one week and Treatment B the next. If the client were seen monthly, alternations would be monthly. Contrast this with the usual A-B-A withdrawal design where, after a baseline, an experimenter would need at least three, and usually more, consecutive data points measuring the effect of Treatment A in order to examine any trends toward improvement. For a client seen weekly, at least 3 weeks would be needed to establish the trend.

Since one is alternating two or more treatments, an experimenter is not interested simply in the trend toward improvement over time. Therefore, one would not plot the data simply by connecting data points for Weeks 1, 2, 3, and so on. Rather, what one is interested in is comparing treatments A and B. Therefore, in order to examine visually the experimental effects, one would connect all the data points measuring the effects of Treatment A and then connect all the data points measuring the effects of Treatment B. If, over time, these two series of data points separated (i.e., Treatment B, for example, produced greater improvement than Treatment A), then one could say with some certainty that Treatment B was the more effective. Naturally, these results would then need replication on additional clients with the same problem. Such hypothetical data are plotted in Figure 8-1 for a client who was treated and assessed weekly.

Of course, one would not want to proceed in a simple A-B-A-B-A-B-A-B fashion. Rather, one would want to randomize the order of introduction of the treatments to control for sequential confounding, or the possibility that introducing Treatment A first, for example, would bias the results in favor of Treatment A. Therefore, notice in the hypothetical data that A and B are introduced in a relatively random fashion. Thus, if one were seeing a client in an office or a child in a school setting, one might administer the treatments in an A-B-B-A-B-A-A-B fashion, as in the hypothetical data. For a client in an office setting, these treatment occasions might be twice a week, with the experiment taking a total of 4 weeks. For a child in a school setting, one might alternate treatments 4 times a day, and the experiment would be completed in a total of 2 days. Randomizing introduction of treatments and

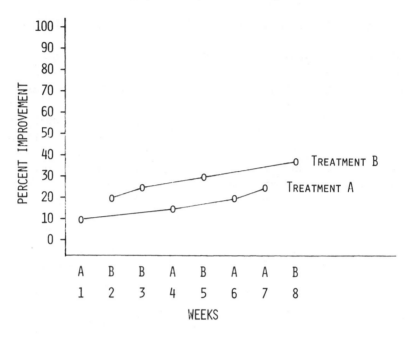

FIGURE 8-1. Hypothetical example of an ATD comparing treatments A and B.

other procedural considerations will be discussed more fully in section 8-2. The basic logic of this design, then, requires the comparison of two separate series of data points. For this reason, this experimental design has also been described as falling within a general strategy referred to as *between-series*, where one is comparing results between two separate series of data points. On the other hand, A-B-A withdrawal designs, described in chapters 5 and 6, look at data *within* the same series of data points, and therefore the strategy has been described as *within-series* (Barlow et al., 1983).

Terminology

While this basic research strategy has been used for years within a number of experimental contexts, a confusing array of terminology has delayed a widespread understanding of the basic logic of this design. In the first edition of this book, we termed this strategy a *multiple schedule design*. Others have termed the same design a *multi-element baseline design* (Sidman, 1960; Ulman & Sulzer-Azaroff, 1973, 1975), *a randomization design* (Edgington, 1967), and *a simultaneous treatment design* (Kazdin & Hartmann, 1978; McCullough, Cornell, McDaniel, & Meuller, 1974). These terms were origina-ted for somewhat different reasons, reflecting the multiple historical origins of single-case research. For example, several proponents of the term *multiple schedule* were associated in Vermont in the late 1960s in an effort to apply operant procedures and methods to clinical problems (e.g., Agras et al., 1969; Leitenberg, 1973). These procedures and terminology were derived directly from operant laboratories.

The term *multiple schedule* implies not only a distinct reinforcement schedule as one of the treatments, but also a distinct stimulus or signal that will allow the subjects to discriminate as to when each of the two or more conditions will be in effect. However, in recent years it has become clear (particularly in applied research with human subjects) that signs or signals functioning as discriminative stimuli (SDs) are either an inherent part of the treatment, and therefore require no further consideration, or are not needed. For example, alternating a pharmacological agent with a placebo, using at ATD design, would be perfectly legitimate, but each drug would not require a discriminative stimulus. In fact, this would be undesirable; hence, the usual double-blind experimental strategies in drug research (see chapter 6). For this reason, the more appropriate analogy within the basic operant laboratories would be a mixed schedule rather than a multiple schedule, since a mixed schedule does not have discriminative stimuli. But the term *schedule* itself implies a distinct reinforcement schedule associated with each treatment, and there is no reason to think that specific treatments under investigation would contain schedules of reinforcement. Thus the terms *multiple schedule* and *mixed schedule* are not really appropriate.

Ulman and Sulzer-Azaroff (1975) used one of Sidman's terms, *multi-element baseline design*, to describe this strategy. Sidman himself (1960) used the term *multi-element manipulation* to describe this particular design. Thus some researchers have settled on the term *multi-element design* (Bittle & Hake, 1977), but these terms also are derived directly out of the basic research laboratories and in their original usage have little applicability to applied situations (Barlow & Hayes, 1979).

Edgington (1966, 1972), from a somewhat different perspective, originated the term *randomization design* to describe his variation of a time series approach amenable to statistical analysis. He was most interested in exploring statistical procedures applicable to randomly alternated treatments. In this respect he continued a tradition begun by R. A. Fisher (1925), who explored the abilities of a lady to discriminate tea prepared in two different ways. Edgington emphasized the randomness of the alternation as well as the number of alternations in developing his statistical arguments. While these and other statistical approaches discussed below are useful and valuable, they are not essential to the logic of the design in our view.

The final alternative mentioned above that is sometimes used to describe alternating treatments designs is the term *simultaneous treatment design*. But this is a bit confusing because there is, in fact, a little-used design in which two or more treatments are actually available simultaneously. Since the treatments are presented simultaneously, what happens is that the subject "chooses" a preferred treatment or condition. Furthermore, this design has also been called the *simultaneous treatment design* (Browning, 1967). In fact, the design has little application in applied research and has not been used since 1967. Therefore, it will be described only briefly at the end of this chapter (see section 8-6).*

The basic feature of this design, under its various names, then, is the "rapid" alternation of two or more different treatments or conditions. For this reason, we suggested in 1979 the term *alternating treatments design* (Barlow & Hayes, 1979), which, most likely because of its descriptive properties, has been widely adopted (see Table 8-1). Although we use the term *alternating treatments*, we pointed out in 1979 that *treatments* refers to the particular condition in force, not necessarily therapy. Baseline conditions can be alternated with specific therapies as easily as two or more distinct therapies can be alternated. Whether or not this is needed, of course, depends on the specific question one is asking. The use of the term *treatment* in this way continues a long tradition in experimental design of referring to various conditions as *treatments*.

8.2. PROCEDURAL CONSIDERATIONS

In a single-case design, most procedures utilized in an ATD are similar to those described earlier for other designs. However, because of the unique purpose of this design (comparing two treatments or conditions in a single subject) and because of the strategy of rapid alternation, some distinct procedural issues arise that the experimenter will want to consider.

Multiple-treatment interference

Multiple-treatment interference (Barlow & Hayes, 1979; Campbell & Stanley, 1963) raises the issue: Will the results of Treatment B, in an ATD where it is alternated with Treatment A, be the same as when Treatment B is the only treatment used? In other words, is Treatment A somehow interfering with Treatment B, so that we are not getting a true picture of the effects of treatment? This notion enjoys much common sense, because at first glance

---

*Kazdin (1982b) has used the term *multiple-treatment designs* very accurately, in our view, to subsume both alternating and simultaneous treatment designs. However, since simultaneous treatment designs are so rare and would seem to have such little applicability in applied research, this book will concentrate on the description and illustration of alternating treatment designs.

there are few strictly "applied" situations where treatments are ever alternated. Thus it is not immediately apparent to practitioners how these results could generalize to their own situations.

On closer analysis, however, we will suggest that this is a relatively small problem, and in some cases not a problem at all, for applied researchers (although it is a major issue in basic research). Also, there are steps applied researchers can take to minimize multiple-treatment interference. After a discussion of the nature of multiple-treatment interference, the remainder of this section will describe procedures for minimizing it.

In a sense, all applied research is fraught with potential multiple-treatment interference. Unlike with the splendid isolation of the experimental animal laboratories where rats are returned to their cages for 23 hours to await the next session, the children and adults who are the subjects of applied research experience a variety of events before and between treatment sessions. A college student on the way to an experiment may have just failed an examination. A subject in a fear-reduction experiment may have been mugged on the way to the session. Another experimental patient may have lost a family member in recent weeks or just had sexual intercourse before the session. It is possible that these subjects respond differently to the treatment than otherwise would have been the case, and it is these historical factors that account for some of the enormous intersubject variability in between-group designs comparing two treatments. ATDs, on the other hand, control for this kind of confounding experience perfectly by "dividing the subject in two" and administering two or more treatments (to the same subjects) within the same time period. Thus, if a family member died during the previous week, that experience would presumably affect each rapidly alternated treatment equally. But the one remaining concern is the possibility that one experimental treatment is interfering with the other within the experiment itself. Essentially, there are three related concerns: sequential confounding, carryover effects, and alternation effects (Barlow & Hayes, 1979; Ulman & Sulzer-Azaroff, 1975).

We earlier discussed sequential confounding as referring to the fact that Treatment B might be different if it always followed Treatment A. Another name for sequential confounding is *order effects*. That is, much of the benefit of Treatment B might be due simply to the order in which it is administered vis-à-vis other treatments. Sequential confounding with A-B-A withdrawal designs has been discussed in section 5.3. The solution, of course, is to arrange for a random (or semirandom) sequencing of treatments. One can view this random order of sequencing treatments in a typical ATD in the hypothetical data presented in Figure 8-1. Such counterbalancing also allows for statistical analyses of ATDs for those who so desire (see chapter 9).

Carryover effects, on the other hand, refer to the influence of one treatment on an adjacent treatment, irrespective of overall sequencing. Terms such as *induction* and, more frequently, *contrast* (Rachlin, 1973; G. S. Reynolds, 1968), are used to describe these phenomena. Several of these terms carry specific theoretical connotations. For our purposes, it will be enough to speak of positive carryover effects and negative carryover effects. To return to the hypothetical data in Figure 8-1 as an example, positive carryover effects would occur if Treatment B were *more* effective, *because it was alternated with Treatment A* than it would be if it were the only treatment administered. Negative carryover effects would occur if Treatment B were *less* effective because it was alternated with Treatment A than if it were administered alone. In other words, Treatment A is somehow interfering with the effects one would see from Treatment B if it were administered in isolation.

Recent basic research has shed more light on the nature and parameters of carryover effects. In basic research laboratories, where the understanding of carryover effects is very important to various theories of behavior, investigators have discovered that such effects are almost always transient and due mostly to the inability of the subject to discriminate among two treatments (Blough, 1983; Hinson & Malone, 1980; Malone, 1976; McLean & White, 1981). Fortunately for us, the types of experimental situations where carryover effects are observed in basic research rarely occur in applied research. In basic research, treatments (schedules of reinforcement in this particular context) are often alternated by the minute. Furthermore, the treatments themselves are almost impossible to discriminate as they are occurring. For this reason, signs or signals (discriminative stimuli), referred to as SDs, are associated with each treatment. As these signals themselves become harder to discriminate (for example, increasingly closer wavelengths of light), carryover effects occur (Blough, 1983). But even with these difficult-to-discriminate treatments and signals, carryover effects eventually disappear as discriminations are learned. Recently, Blough (1983) has proposed that in situations where carryover effects are more permanent within this context, individual differences in ability to learn discrimination may be the reason. That is, those subjects (pigeons or rats) that are slower in learning the discriminations are associated with longer periods of carryover effects, whereas subjects learning the discriminations quickly evidence very short and transient carryover effects.

When carryover effects have been noticed in humans (e.g., Waite & Osborne, 1972), experimental operations similar to those employed in the laboratories of basic research were in operation. Presumably the same lack of discriminability was occurring.

In applied research, this would imply that carryover effects of the type discussed here are a possibility only when learning is occurring. This would exclude most biological treatments, such as pharmacotherapy, where no real learning occurs (although biological multiple-treatment interference will occur if drugs are alternated too quickly, depending on the half-life of the particular drug, see chapter 6). On the other hand, almost all psychosocial interventions do involve some learning. But treatments are usually so distinct that they are very easily discriminated even without any sign or signal. In fact, in the examples to be described below, adults are usually told which treatment is in effect from session to session, and therefore discriminations are perfect. Similarly, children of all ages are certainly capable of discriminating different treatments (e.g., time-out versus praise in the classroom) very quickly.

Nevertheless, until we know even more about carryover effects, it would be prudent to consider the following procedures when implementing an ATD. *First*, counterbalancing the order of treatments should minimize carryover effects and control for order effects. The remaining steps involve ensuring that treatments are discriminable. *Second*, for example, separating treatment sessions with a time interval should reduce carryover effects. Powell and Hake (1971) minimized carryover effects in this way in a study comparing two reinforcement conditions by presenting only one condition per session. Fortunately, in applied research it is the usual case that only one treatment per session is administered even if several sessions are held each day (e.g., Agras et al., 1969; McCullough et al., 1974). Similar procedures have been suggested to minimize carryover effects in the traditional, within-subjects, group comparison approaches (Greenwald, 1976). *Third*, the speed of alternations seems to increase carryover effects, at least until discriminations are formed. This is particularly true in basic research, as noted above, where treatments may be alternated by the minute. Slower and, once again, more discriminable alternations should minimize carryover effects (Powell & Hake, 1971; Waite & Osborne, 1972). In summary, based on what we now know about carryover effects, counterbalancing and insuring discriminability of treatments will minimize this problem. In applied research, where possible, simply telling the subjects which treatment they are getting should be sufficient.

Finally, in the event that some carryover effects may be occurring even with the procedural cautions mentioned above in place, there is no reason to think that these carryover effects would reverse the relative positions of the two treatments. Returning to the hypothetical data in Figure 8-1, Treatment B is seen as better than Treatment A. In this particular ATD, B may not be as effective as it would be if it were the only treatment administered, and A may be more effective, but it is extremely unlikely that carryover effects would make A better than B. Thus, even if carryover effects were observed in the major comparison of treatments, the experimenter would have clear evidence concerning the effectiveness of Treatment B, but would have to emphasize caution in determining exactly how effective Treatment B would be if it were not alternated with Treatment A.

Assessing multiple-treatment interference. For those investigators who are interested, it is possible and sometimes desirable to assess directly the extent to which carryover effects are present. Sidman (1960) suggested two methods. One is termed *independent verification* and essentially entails conducting a controlled experiment in which one or another of the component treatments in the ATD is administered independently. For example, returning to Figure 8-1 once again, Treatments A and B would be compared using an ATD in the manner presented in Figure 8-1, and this experiment would be replicated across two subjects. The investigator could then recruit 3 more closely matched subjects to receive a baseline condition, followed by Treatment A in an A-B fashion. Treatment B could be administered to a third trio of subjects in the same manner. Any differences that occur between the treatment administered in an ATD or independently could be due to carryover effects. Alternatively, these subjects could receive treatment A alone, followed by the ATD which alternated Treatments A and B, returning to Treatment A alone. An additional 3 subjects could receive Treatment B in the same manner. Trends and levels of behavior during either treatment alone could be compared with the same treatment in the ATD. Obviously, this type of strategy would also be very valuable for purposes of replication and for estimating the generalizability or external validity of either treatment.

A more elegant method was termed *functional manipulation* by Sidman (1960). In this procedure the strength of one of the components is altered. For example, if comparing imaginal flooding versus reinforced practice in the treatment of fear, the amount of time in flooding could be doubled at one point. Changes in fear behavior occurring during the second unchanged treatment (reinforced practice) could be attributed to carryover effects.

In an important, more recent example using these types of strategies, E. S. Shapiro, Kazdin, and McGonigle (1982) examined the possible multiple-treatment interference in an experiment with five retarded, behaviorally disturbed children. The target behavior in this particular experiment was on-task behavior in a classroom located in a children's psychiatric unit. With a very clever and elegant variant of the method of independent verification, the effects of two treatments and a baseline condition were examined within the context of an ATD for increasing on-task behavior. One treatment was token reinforcement for on-task behavior, the second treatment was response cost where tokens were removed for off-task behavior. Two 25-minute sessions were held per day: one in the morning and one in the afternoon. On any one day, two treatments would be administered, and these would be counterbalanced over a number of days. After a 4-day phase in which baseline conditions were in effect during both time periods, baseline and token reinforcement were alternated over a 6-day phase. This was followed by the alternation of token reinforcement and response cost over a 10-day period. The investigators then returned to the baseline versus token reinforcement phase for 6 more days, followed by a return to the token reinforcement versus response cost phase for yet another 6-day period. Finally, this was followed by a phase where token reinforcement was administered during both time periods.

The experimental design and the results are represented in Figure 8-2, where the average responses of the five subjects are presented. (Individual data were also presented, but this figure will suffice for purposes of illustration.) Thus this experiment really consisted of four separate ATDs after the baseline condition, in which token reinforcement was alternated with either baseline or response costs. Each of these ATDs was repeated twice. The elegance of this design for examining multiple-treatment interference is found in the fact that one can examine the effects of token reinforcement when alternated with either another treatment or baseline. If multiple-treatment interference is evident when token reinforcement is alternated with the other treatment, response cost, then the effects of token reinforcement should be different during that part of the experiment from when token reinforcement is alternated with baseline.

First, it is important to note here that both token reinforcement and response costs produced strong and comparable effects in increasing on-task behavior, and that token reinforcement was clearly effective when compared to baseline. The investigators decided, however, that token reinforcement was the preferable treatment because they noticed that more disruptive behavior occurred during the response-cost procedure than during the token reinforcement procedure. Thus token procedures were continued during both sessions in the last phase.

The investigators reported three different sets of findings from their examination of potential multiple-treatment interference. First, no evidence was

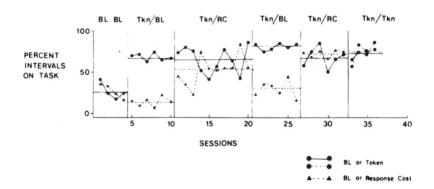

FIGURE 8-2. Group mean percentages of on-task behavior. Paired interventions in each phase consisted of Baseline/Baseline; Token Reinforcement/Baseline; Token Reinforcement/Response Cost; Token Reinforcement/Baseline; Token Reinforcement/Response Cost; Token Reinforcement/Token Reinforcement. (Figure 1, p. 110, from: Shapiro, E. S., Kazdin, A. E., & McGonigle, J. J. (1982). Multiple-treatment interference in the simultaneous- or alternating-treatments design. *Behavioral Assessment, 4,* 105–115. Copyright 1982 by Association for Advancement of Behavior Therapy. Reproduced by permission.)

found that the overall level of on-task behavior was different when it was alternated with either baseline or response cost. This, of course, is an extremely important finding, particularly in terms of estimating what the effects of token reinforcement in this context would be when applied in isolation; that is, without the potentially interfering effects of another treatment. In other words, the investigator or clinician can feel somewhat safe in determining that the effects of token reinforcement, when alternated with response costs, are about what they would be if response cost were not present. Of course, this still is not a "pure" test because it is possible that alternating token reinforcement with baseline in an ATD yields a somewhat different effect from token reinforcement administered in isolation. Strict adherence to Sidman's method of independent verification would be necessary to estimate if any carryover effects were present when a treatment was alternated with a baseline condition.

Nevertheless, the investigators do point out that on-task behavior was more variable during token reinforcement when alternated with response cost than when alternated with baseline. Visual inspection of the data indicates that this was particularly true in 3 out of 5 subjects. While this finding in no way effects the interpretation of the results, it is an interesting observation in itself that could be followed up in a number of ways. It is possible, for example, that "disruptiveness" noted during response cost temporarily carried over into the next token phase, thereby causing some of the variability. A greater spacing of sessions and subsequent sharpening of stimulus control might have decreased this variability.

Also, the investigators observed a sequence effect, in that token reinforcement was more effective when applied in the morning session than in the afternoon session. Once again, this demonstrates the importance of counterbalancing. Finally, the investigators observed another possible example of multiple-treatment interference not directly connected with the comparison of the two treatments. In the first phase, where token reinforcement and baseline were alternated, on-task behavior averaged 14 percent during the baseline condition. In the second phase, where this same alternation occurred, however, on-task behavior averaged approximately 30 percent during the baseline session. Inspection of individual data revealed that this trend occurred in four out of five children. This may represent a positive carryover or a generalization of treatment effects to the baseline condition; thus, the first phase probably presents a truer picture of baseline responding. Studies of this type will be very critical in the future in mapping out the exact nature of multiple-treatment interference and improving our ability to draw causal inferences from ATDs.

The study of carryover effects, or treatment interactions, when they occur, can be interesting in its own right (Barlow & Hayes, 1979; Sidman, 1960). For example, it is possible that carryover effects might increase the efficacy of some treatments. In an early study of fantasy alteration in a sadistic rapist, Abel, Blanchard, Barlow, and Flanagan (1975) alternated orgasmic reconditioning daily, first using a sadistic fantasy and then a desired heterosexual fantasy. It is important to note that treatments were not counterbalanced and alternations were rather rapid. Sexual arousal to the heterosexual fantasy increased more quickly during the fast alternation than during orgasmic reconditioning to the appropriate fantasy alone. More recently, Leonard and Hayes (in press) have also demonstrated that fantasy alteration produces stronger changes in sexual arousal patterns when alternations are fast rather than when alternations are slow. This may represent a carryover effect or simply a sharpening of stimulus control.

Counterbalancing relevant experimental factors

If certain factors extraneous to the treatments themselves might influence treatment, then these factors should be counterbalanced. Actually, this should be quite obvious to any investigator designing an experiment. For example, if Treatments A and B in Figure 8-1 referred to two distinct manipulations within a classroom, and two classrooms were involved, then it would be important that one treatment did not always occur in the same classroom. For example, in McCullough et al (1974) ATD examining the effects of two treatments on disruptive behavior in a 6-year-old boy, two factors were counterbalanced (see Table 8-1). In this particular experiment the first treatment was social reinforcement for cooperative behavior and ignoring of uncooperative behavior. The second treatment was social reinforcement for cooperative behavior plus time-out for uncooperative behavior, in this case removal from the classroom for 2 minutes. A teacher and a teacher's aide administered the treatments, with the teacher administering Treatment A the first two days and Treatment B the last two days. Thus the two people

Table 8-1

TIME	TREATMENT			
	DAY 1	DAY 2	DAY 3	DAY 4
AM	A T-1	B T-2	A T-2	B T-1
PM	B T-2	A T-1	B T-1	A T-2

NOTE: T-1 = teacher, T-2 = teacher's aide

administering treatments were counterbalanced because, of course, differential effectiveness might have something to do with the person administering the treatments. In addition, treatments were administered during both a morning session and an afternoon session. Once again, rather than the experimenters offering Treatment A only in the morning and Treatment B only in the afternoon, treatments were alternated such that administration of them was counterbalanced across morning and afternoon. In the example described above (E. S. Shapiro et al., 1982), the investigators observed greater effectiveness of token reinforcement sessions in the morning than with afternoon sessions, underscoring once again the need for counterbalancing.

Of course, what should and should not be counterbalanced will be up to the investigator. Naturally, if different therapists, teachers, or other practitioners are involved in administering the treatments, then they must be counterbalanced. Some investigators may also want to counterbalance times of day if these differ, whereas others may not consider this important, depending on the question asked. Most investigators will have a good feel for this.

Number and sequencing of alternations

The major question one must consider in determining the number of alternations is the potential for determining differences among two or more treatments. In determining behavior trends within a baseline phase or one of the phases of an A-B-A withdrawal design, we suggested that three data points were the minimum necessary to determine a trend. In the ATD, however, when one is comparing two treatments, a minimum number of two data points for each treatment would be necessary, although a higher number would, of course, be much more desirable. Two data points per treatment would allow an examination of the relative position of each treatment and some tentative conclusions on treatment efficacy. However, returning to Figure 8-1 once again, few investigators would be convinced of the superiority of Treatment B if the experiment were stopped after Week 4. Nevertheless, if other practical considerations prevented continuation, the findings might be potentially important, pending replication.

Naturally, frequency of alternations will be limited by practical and other considerations. It is possible, for example, that treatment and meaningful measurement opportunities would occur only once a month. Once again, one could conceive of this situation occurring in the alternation of two drugs with long half-lives, where a meaningful measurement of behavioral or mood changes could occur only after one month; this might consist of two weeks of treatment with the drug and two weeks of consolidation of drug effects. Similar situations might obtain for two different physical interventions in a rehabilitation setting.

Finally, in arranging for random alternation of treatments to avoid order effects, one must be careful not to bunch too many administrations of the same treatment together in a row. For example, in determining the random order of two treatments by coin toss or a random-numbers table, it is conceivable that one might arrive by chance at an order that dictates four administrations of Treatment A in a row. If only one has time for only eight alternations altogether, then this would not be desirable. Thus the investigator must move to a "semirandom" order with an upper limit on the number of times a treatment could be administered consecutively. The investigator will make this determination based on the total number of alternations available. For example, if eight alternations were available, as in the hypothetical data in Figure 8-1, then the investigator might want to set an upper limit of three consecutive administrations of one treatment.

8.3. EXAMPLES OF ALTERNATING TREATMENTS DESIGNS

ATDs have been used in at least two ways: to compare the effect of treatment and no treatment (baseline) and to compare two distinct treatments. Some examples of ATDs with specification of the experimental comparison are presented in Table 8-2.

Comparing treatment and no-treatment conditions

Several investigators have compared treatment and no treatment in an ATD. Among early examples, O'Brien, Azrin, and Henson (1969) compared the effect of following and not following suggestions made by chronic mental patients in a group setting on the number of suggestions made by these patients. Doke and Risley (1972) alternated daily the presence of three teachers versus the usual one teacher and noted the effect on planned activities in the classroom (contingencies on individual versus groups were also compared in an ATD later in the experiment). Redd and Birnbrauer (1969), J. Zimmerman, Overpeck, Eisenberg, and Garlick (1969), and Ulman and Sulzer-Azaroff (1975) also reported early examples comparing treatment and no treatment in an ATD.

A particularly good example of this strategy was reported by Ollendick, Shapiro, and Barrett (1981). In this experiment the effects of two treatments (physical restraint and positive-practice overcorrection) were compared to no treatment in the reduction of stereotypic behavior in three mentally retarded emotionally disturbed children. The investigators targeted stereotypic behaviors for reduction involving bizarre hand movements, such as repetitive hair twirling and repetitive hand posturing. In a very important consideration

TABLE 8-2. Examples of Alternating Treatment Designs

AUTHORS	CLIENTS	BEHAVIOR	TREATMENTS
C. M. Smith (1963)	A narcoleptic	Narcolepsy	a. Methyl amphetamine b. Dextroamphetamine c. Adrenaline methyl amphetamine
O'Brien, Azrin, & Henson (1969)	13 chronic schizophrenic outpatients	Increase patient suggestions for improved environment	a. Response priming b. No response priming
Redd & Birnbrauer (1969)	2 severely retarded boys	Play behavior	a. Reinforced play b. Noncontingent reinforcement
J. Zimmerman, Overpeck, Eisenberg, & Garlick (1969)	13 multiply handicapped clients in a prevocational workshop	Work rate	a. No treatment b. Isolation-avoidance procedure
Steinman (1970)	6 normal girls	Imitation behavior	a. Reinforced imitation b. Nonreinforced imitation
Corte, Wolf, & Locke (1971)	1 institutionalized profoundly retarded resident	Self-injurious behavior	a. DRO b. Noncontingent condition
A. S. Kircher, Pear, & Martin (1971)	2 retarded children	Picture naming	a. Ignoring of incorrect responses b. Shock as punishment for incorrect responses
Mann & Baer (1971)	4 normal 4-year-olds	Language skills	a. Articulation training b. No training
Doke & Risley (1972)	14 normal children	Group participation	a. Scheduled activities b. Optional activities
Johnson & Lobitz (1974)	12 families	Children's disruptive behavior	a. Instruction to parents to make their child look "bad" b. Instruction to parents to make their child look "good"
Ulman & Sulzer-Azaroff (1975)	6 retarded adults	Academic behavior	a. Group reinforcement contingencies b. Individual reinforcement contingencies
Bittle & Hake (1977)	8-year-old autistic boy	Self-stimulatory behavior	Treatment procedures applied in 4 different settings
Kazdin & Geesey (1977)	2 mentally retarded boys aged 7 and 9	Disruptive and inattentive behavior	a. Earning tokens for oneself b. Earning tokens for the entire class
Rojahn, Mulick, McCoy, & Schroeder (1978)	2 blind, profoundly retarded men	Self-injurious behavior	a. Adaptive clothing b. Adaptive clothing and time-out

TABLE 8-2. Examples of Alternating Treatment Designs *(Continued)*

AUTHORS	CLIENTS	BEHAVIOR	TREATMENTS
Weinrott, Garrett, & Todd (1978)	6 boys in kindergarten through 3rd grade	Social aggression	a. Observer present b. Observer absent
E. B. Fisher (1979)	13 chronic psychiatric patients	Toothbrushing	a. Reward with 5 tokens b. Reward with 1 token c. No token reward
G. Martin, Palotta-Cornick, Johnstone & Celso-Goyos (1980)	16 retarded clients in institutionalized sheltered workshop	Work performance	a. Multiple component strategy to increase work production b. "Normal" procedure
Neef, Iwata & Page (1980)	3 mentally retarded students	Spelling acquisition and retention	a. High-density reinforcement b. Interspersal training
Ollendick, Matson, Elsveldt-Dawson, & Shapiro (1980)	Exp. 1: 2 emotionally disturbed, hospitalized children aged 8 and 10 Exp. 2: 2 emotionally disturbed, hospitalized children aged 12 and 13	Increase spelling achievement	Exp. 1: a. Positive practice overcorrection plus positive reinforcement b. Positive practice alone c. No-remediation control condition Exp. 2: a. Positive practice plus positive reinforcement b. Traditional corrective procedure plus positive reinforcement c. Traditional procedures alone
E. S. Shapiro, Barrett & Ollendick (1980)	3 female mentally retarded children aged 6, 7, and 8	Stereotypic mouthing or face-patting behavior	a. Physical restraint b. Positive practice overcorrection
Barrett, Matson, Shapiro & Ollendick (1981)	2 mentally retarded children aged 5 and 9	Stereotypic behavior	a. Punishment b. Differential reinforcement of other behavior c. No treatment
Ollendick, Shapiro, & Barrett (1981)	3 mentally retarded, emotionally disturbed children aged 7 and 8	Stereotypic behavior	a. Physical restraint b. Positive practice overcorrection c. No treatment
Hallahan, Lloyd, Kneedler & Marshall (1982)	8-year-old learning disabled boy	Difficulty attending to task	a. Self-assessment b. Teacher assessment
E. S. Shapiro, Kazdin, & McGonigle (1982)	5 mentally retarded, behaviorally disturbed children	On-task behavior	a. Baseline b. Token reinforcement c. Response cost

267

TABLE 8-2. Examples of Alternating Treatment Designs *(Continued)*

AUTHORS	CLIENTS	BEHAVIOR	TREATMENTS
VanHouten, Nau, Mackenzie-Keating, Sameoto, & Colavecchia (1982)	Exp. 1: 2 elementary school boys aged 9 and 12 Exp. 2: 2 elementary school boys aged 9	Disruptive behavior	Exp. 1 a. Verbal reprimands with eye contact and grasp b. Verbal reprimands without eye contact and grasp Exp. 2: a. Reprimands delivered from 1 m away b. Reprimands delivered from 7 m away
Hurlbut, Iwata & Green (1982)	3 severely handicapped, nonvocal adolescents	Language acquisition	a. Bliss symbol system b. Iconic picture system
Last, Barlow, & O'Brien (1983)	32-year-old married female	Generalized Anxiety Disorder	a. Coping self-statements b. Paradoxical intention
Singh, Winton, & Dawson (1982)	2-year-old develomentally normal girl	Screaming behavior	a. 1-minute facial screening b. 30-sec (phase 1), 3-sec (phase 2) facial screening
Carey & Bucher (1983)	5 institutionalized retarded children aged 10–13	Off-task behavior in object placement task	a. Short (30 sec) positive practice duration b. Long (3 mins) positive practice duration
Barrera & Sulzer-Azaroff (1983)	3 echolalic autistic children aged 6–9	Teaching expressive labeling skills	a. Oral communication training program b. Total communication training program
McKnight, Nelson, Hayes & Jarrett (1983)	9 depressed women	Depression	a. Cognitive therapy b. Social skills training

NOTE: In some cases these designs were mislabeled in the original article. In other cases the data were misanalyzed.

268

before beginning the experiment, the investigators ruled out the use of an A-B-A withdrawal design because even temporary increases in stereotypic behavior during withdrawal phases were unacceptable in this setting. Furthermore, previous experience of these investigators suggested that there was a chance the two treatments might be equally effective. Thus a no-treatment condition might be necessary to determine if these treatments were effective at all. Of course, this problem also arises in between-group research because, if two treatments were equally effective (on the average) in two groups, a control group would be necessary to determine if any clinical effects occurred over and above no treatment.

In this procedure, three 15-minute sessions were administered by the same experimenter each day. Individual sessions were separated by at least one hour. Following baseline conditions for all three time periods, the two treatments and the no-treatment conditions were administered in a counterbalanced order across sessions. When one of the treatments produced a zero or near-zero rate of stereotypic behavior, that treatment was then selected and implemented across all three time periods during the remainder of the study. During sessions, each child was escorted to a small table in a classroom and instructed to work on one of several visual motor tasks. One treatment was physical restraint, consisting of a verbal warning and manual restraint of the child's hand on the tabletop for 30 seconds contingent on each occurrence of stereotypic behavior. The second treatment, positive-practice overcorrection, involved the same verbal warning but was followed by manual guidance in appropriate manipulation of the task materials for 30 seconds. Measures taken included number of stereotypic behaviors during each session and performance on the task.

The results for two of the three subjects are presented in Figures 8-3 and 8-4. In Figure 8-3 it is apparent during the ATD phase of this experiment that physical retraint was the superior treatment for John. Therefore, this treatment was chosen for the remainder of the experiment. Task performance increased rather steadily throughout the experiment, but was greatest during physical restraint. On the other hand, Figure 8-4 shows that positive practice intervention was the superior treatment for Tim.

Several features of this noteworthy experiment are worth mentioning. *First*, the ATD part of this experiment was concluded in 3 or 4 days (three sessions per day), and proper determinations of the effective treatment in each case were made. This is a relatively brief amount of time for an experiment in applied research, and yet it is typical of ATDs, particularly in this context (e.g., McCullough et al., 1974). *Second*, the addition of a baseline phase prior to introduction of the ATD allowed further identification of the naturally occurring frequencies of the target problem and the absolute amount of reduction in the target problem when treatments were instigated. Of course, this is not necessary in order to determine which of three condi-

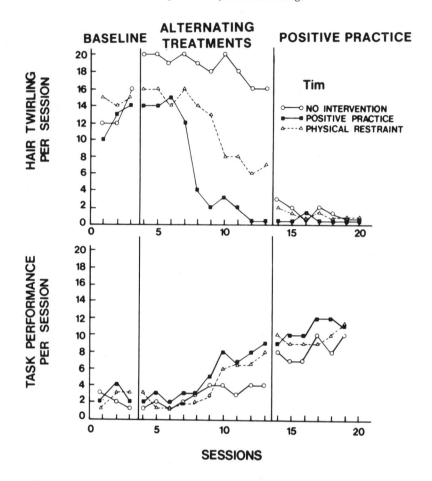

FIGURE 8-3. Stereotypic hair twirling and accurate task performance for John across experimental conditions. The data are plotted across the three alternating time periods according to the schedule that the treatments were in effect. The three treatments were presented only during the alternating-treatments phase. During the last phase, physical restraint was used during all three time periods. (Figure 1, p. 573, from Ollendick, T. H., Shapiro, E. S., & Barrett, R. P. (1981). Reducing stereotypic behaviors: An analysis of treatment procedures utilizing an alternating treatments design. *Behavior Therapy, 12*, 570–577. Copyright 1981 by Association for Advancement of Behavior Therapy. Reproduced by permission.)

tions was more effective, but it provides important additional information to the investigator. *Third*, The ATD in this case also served as a clinical assessment procedure for each client, since the most effective treatment was immediately applied to eliminate the problem behavior. The rapidity with which the ATD can be implemented makes this design very useful as a clinical assess-

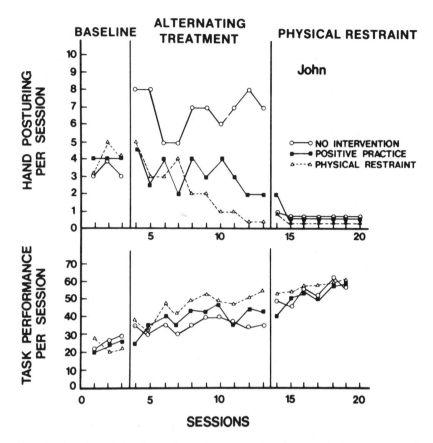

FIGURE 8-4. Stereotypic hand posturing and accurate task performance for Tim across experimental conditions. The data are plotted across the three alternating time periods according to the schedule that the treatments were in effect. The three treatments were presented only during the alternating-treatments phase. During the last phase, positive practice overcorrection was used during all three time periods. (Figure 2, p. 574, from Ollendick, T. H., Shapiro, E. S., & Barrett, R. P. (1981). Reducing stereotypic behaviors: An analysis of treatment procedures utilizing an alternating treatments design. *Behavior Therapy, 12*, 570–577. Copyright 1981 by Association for Advancement of Behavior therapy. Reproduced by permission.)

ment tool as well as an experimental strategy (see Barlow et al., 1983). *Fourth,* John did better with physical restraint, whereas Tim did better with positive practice intervention. The third subject also did better with positive practice intervention. This is a good example of the handling of intersubject variability in an ATD design. As discussed in chapter 2, a between-group strategy would average out, rather than highlight, these individual differences in response to treatment. By demonstrating this intersubject variability, however, the investigators were in a position to speculate on the reasons for these differences, which in fact they did. Because of this, they were in a position to examine more carefully client-treatment interactions that would predict which treatment would be successful in an individual case. Once again, highlighting intersubject variability in this way can only increase the precision with which one can generalize the effects of these specific treatments to other individual clients (see chapter 2).

Finally, the discerning reader will notice that posturing during the no-treatment condition of the ATD is somewhat higher with John and Tim than during baseline, where the same condition was in effect across all three time periods (but this increased response during no treatment was not true for the third subject). It is possible that this is an example of negative carryover effects, because responding during no treatment was worse when it was alternated with treatment than it was alone; that is, in baseline. In this experiment the authors purposefully blurred the discriminability of the three conditions as part of their experimental strategy, which may account, in part, for the carryover effects. This finding, once again, occurred in baseline and did not affect the ability of the investigators to determine the most effective treatment and then to apply it successfully during the last phase.

Of course, determination of the effectiveness of a single treatment compared to no treatment can also be examined via the most common A-B-A-B withdrawal design (see chapter 6, section 6-3). In this particular experiment, however, the authors were interested in comparing the effects of two treatments with each other as well as the effects of each compared to no treatment, and thus the ATD was the only choice. Furthermore, they had determined clinically that it was not possible to allow an increase in stereotypic responding in the absence of treatment, a condition that would obtain during the withdrawal phase of any A-B-A design. Nevertheless, when one wishes to compare treatment with no treatment, one has a choice between a more standard withdrawal design and an ATD. The advantages of the ATD have already been mentioned. In addition to not requiring a withdrawal of treatment for a period of time, the comparison within the ATD can usually be made more quickly, and it can proceed without a formal baseline if this is necessary. On the other hand, there is no single phase in the ATD where treatment is applied in isolation as it would be in a clinical situation. Therefore, estimating the generalizability of any given treatment is less certain if one has any reason to worry about multiple-treatment interference effects. Investigators will have to weigh these advantages and disadvantages in choosing a particular design to compare treatment and no treatment.

Ollendick and his colleagues have also produced two other excellent examples of ATDs comparing three conditions. In each case two treatments were compared to no treatment (Barrett, Matson, Shapiro, & Ollendick, 1981; Ollendick, Matson, Esveldt-Dawson, & Shapiro, 1980). In the Barrett et al. study, punishment and DRO procedures were compared to no treatment in dealing with stereotypic behavior of mentally retarded children. In the Ollendick et al. (1980) study, two spelling remediation procedures were compared to no treatment. Unlike the Ollendick et al. (1981) study reported earlier, the investigators chose to make each condition clearly discriminable through either instructions at the beginning of each session or other clear signs and signals. There is little or no evidence of multiple-treatment interference in either of these experiments. Once again, if one wants to eliminate the possibility of multiple-treatment interference, it would seem advisable to make conditions as discriminable as possible. The easiest method is to use simple instructions announcing what condition the subject is in.

Comparing multiple treatments

The majority of ATDs compare the effects of two treatments rather than the effects of treatment with no treatment. An early example in an adult clinical situation examined the effects of two fear-reduction procedures (Agras et al., 1969, see Figure 8-5). This study examined the effects of two forms of exposure-based therapy. The subject was a 50-year-old female with severe claustrophobia. Her fears had intensified following the death of her husband some 7 years before admission to the treatment program. When admitted, the patient was unable to remain in a closed room for longer than one minute without experiencing considerable anxiety. As a consequence of this phobia, her activities were seriously restricted. During the study she was asked four times daily to remain inside a small room until she felt she had to come out. Time in the room was the dependent measure. During the first four data points, representing treatment, she kept her hand on the doorknob. Before the fifth treatment data point (sixth block of session), she took her hand off the doorknob, resulting in a considerable drop in times. During one treatment she was simply exposed to the closet, with the therapist nearby (outside the door). In the second treatment the therapist administered social praise contingent on her remaining in the room for an increasing period of time. The two therapists alternated sessions with one another. In the original experimental phase the therapists switched roles, but they returned to their original reinforcing or nonreinforcing roles in the third phase. The data indicate that reinforced sessions were consistently superior to nonreinforced sessions.

Several procedural considerations deserve comment. *First*, the counterbalancing was rather weak because the therapists switched roles only twice during the whole experiment. Ideally, a more systematic counterbalancing strategy would have been planned. *Second*, the treatments were not administered randomly. Sessions involving exposure without contingent praise always preceded exposure with contingent praise. Despite this fact, a clear superiority of one treatment over the other emerged. Nevertheless, the experiment

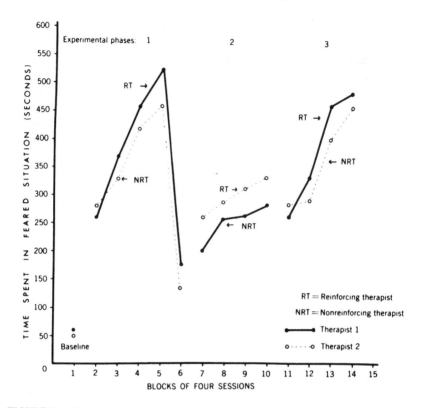

FIGURE 8-5. Comparison of effects of reinforcing and nonreinforcing therapists on the modification of claustrophobic behavior. (Figure 3, p. 1438, from: Agras, W. S., Leitenberg, H., Barlow, D. H., & Thomson, L. E. (1969). Instructions and reinforcement in the modification of neurotic behavior. *American Journal of Psychiatry, 125,* 1435–1439. Copyright 1969 by the American Psychiatric Association. Reproduced by permission.)

would be stronger with counterbalancing. *Finally,* one data point representing a block of four sessions served as a baseline comparison. While formal baseline phases are not necessary for ATD comparisons, and one baseline point is perhaps better than none, the examination of trends is always more informative than having simply a one-point pretest (or posttest).

The one indication of how far we have come in using the ATD to its fullest potential can be found in the next illustration, comparing the effectiveness of two treatments for depression in an adult clinical population (McKnight, Nelson, Hayes, & Jarrett, in press). Nine women diagnosed as depressed, based on a Schedule for Affective Disorders and Schizophrenia (SADS) interview, were included in this project. Subjects with strong suicidal tendencies or on medication at the time of the initial interview were excluded from the project, but all who eventually participated were severely depressed.

While depression is a problem with multiple components, two components that play a prominent role in many depressed cases are irrational cognitions and deficient social skills. In fact, treatment modalities with proven effectiveness have concentrated on one or another of these problem areas. For example, Beck's approach (A. T. Beck, Rush, Shaw, & Emery, 1979) concentrated on cognitive aspects of depression, and Lewinsohn, Mischel, Chaplin, and Barton's (1960) concentrated on deficient social skills.

Careful assessment revealed that 3 depressive subjects were primarily deficient in social skills, with few if any problems with irrational cognitions. Another 3 subjects presented with clear difficulties with irrational cognitions but few, if any, problems with social skills, while yet a third trio of subjects had difficulties in both areas.

An ATD was used to compare social skills training and cognitive therapy in each of the three sets of 3 subjects. The two therapies were randomly assigned to 8 weeks of therapy such that each subject received four sessions of cognitive therapy and four sessions of social skills therapy. Appropriate counterbalancing was employed. The results for the first 2 trios of subjects displaying either difficulties with irrational cognitions or difficulties with social skills are presented in Figures 8-6 and 8-7.

One will notice, upon examining these figures, another experimental design feature that adds to the elegance of this experiment. Not only were treatments compared in individual subjects with an ATD, but in each trio of three subjects a multiple baseline across subjects design was implemented in order to observe the effects of treatment, compared to the initial baseline, and to insure that the effects of any treatment occurred only when that treatment was introduced. This strategy, of course, controls for potential confounds that are a function of multiple meaures and other conditions present during baseline (see chapter 7). Thus this experimental design allows a determination of the effects of treatment over baseline by means of a multiple baseline across subjects design as well as a comparison of two treatments within the ATD portion of the experiment.

Examining Figure 8-6, one can see that social skills training was the more effective treatment for depression in each of the 3 subjects presenting with social skills deficits, as indicted by scores on the Lubin Depression Adjective Checklist. Social skills training was also significantly better on a measure of social skills, the Interpersonal Events Schedule, than was cognitive therapy, as would be expected. These findings were statistically significant. No significant differences emerged on measures of irrational cognitions as assessed by the Personal Beliefs Inventory.

In Figure 8-7, on the other hand, which presents data for the 3 subjects experiencing primarily cognitive deficits, cognitive therapy was clearly superior to social skills training, on both measures of depression and measures of irrational cognitions. These findings were also statistically significant. No

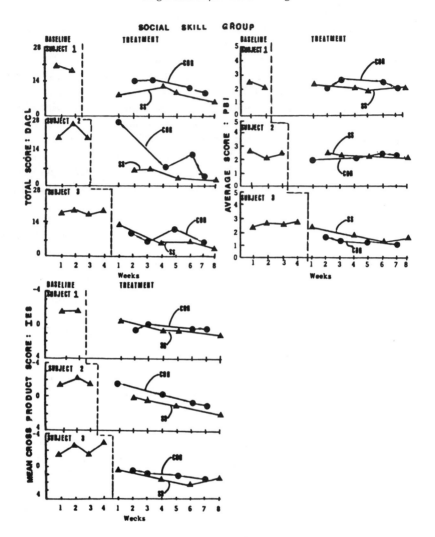

FIGURE 8-6. The effects of each treatment (COG = cognitive treatment; SS = social skill treatment) in a multiple baseline design across the 3 subjects experiencing difficulties in social skills on the weekly dependent measures administered. (Total score on the Lubin Depression Adjective Checklist; Average score on the Personal Beliefs Inventory; Mean cross-product score on the Interpersonal Events Schedule.) (Figure 2 from: McNight, D. L., Nelson, R. O., Hayes, S. C., & Jarrett, R. B. (in press). Importance of treating individually assessed response classes in the amelioration of depression. *Behavior Therapy*. Copyright 1984 by Association for Advancement of Behavioral Therapy. Reproduced by permission.)

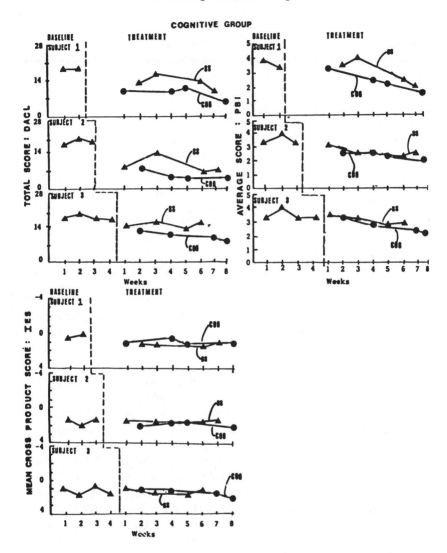

FIGURE 8-7. The effects of each treatment (COG = cognitive treatment; SS = social skill treatment) in a multiple baseline design across the 3 subjects experiencing difficulties in irrational cognitions on the weekly dependent measures administered. (Total score on the Lubin Depression Adjective Checklist; Average score on the Personal Beliefs Inventory; Mean cross-product score on the Interpersonal Events Schedule.) (Figure 4, from: McKnight, D. L., Nelson, R. O., Hayes, S. C., & Jarrett, R. B. (in press). Importance of treating individually assessed response classes in the amelioration of depression. *Behavior Therapy.*

statistically significant differences emerged on the measure of social skills, however, for people with primarily cognitive deficits.

This very elegant experiment is a model in many ways for the use of the ATD in adult clinical situations. The major conclusions derived from these data concern the importance of carefully and specifically assessing depression and all of its multiple components in order to tailor appropriate treatments to the individual. While these data were not necessary for this presentation, the third trio of subjects, displaying both irrational cognitions and social skill deficits, benefited from both treatments. Furthermore, consistent with the advantages of ATDs in investigating other problems, the results were apparent rather quickly after a total of eight treatment sessions. Also, the two treatments require the presentation of somewhat different therapeutic rationales to the patients, but this does not present a problem in our experience, and it did not in this experiment. Usually clients are simply told, correctly, that each treatment is directed at a somewhat different aspect of their problem and/or that the experimenters are trying to determine which of two treatments might be best for them. Contrast this experiment with the early example of an ATD with adult clinical problems described earlier (Agras et al., 1969), and one can see how far we have advanced our methodology. The elegant experimental manipulations and the wealth of information available due to combining the ATD with a multiple baseline across subjects make these data very useful indeed.

In one final, good example of an alternating treatment design comparing two treatments, Kazdin and Geesey (1977) investigated two different forms of token reinforcement in a special education classroom. Two mentally retarded children could earn tokens exchangeable for backup events for themselves or for the entire class. Tokens were contingent on attentive behavior in the classroom. Data from one of the children are presented in Figure 8-8. Data on attentive behavior were collected in the classroom during two different time periods each day. The two different conditions, earning tokens for oneself or for the entire class, were counterbalanced across these time periods. Data from the lower panel illustrate the ATD. During baseline, rates of attending behavior were essentially equal across time periods. During the ATD, attentive behavior was higher when the subject could earn backup reinforcers for the whole class. This condition was then implemented in the final phase across both time periods. As indicated in the figure caption, data were averaged in the upper panel to convey an overall level of attending behavior during these phases. As in the Ollendick et al. (1981) experiment described above, the baseline phase of this experiment provides the investigator with information on the naturally occurring frequency of the behavior and therefore allows an estimate of the absolute extent of improvement, as well as the relative effectiveness of the two conditions. In this experiment, the

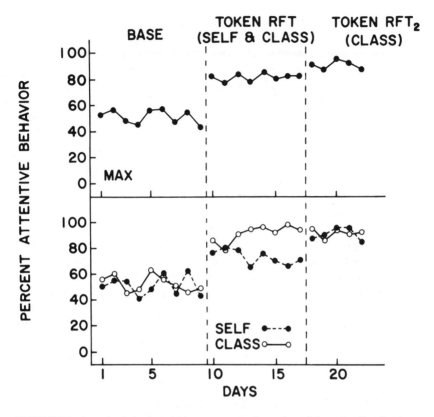

FIGURE 8-8. Attentive behavior of Max across experimental conditions. Baseline (base)—no experimental intervention. Token reinforcement (token rft)—implementation of the token program where tokens earned could purchase events for himself (self) or the entire class (class). Second phase of token reinforcement (token rft₂)—implementation of the class-exchange intervention across both time periods. The upper panel presents the overall data collapsed across time periods and interventions. The lower panel presents the data according to the time periods across which the interventions were balanced, although the interventions were presented only in the last two phases. (Figure 2, p. 690, from: Kazdin, A. E., & Geesey, S. (1977). Simultaneous-treatment design comparisons of the effects of earning reinforcers for one's peers versus for oneself. *Behavior Therapy, 8,* 682–693. Copyright 1977 by Association for Advancement of Behavior Therapy. Reproduced by permission.)

ATD also served as a clinical assessment procedure, in that the investigators were then able to implement the most successful treatment during the last phase. Finally, the ATD phase of this experiment took only 8 days, demonstrating once again the relative rapidity with which conclusions can be drawn using this design. Naturally, this feature depends on the frequency of potential measurement occasions. With institutionalized patients or subjects in a classroom, several experimental periods per day are possible. In outpatient settings, however, measurement occasions might be limited to once a week, or perhaps even once a month. Of course, the frequency of measurement occasions is also the function of the particular behavior under study.

In the examples provided thus far, times of treatment administration and, in some cases, therapists, have been counterbalanced so that the effects of the treatments themselves become clear. Naturally, the ATD also makes it very easy to examine directly the effects of different therapists, times of treatment administration, or settings on a particular intervention. For example, two therapists could alternately (and randomly) administer a treatment for generalized anxiety disorder from a relatively fixed treatment protocol. Weinrott, Garrett, and Todd (1978) examined the effects of the presence or absence of an observer on social aggression in six elementary schoolchildren. The results of the ATD demonstrated minimal observer reactivity in the situation. Finally, as mentioned above, E. S. Shapiro et al. (1982) discovered that token reinforcement was more effective in the morning than in the afternoon.

In some cases the setting in which treatment is administered becomes an important question. Bittle and Hake (1977) discovered comparable rates of reduction of self-stimulatory behavior in both experimental and natural settings during the administration of a given treatment. In other contexts, the implication of this work is that treatment can then be administered in the natural setting, where less experimental or therapeutic control exists.

8.4. ADVANTAGES OF THE ALTERNATING TREATMENTS DESIGN

The various strengths and weaknesses of the ATD have been reviewed before (Barlow & Hayes, 1979; Barlow et al., 1983; Ulman & Sulzer-Azaroff, 1975) and mentioned throughout this chapter. The major advantages and disadvantages will be listed briefly once again. *First*, the ATD does not require withdrawal of treatment. If two or more therapies are being compared, questions on relative effectiveness can be answered without a withdrawal phase at all. If one is comparing treatment with no treatment, then one still would not require a lengthy phase where no treatment was administered. Rather, no-treatment sessions are alternated with treatment sessions, usually within a relatively brief period of time.

Second, an ATD will produce useful data more quickly than a withdrawal design, all things being equal. This is because the relatively lengthy baseline, treatment, and withdrawal phases necessary to establish trends in A-B-A withdrawal designs are not important in an ATD design. The examples provided in this chapter illustrate this point. In fact, the relative rapidity of an ATD will often make it more suitable in situations where measures can be taken only infrequently. For example, if it is only practical to take measures infrequently, such as monthly, then an ATD will also result in a considerable saving of time. In an example provided in Barlow et al., (1983), it was noted that it often requires several hours and careful testing by two professional staff in a physical rehabilitation center to work up a stroke patient's muscular functioning. Obviously these measures cannot be taken frequently. If one were testing a rehabilitation treatment program using an A-B-A-B design, with at least three data points in each phase, then 12 months would be required to evaluate the treatment, assuming that measures could be taken no more frequently than monthly. On the other hand, if one month of treatment were alternated with one month of maintenance, then useful data within the ATD format would begin to emerge after four months.

Third, trends that are extremely variable or rapidly rising or falling present some problems for other single-case designs where interpretation of results is based on levels and trends in behavior. But the ATD design is relatively insensitive to background trends in behavior because one is comparing the results of two treatments or conditions in the context of whatever background trend is occurring. For example, if a specific behavioral problem is rapidly improving during baseline, it would be problematic to introduce a treatment. But in an ATD, two treatments could be alternated in the context of this improving behavior, with the potential for useful differences emerging. *Finally*, no formal baseline phase is required.

Naturally, these advantages vis-à-vis other design choices, apply only to situations where other design choices are indeed possible. There are many situations where other experimental designs are more appropriate for addressing the question at hand. Furthermore, the ATD suffers from the, as yet, unknown effects of multiple-treatment interference, and although recent research indicates that this problem may not be a great as once feared, we must still await systematic investigation of this issue to proceed with certainty. In any case, when it comes to generalizing the results of single-case experimental investigations to applied situations, there seems little question that the first treatment phase of an A-B-A-B design (or a multiple baseline design) is closer to the applied situation than is a treatment that is rapidly alternated with another treatment or with no treatment. These are only a few of the many factors the investigator must consider when choosing an appropriate experimental design.

8.5. VISUAL ANALYSIS OF THE ALTERNATING TREATMENTS DESIGNS

If enough data points have been collected for each treatment, and if one is so inclined, a variety of statistical procedures are appropriate for analyzing alternating treatment designs (see chapter 9). However, visual analysis should suffice for most ATDs. Throughout this book, the visual analysis of single-case designs is discussed in terms of observation of both levels of behavior and trends in behavior across a phase. Within at ATD, as noted above, levels and trends in behavior are not necessarily relevant because the major comparison is between two or more series of data points representing two or more treatments or conditions. To date, most investigators have been relatively conservative, in that very clear divergence among the treatments has been required. In most cases the series have been nonoverlapping. For example, with the exceptions of Points 1 and Points 11, which represented data points immediately following the switch in therapists, the Agras et al. (1969) ATD presented nonoverlapping series (see Figure 8-5).

Kazdin and Geesey (1977) also presented two series of data from the two treatments tested in their experiment which do not overlap, with the exception of one point very early in the ATD experiment (see Figure 8-8). Also, these data diverge increasingly as the ATD proceeds. Finally, Ollendick, Shapiro, and Barrett (1981) demonstrated a clear divergence between treatment and no treatment (see Figures 8-3 and 8-4). When one examines the effects of the two treatments, several data points overlap initially, but the two series increasingly diverge as the ATD proceeds. One must also remember that in this particular experiment (Ollendick et al., 1981) there were no clear signs or signals discriminating the treatments, and therefore this overlap may reflect some confusion about which treatment was in effect early in the experiment.

If overlap among the series occurs, then there is little to choose among the treatments or conditions, and most investigators say so. For example, Weinrott et al. (1978) observed considerable overlap between observer-present and observer-absent conditions in their experiment and concluded that observer reactivity was not a factor. Last, Barlow and O'Brien (1983) also observed overlap between two cognitive therapies and concluded that each was effective. Of course, when some overlap does exist, it is possible to utilize statistical procedures to estimate if any differences that do exist are due to chance or not (e.g., McKnight et al., 1983, Figure 8-7; E. S. Shapiro et al., 1982, Figure 8-2). However, as discussed in chapter 9, one must then decide if these rather small effects, even if statistically significant, are clinically useful. Our recommendation for these designs, and throughout this book, is to be conservative and to look for large visually clear, clinically significant effects. On the other hand, the ATD lends itself to a wide number of statistical tests, as outlined by Edgington (1984) and reviewed in chapter 9. Many of these tests require relatively few data points in each series. For example, using some of the examples presented in this chapter, Edgington (1984) has demonstrated how a variety of tests would be applicable to these data sets.

8.6. SIMULTANEOUS TREATMENT DESIGN

In the beginning of the chapter we noted the existance of a little-used design that actually presents two or more treatments simultaneously to an individual subject. In the first edition of this book, this design was referred to as a

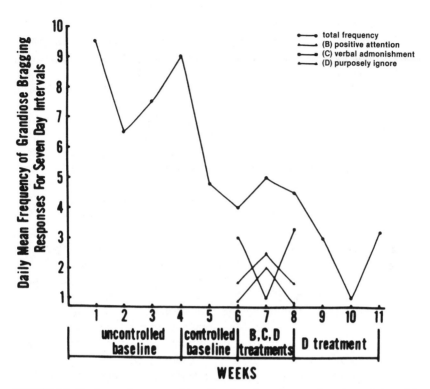

FIGURE 8-9. Total mean frequency of grandiose bragging responses throughout study and for each reinforcement contingency during experimental period. (Figure 3, p. 241, from: Browning, R. M. (1967). A same-subject design for simultaneous comparison of three reinforcement contingencies. *Behaviour Research and Therapy*, *5*, 237–243. Copyright 1967 by Pergamon Press. Reproduced by permission.)

concurrent schedule design. But the implication that a distinct schedule of reinforcement is attached to each treatment produces the same unnecessary narrowness as calling an alternating treatments design a multiple schedule design. Browning's (1967) term, *simultaneous treatment design*, seems both more descriptive and more suitable. Nevertheless, both terms adequately describe the fundamental characteristic of this design—the concurrent or simultaneous application of two or more treatments in a single-case. This contrasts with the fast alternation of two or more treatments in the ATD. The only example of the use of this design in applied research of which we are aware is the original Browning (1967) experiment, also described in Browning and Stover (1971). In this experiment, Browning (1967) obtained a baseline on incidences of grandiose bragging in a 9-year-old child. After 4 weeks, three treatments were used simultaneously: (1) positive interest and praise contingent on bragging, (2) verbal admonishment, and (3) ignoring. Each treatment was administered by a team of two therapists who were staff in a residential college for emotionally disturbed children. To control for possible differential effects with individual staff, each team administered each treatment for one week in a counterbalanced order. For example, the second group of two therapists admonished the first week, ignored the second week, and praised the third week. All six of the staff involved in the study were present simultaneously to administer the treatment. Browning hypothesized that the boy ". . . would seek out and brag to the most reinforcing staff, and shift to different staff on successive weeks as they switched to S's preferred reinforcement contingency" (p. 241). The data from Browning's subject (see Figure 8-9) indicate a preference for verbal admonishment, as indicated by frequency and duration of bragging, and a lack of preference for ignoring. Thus ignoring became the treatment of choice and was continued by all staff.

In this experiment the effects of three treatments were observed, but it is unlikely that a subject would be equally exposed to each treatment. In fact, the very structure of the design ensures that the subject won't be equally exposed to all treatments because a choice is forced (except in the unlikely event that all treatments are equally preferred). Thus this design is unsuitable for studying differential effects of treatments or conditions.

The STD might be useful anytime a question of individual preferences is important. Of course, in some cases preferences for a treatment may be an important component of its overall effectiveness. For example, if one is treating a phobia, and either one of two cognitive procedures combined with exposure-based therapy is equally effective, the client's preference becomes very important. Presumably a client would be less likely to continue using, after treatment is terminated, a fear-reduction strategy that is less preferred or even mildly aversive. But the more preferred or least aversive treatment procedure would be likely to be used, resulting most likely in a more favorable response during follow-up. Similarly, one could use an STD to determine the reinforcing value of a variety of potential consequences before introducing a program based on selective positive reinforcement. But it is also possible that a particular subject might prefer reinforcing consequences or treatments that are less effective in the long run. The investigator must remember that preference does not always equal effectiveness. The STD, then, awaits implementation by creative investigators studying areas of behavior change or psychopathology where strong experimental determinations of behavioral preference are desired. Presumably, these situations will be such that the self-report resulting from asking a subject about his or her preference will not be sufficient, for a variety of reasons. When these questions arise, the STD can be a very powerful tool for studying preference in the individual subject. But the STD is not well suited to an evaluation of the effectiveness of behavior change procedures.

CHAPTER 9

Statistical Analyses for Single-case Experimental Designs

by Alan E. Kazdin*

9.1. INTRODUCTION

Data evaluation consists of methods that are used to draw conclusions about behavior change. In applied research where single-case designs are used, experimental and therapeutic criteria are invoked to evaluate data (Risley, 1970). The *experimental criterion* refers to the way in which data are evaluated to determine if an intervention has had a reliable or veridical effect on behavior. The experimental criterion is based on a comparison of behavior under different conditions, usually during intervention and nonintervention (baseline) phases. To the extent that performance reliably varies under these separate conditions, the experimental criterion has been met.

The *therapeutic criterion* refers to whether the effects of the intervention are important. This criterion entails a comparison between behavior change that has been accomplished and the level of change required for the client's adequate functioning in society. Even if behavior change is reliable and clearly related to the experimental intervention, the change may not be of clinical or applied significance. To achieve the therapeutic criterion, the intervention needs to make an important change in the client's everyday functioning.

Completion of this chapter was facilitated by a Research Scientist Development Award (MH00353) from the National Institute of Mental Health.

*Please address all correspondence to: Alan E. Kazdin, Department of Psychiatry, University of Pittsburgh School of Medicine, Western Psychiatric Institute and Clinic, 3811 O'Hara Street, Pittsburgh, PA 15213.

Within single-case research, data can be evaluated in different ways to address the experimental and therapeutic criteria. Visual inspection is the most commonly used method of evaluating the experimental criterion and consists of examining a graphic display of the data (see Baer, 1977a; Michael, 1974). The data are plotted across separate phases of the single-case design. A judgment is made about whether the requirements of the design have been met, to draw a causal relationship between the intervention and behavior change. To those unfamiliar with the method, visual inspection seems to be completely subjective and free from specifiable criteria that guide decision making. Yet for visual inspection to be applied, special data requirements need to be met. Also, the data are visually inspected according to specific criteria (e.g., changes in trend, latency of the change at the point of intervention) to indicate whether the changes are reliable (see Kazdin, 1982b; Parsonson & Baer, 1978).

Statistical analysis represents another method of data evaluation in single-case research. Statistical tests provide a quantitative method and a set of rules to determine if a particular experimental effect is reliable. Statistical tests do not eliminate judgment from data evaluation. Rather, they provide replicable methods of evaluating information and reaching a conclusion about the experimental criterion. For statistical evaluation, a level of confidence (significance), decided by consensus, is used as a criterion to define whether a change in behavior is reliable (i.e., meets the experimental criterion). Judgment still enters into data analysis in terms of defining the datum, selecting the unit of analysis, identifying the statistical test, and so on. But the analyses themselves consist of replicable computational methods and rules for making decisions about the data.

Visual inspection and statistical data evaluation address the experimental criterion for single-case research. The applied, or clinical, significance of the change also is important. The therapeutic criterion has been addressed in different ways (Kazdin, 1977; Wolf, 1978). One method is to evaluate if the changes in the client's behaviors bring him or her within the level of his or her peers who are functioning adequately in society. For example, in the case of treatment for deviant behavior, a clinically significant change is achieved if the client's behavior after treatment falls within the range of persons who have not been identified as having problems. Another method is to have various persons (the client, relatives, experts, and other people in everyday life) evaluate the magnitude of change achieved by the client. If such persons perceive a distinct improvement in behavior or qualitative differences before and after treatment, the results suggest that the change is of applied significance.

The purpose of the present chapter is to detail statistical analyses for single-case experimental designs. The statistical analyses need to be viewed in the context of other methods of data evaluation to which they are compared. In between-group research, statistical analysis obviously has been widely adopted and accepted as the method of data evaluation. Even though questions are occasionally raised about whether statistical significance is an appropriate criterion, whether certain types of tests should be used, and so on, they remain in the background in terms of the actual conduct of research. Within single-case research, application of statistical tests is far less well developed or established. The types of statistical tests available are not widely familiar, and their appropriate application has relatively few exemplars (Kratochwill, 1978b; Kratochwill & Brody, 1978). More basic than the application of the tests is the question of whether such tests should be used at all in single-case research. The present chapter discusses issues regarding the use of statistical analyses in single-case research. However, major emphasis will be given to various tests themselves and how they are applied. Advantages and limitations in applying particular tests will be presented as well.

9.2. SPECIAL DATA CHARACTERISTICS

Most research in the behavioral sciences utilizes between-group designs, where multiple subjects are observed at one or a few points in time. Parametric statistical analyses are applied that invoke several assumptions about the nature of the data and the population from which subjects are drawn. In single-case research, one or a few individuals are observed at several different points in time. Statistical tests applicable to group studies may not be appropriate for single cases where data are collected over time.

Serial dependency

In applications of analyses of variance in group research, researchers are familiar with the fact that the tests are "robust" and can handle the violation of various assumptions (e.g., Atiqullah, 1967; G. V. Glass, Peckham, & Sanders, 1972; Scheffé, 1959). There is one assumption which, if violated, seriously affects analysis of variance and makes t or F tests inappropriate. The assumption is the independence-of-error components. The assumption refers to the correlation between the error (e) components of pairs of observations (within and across conditions) for i and j subjects. The expected value of the correlation for pairs of observations is assumed to be zero (i.e., $r_{e_i e_j} = 0$). Typically, in between-group designs, independence-of-error components are assured by randomly assigning subjects to conditions. In the case of continuous or repeated measures over time, the assumption of independence-of-observations often is not met. Successive observations in a time series tend to be correlated, in which case the data are said to be *serially dependent*. The correlation among successive data points means that knowing the level of performance of a subject at a given time allows one to predict subsequent points in the series.

The extent to which there is dependency among successive observations can be assessed by examining *autocorrelation* in the data. Autocorrelation refers to a correlation (r) between data points separated by different time intervals (lags) in the series. An autocorrelation of lag 1 (or r_1) is computed by pairing the initial observation with the second observation, the second with the third, the third with the fourth, and so on throughout the time series. Autocorrelation of lag 1 yields the correlation coefficient that reflects serial dependency. If the correlation is significantly different from zero, this indicates that performance at a given point in time can be predicted from performance on the previous occasion (the direction of the prediction determined by the sign of the autocorrelation).

Generally, autocorrelation of lag 1 is sufficient to reveal serial dependency in the data. However, a finer analysis of dependency may be obtained by computing several autocorrelations with different time lags (e.g., autocorrelations of lags of 2, 3, 4, and so on). For the general case, an autocorrelation of the lag t is computed by pairing observations t data points apart. For example, autocorrelation of lag 2 is computed by pairing the initial observation in the series with the third, the second with the fourth, the third with the fifth, and so on.

Serial dependency throughout the time series is clarified by computing and plotting correlations of different *lags*.[1] The plot of the autocorrelations is referred to as a *correlogram*. Figure 9-1 provides correlograms (i.e., autocorrelations plotted as a function of different lags) for two hypothetical sets of data. In each correlogram, the point that is plotted reflects the correlation coefficient for observations of a given lag. As can be seen for the data in the upper portion of the figure, the correlations with short lags are positive and relatively high. As the lag (i.e., the distance between the data points) increases, the autocorrelation approaches zero and eventually becomes negative. The hypothetical data in the upper portion of Figure 9-1 reflect serial dependency because the autocorrelation of lag 1 is likely to be significantly different from 0.[2] Moreover, the correlogram reveals that the dependency continues beyond lag 1 until the autocorrelation approaches 0. In contrast, the lower portion of Figure 9-1 reveals a hypothetical correlogram where the observations in the time series are not dependent. The autocorrelations do not significantly deviate from 0. The lack of dependence signifies that the errors of successive observations are "random," that is, a data point below the "average" value is just as likely to be followed by a high value as by another low value. Time series data that reveal this latter pattern can be treated as independent observations and can be subjected to conventional statistical analyses.

When autocorrelation is significant, serious problems occur if conventional analyses are used (Scheffé, 1959). Initially, serial dependency reduces the number of independent sources of information in the data. The degrees of freedom based upon the actual number of observations is inappropriate because it assumes that the observations are independent. Any F test is likely to overestimate the true F value because of an inappropriate estimate of the

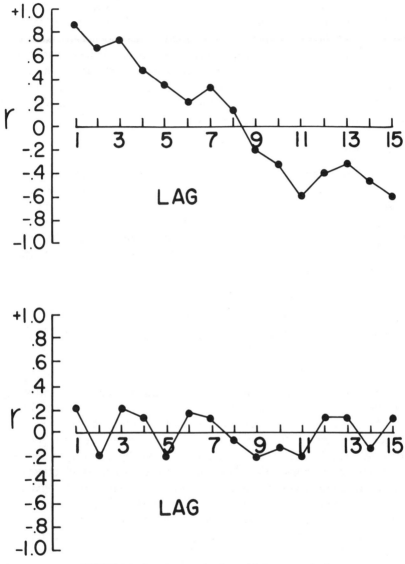

FIGURE 9-1. Correlograms for data with (upper portion)
and without serial dependency (lower portion).

degrees of freedom. For the appropriate application of t and F tests, the
degrees of freedom must be independent (uncorrelated) sources of information. A second and related problem associated with dependency is that the
autocorrelation spuriously reduces the variability of the time series data. Thus,
error terms derived from the data underestimate the variability that would

result from independent observations. The smaller error term inflates or positively biases F. In general, significant autocorrelation can greatly bias t and F tests. Use of these tests when the data are serially dependent can lead to Type I and Type II errors, and simple corrections to avoid these biases (e.g., adjustment of probability level) do not address the problem. (In passing, it may be important to note as well that serial dependency in the data can also bias the conclusions reached through visual inspection as well as statistical analyses [see R. R. Jones, Weinrott, & Vaught, 1978].)

General comments

Serial dependency is not a necessary characteristic of single-case data or observations over time. However, significant autocorrelation is a likely characteristic of continuous data and is a central consideration in deciding if particular statistical tests should be applied to single-case data. Several statistical tests for single-case data, including variations of t and F, are presented below. The tests vary as to whether they acknowledge, take into account, or are influenced by serial dependency in the data.

9.3. THE ROLE OF STATISTICAL EVALUATION IN SINGLE-CASE RESEARCH

Sources of controversy

The use of statistical analyses has been a major source of controversy because the approach embraced by such analyses appears to conflict with the purposes of single-case research and the criteria for identifying effective interventions. To begin with, identifying reliable intervention effects does not necessarily require statistical evaluation, as implicitly assumed in between-group research. In single-case research, demonstration of a reliable effect (i.e., meeting the experimental criterion) is determined by replication of intervention and baseline levels of performance over the course of an experiment, as is commonly illustrated in A-B-A-B designs. Other single-case experimental designs replicate intervention effects in different ways and permit comparisons to be made between what performance would be with and without treatment. In practice, whether the results clearly meet the experimental criterion depends upon the pattern of the data in light of the requirements of the specific design. Several characteristics such as changes in means or slope across phases, abrupt shifts or repeated changes in performance as an intervention is presented and withdrawn, and similar characteristics can be used to evaluate intervention effects without inferential statistics (Kazdin, 1982b).

Statistical criteria are objected to in part because of the goal of applied single-case research. The goal is to identify and evaluate potent interventions (Baer, 1977a; Michael, 1974). Visual inspection, the method commonly used to evaluate single-case data, is viewed as a relatively *in*sensitive method for determining if an intervention has been effective. Only marked effects are likely to be regarded as reliable through visual inspection. In contrast, statistical analyses may identify as significant subtle changes in performance. The tests may detect changes in performance that are not replicable. Indeed, within statistical evaluation, the possibility exists that the findings were obtained by "chance."

Single-case research designs do not necessarily require visual inspection or statistical analysis as a method of data evaluation. However, applied research where single-case designs are used (applied behavior analysis) has emphasized the importance of searching for potent intervention effects and subjecting the data to visual inspection rather than statistical evaluation. The two different methods are not fundamentally different, but they do vary in the sorts of effects that are sought and the manner in which decisions are reached about intervention effects.[3]

Some of the objections to statistics in single-case research have stemmed from the focus on groups of subjects in between-group research. Within-group variability is often a basis for evaluating the effect of interventions in group research. Yet, within-group variability is not part of the behavioral processes of individual subjects and perhaps should not be included in the evaluation of performance (Sidman, 1960; also see chapter 2). Related group research often obscures the performance of the individual subject. Statistical analyses usually reflect the performance of the group as a whole with data characteristics (means, variances) that do not bear on the performance of any single subject. It remains unclear how the intervention affects individuals and the extent to which group performance represents individual subjects. As these objections illustrate, concerns over statistical analyses extend beyond the manner in which data are evaluated. The objections pertain to fundamental issues about experimental design and the approach toward research more generally (J. M. Johnston & Pennypacker, 1981; Kazdin, 1978).

Potential contributions

Statistical analyses in single-case research may provide a valuable supplement rather than an alternative to visual inspection. In many applications, inferences about the effects of the intervention can be readily drawn through visual inspection. Statistical analyses in such situations may not add an increment of useful information unless a specific question arises about a particular facet of the data at a given point in time. In many situations, the pattern of data required for visual inspection may not be met, and statistical tests may provide important advantages.

Evaluation of intervention effects can be difficult when performance during baseline is systematically improving. An intervention may still be required to accelerate the rate of change. For example, self-destructive behavior of an autistic child might be decreasing gradually during baseline but an intervention may be required to achieve more rapid progress. Visual inspection is often difficult to invoke with a baseline trend reflecting improvement. Selected statistical analyses (discussed later in the chapter) can readily examine whether a reliable intervention effect has been achieved over and above what would be expected by continuation of the initial trend. Thus statistical analyses provide an evaluative tool in cases where visual inspection may be difficult to invoke.

Apart from trend in baseline, visual inspection is also difficult to invoke if data show relatively high variability within and across phases. Single-case research designs have been applied in a variety of settings such as psychiatric hospitals, institutions, classrooms, and others. In such settings, investigators have frequently been able to control several features of the environment such as staff behavior and activities of the clients, in addition to the intervention. Because extraneous factors are held relatively constant for purposes of experimental control, variability in subject performance can be held to a minimum. Visual inspection is more readily applied to single-case data when variability is small.

Over the years, single-case research has been extended to several community or open-field settings (Geller, Winett, & Everett, 1982; Kazdin, in press). In such extensions, control over extraneous factors in the situation may be minimal. Moreover, the persons who serve as subjects may change over the course of the project, so that the effect of the intervention is evaluated against the backdrop of intrasubject and intersubject variability. Increased variability in performance decreases the likelihood of demonstrating marked effects in performance and the ability of visual inspection to detect reliable changes. Statistical evaluation may provide a useful aid in detecting if the intervention has produced a reliable effect.

Proponents of applied single-case research have stressed the need to investigate interventions that produce potent effects. Yet there may be different situations where it is important to detect reliable intervention effects, even if relatively small. To begin with, investigators may embark on new lines of research where the interventions are not well developed. The interventions may not be potent at this stage because of lack of information about the intervention or the conditions that maximize its efficacy. Statistical analyses at this initial stage of research may help identify interventions and variables that produce reliable effects. More stringent criteria of visual inspection might lead to abandonment of interventions that do not produce marked effects at the outset. Yet identification of procedures through statistical analyses may help screen among variables that warrant further pursuit. Interventions identified in this fashion might be developed further through subsequent research and perhaps eventually produce large effects that meet the criteria of visual inspection. But, at the initial stage of research, statistical analyses may serve a useful purpose in identifying variables that warrant further scrutiny and development.

It may be important to detect small effects in other situations. As applied research has been extended to community settings, small changes in the behaviors of individual subjects have become increasingly important. These changes, when accrued across many persons, become highly significant. For example, small changes in energy consumption within individuals are important because such effects become socially significant when extended on a larger scale. Also, in community applications, small changes in performance may be important to detect because of the significance of the behaviors. For example, interventions designed to reduce violent crimes in the community may produce minute effects that do not pass the test of visual inspection. Yet small but reliable changes are important to detect because of the significance of any change in such behaviors.

General comments

The controversy over statistical analyses is not whether all data in single-case research should be evaluated statistically. Single-case research designs, the tradition from which they derive, and the dual concerns in applied work for experimental and therapeutic criteria for evaluating change all place limits on the role of statistical analysis. Within the approach of single-case research, the question is whether statistical tests can be of use in situations where visual inspection might be difficult to apply. There are different reasons for posing an affirmative answer. Although visual inspection can be readily applied to many investigations, the method has its own weaknesses. In a variety of circumstances, researchers often have difficulty in judging (via visual inspection) whether reliable effects have been produced and disagree in their interpretations of the data (DeProspero & Cohen, 1979; Gottman & Glass, 1978; R. R. Jones et al., 1978). Also, systematic biases may operate when invoking visual inspection criteria, such as ignoring the impact of autocorrelation and being influenced by the metric by which data are graphed (R. R. Jones et al., 1978; Knapp, 1983; Wampold & Furlong, 1981a). An attractive feature of statistical analyses is that once the statistic is decided, the results are (or should be) consistent among different investigators. Judgment plays less of a role in applying a statistical analysis to the data. Thus statistical analyses can be a useful tool in cases where the idealized data patterns required for visual inspection are not obtained.

9.4. SPECIFIC STATISTICAL TESTS

There are a large number of statistical tests that can be applied to data obtained from a single subject over time. The range of available tests has not been conveniently codified or illustrated. Indeed, the task is rather large because a given test might be applied in a variety of different ways depending on the specific variant of single-subject designs and the statement the investigator wishes to make about the intervention. Several tests discussed below illustrate major variants currently available but do not exhaust the range of appropriate tests.

Conventional *t* and *F* tests

Although many different statistical tests are available for single-case designs, certainly the most familiar are *t* and *F* tests. Each single-case design includes two or more phases that can be compared with a *t* or *F* test depending, of course, on the number of different conditions or phases. For example, in an A-B-A-B design, comparisons can be made over baseline (A) and intervention (B) phases. An obvious test would be to compare A and B phases (*t* test) or to compare the four A-B-A-B phases (analysis of variance). The test would evaluate whether the difference(s) between (or among) means is statistically significant.

If the single-case design is applied to a group of subjects, correlated *t*-test or repeated-measures analyses of variance can be performed. For data from an individual subject, *t* and *F* tests may *not* be appropriate if the data are serially dependent. A test is appropriate if autocorrelation is computed and shown to be nonsignificant.

Consider, as an example, hypothetical data for a socially withdrawn child who received reinforcing consequences at school for interacting with peers. Consider data from the first two (AB) phases of an A-B-A-B design. The change from baseline to intervention phases can be evaluated with a *t* test. Table 9-1 presents the data for each day, where the numbers reflect the percentage of intervals of appropriate social interaction. The baseline phase tends to show lower rates of performance than the intervention phase, but are the differences statistically significant?

To first assess if the data are serially dependent, autocorrelations are computed for the separate phases. The autocorrelations are computed within each phase rather than for the data across both phases, because the intervention may influence the relation of data points to each other (i.e., their dependency). As shown in the table, neither autocorrelation is statistically significant. The data appear to meet the independence-of-error assumption and can be subjected to conventional *t* testing. The results of a *t* test for independent observations (or groups) and for unequal sample sizes indicate that A and B phases were significantly different ($t(25) = 6.86$, $p < .01$). Thus the differences in social behavior between the two phases are reliable.

Variations of *t* and *F* tests

Variations of *t* and *F* have been suggested for situations where autocorrelation is significant and the data are dependent. Prominent among the sugges-

TABLE 9-1. *t* test Comparing Hypothetical Data
for A and B Phases for One Subject

BASELINE (A)		INTERVENTION (B)	
DAYS	DATA	DAYS	DATA
1	12	13	88
2	10	14	28
3	12	15	40
4	22	16	63
5	19	17	86
6	10	18	90
7	14	19	82
8	29	20	95
9	26	21	39
10	5	22	51
11	11	23	56
12	34	24	86
		25	31
		26	77
		27	76
Mean (A) = 17.00		Mean (B) = 65.87	
Autocorrelation r = .005 (lag 1)		Autocorrelation r = .010 (lag 1)	

tions is the analysis proposed by Gentile, Roden, and Klein (1972). When autocorrelation exists, these investigators suggested that nonadjacent phases that employed the same treatment can be combined and will reduce the effect of serial dependency. For example, in an A-B-A-B design, the two A phases are not adjacent and could be combined and compared with the two B phases. The rationale for combining phases is based on the fact that autocorrelations tend to decrease as the lag between observations increases. Assuming serial dependency in the data, Observation 1 in phase A_1 would be more highly correlated with Observation 1 in Phase B_1 (i.e., the immediately adjacent phase) than with Observation 1 in phase A_2 (i.e., a nonadjacent phase). Since the error components of all observations in A_1 are more like the components for the observations in B_1 than in A_2, it is assumed that combining treatments separated in time will reduce the dependency. Combining phases that are not adjacent should make A and B treatments more dissimilar, due to dependency in the data. The resulting t (or F) should be reduced because the dependency of adjacent observations will minimize treatment differences. Additional variations of t and F have been proposed, some of which attempt to address the issue of serial dependency by developing special error terms to make statistical comparisons of treatment effects (see Gentile et al., 1972; Shine & Bower, 1971).

Considerations and limitations of *t* and *F* tests

Appropriateness of the Tests. There is considerable agreement that *t* and *F* tests are not appropriate if the data from a single subject are serially dependent (Hartmann, 1974; Kratochwill et al., 1974; Thoresen & Elashoff, 1974). The variations alluded to above do not clearly resolve the issues. The effects of trying to compensate for serial dependency (e.g., by combining phases) are not easily estimated and no doubt vary with different patterns of autocorrelation. The safest approach is to precede *t* and *F* tests with an analysis of serial dependency. If significant autocorrelation exists, alternative statistical tests should be considered.

Evaluation of Means. Another issue may influence selection of *t* or *F* tests. Typically, these analyses, when appropriate, are applied to test whether or not there are significant changes in means between or among phases. Trends in the data are ignored. It is possible, for example, that an accelerated slope in baseline and intervention phases is apparent, in which case each data point may exceed the value of the preceding point. A simple test of means across A and B phases could reflect a statistically significant effect, but the effect might be accounted for by the trend. Alternatively, the data might show an increasing slope in baseline and a decreasing slope in treatment, with no overall mean differences. A test of means in both the above instances would lead to interpretive problems if the trends were ignored. The need to consider trend and mean changes as well as other data parameters is clarified in the discussion of time series analysis.

9.5. TIME SERIES ANALYSIS

Time series analysis compares data over time for separate phases for an individual subject or group of subjects (see G. V. Glass et al., 1974; Gottman, 1981; Hartmann et al., 1980; R. R. Jones, Vaught, & Weinrott, 1977). The analysis can be used in single-case designs in which alternative phases (e.g., baseline and intervention) are compared. There are two important features of time series analysis for single-case research. *First*, the analysis provides a *t* test that is appropriate when there is serial dependency in the data. *Second*, the analysis provides important information about different characteristics of behavior change across phases. The notion of serial dependency has been addressed already. The different features of the data that time series analysis reveals require a brief digression.

Patterns of change in time-series data

Continuous observations across separate phases may indicate change along several dimensions. Three dimensions that are especially relevant in understanding time series analysis include change in level, change in slope, and presence or absence of slope in a given phase (R. R. Jones et al., 1977). A *change in level* refers to a change at the point in which the intervention is made. If data at the end of baseline and the beginning of intervention phases show an abrupt departure or discontinuity, this would reflect a change in level. A *change in slope* refers to a change in trend between or among phases.

The notion of a change in level warrants further mention because it differs from the more familiar concern of a change in mean across phases. A change in mean across phases refers to differences in the average performance. A change in level does not necessarily entail a change in mean, and vice versa. However, a change in one does entail a change in the other when there is no slope in the data in either baseline or intervention phases. Applied researchers are concerned primarily with a change in means. Whether or not there is a change in the precise point of intervention (i.e., beginning of the B phase) is not necessarily crucial as long as behavior shows a marked overall increase or decrease.

Time series analysis provides separate tests of a change in level and a change in slope. A change in mean can be inferred from these other parameters. For example, a very gradual change in behavior after the intervention is applied might be detected as a significant change in slope but no change in level. The absence of change in level indicates that behavior did not change abruptly at the point of intervention. The significant change in the slope would imply a change in the means across phases. An advantage of time series analysis is that the nature of the change across phases is examined in a more analytic fashion than by merely evaluating overall means. Because separate tests are provided for changes in slope and level, there is no requirement that baseline phases show little or no trend in the data. The test allows one to evaluate whether any trend in an intervention phase departs from the slope in baseline, if one exists.

To convey how changes in level and slope can appear in single-case data, several different data patterns are illustrated in Figure 9-2. The figure provides hypothetical data over two phases (AB) of a larger design. The data patterns illustrate some of the relationships among changes in level and slope and in means across phases. Also, some of the data patterns (e.g., Figures 9-2a, 9-2b, and 9-2c) represent instances where visual inspection presents problems because of the presence of an overall trend across baseline and intervention phases. Conventional *t* and *F* tests that examine changes in means might overlook important changes when means do not change (as in Figure 9-2d), or they may indicate a significant change when in fact level or slope have not changed (e.g., as in Figure 9-2b).

Data analysis

The actual analysis itself cannot be outlined in a fashion that permits simple computation. Time series analysis depends upon more than entering raw data into a single formula. Several models of time series analysis exist that make different assumptions about the data and require different equations to

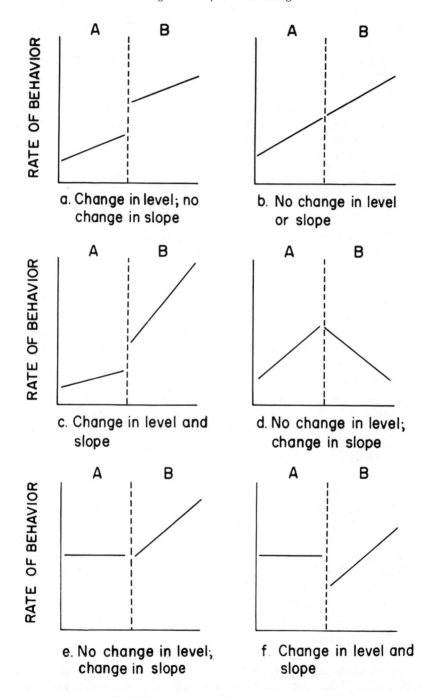

FIGURE 9-2. Examples of selected patterns of data
over two phases (AB), illustrating changes in level and/or trend.

achieve the final statistics. The analysis begins by evaluating serial dependency in the data. Different patterns of dependency may emerge that depend upon the pattern of autocorrelations, which are computed with different lags or intervals, as noted earlier. Once the pattern of serial dependency is identified, a model is applied to the data. The analysis consists of several steps, including adoption of a model that best fits the data, evaluation of the model, estimation of parameters for the statistic, and generation of t for level and slope changes (G. V. Glass et al., 1974; Gorsuch, 1983; Gottman, 1981; Horne, Yang, & Ware, 1982; Stoline, Huitema, & Mitchell, 1980). Computer programs are available to handle these steps (see Gottman, 1981; Hartmann et al., 1980).

It is useful to examine the results of a time series analysis for illustrative purposes and to evaluate the results in light of the characteristics of the data that might be inferred from visual inspection. As an illustration, one program focused on the frequency of inappropriate talking in a second-grade classroom (C. Hall et al., 1971, Exp. 6). Although there were many children in class, the class as a whole was treated as a single subject. The intervention consisted of praise and other reinforcers provided to children for their appropriate class-room behavior. The effects of the intervention, evaluated in an A-B-A-B design, are plotted in Figure 9-3. The results suggest that inappropriate talking out was generally high during the two different baseline phases and was much lower during the different reinforcement phases (praise, tokens plus a sur-prise). The first two phases (AB) have been analyzed using time series analysis (R. R. Jones, Vaught, & Reid, 1975). Through a computer program, the analyses revealed that the data were serially dependent, that is, the adjacent points were significantly correlated. Indeed, autocorrelation for lag 1 was .96 ($p < .01$). Thus conventional t and F test analyses would be inappropriate. Time series analyses revealed a significant change in level across the first two phases (AB) ($t(39) = 3.90, p < .01$) but no significant change in slope. A change in level with no change in slope suggests also a change in mean performance, obvious from visual inspection of the graphical display of the data. The data analysis only addresses the changes in the first two phases of the design. In principle, comparisons could be made across the other phases as well, although restrictions on the number of data points in this particular study present a limiting condition, discussed later.

The analysis is not restricted to variations of an A-B-A-B design. In any design where there is a change across phases, time series analysis provides a potentially useful tool. For example, in multiple baseline designs, time series analysis can evaluate change from baseline to intervention phases for each of the responses, persons, or situations, depending upon the precise design.

Considerations and limitations

Among the available statistical analyses, time series analysis is recom-mended because of the manner in which serial dependency is handled. With conventional t and F tests and many variations, dependency in the data is either

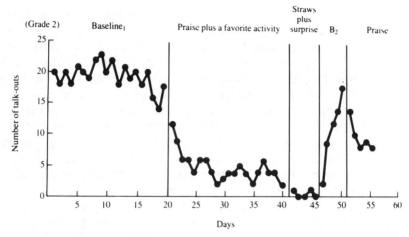

FIGURE 9-3. Daily number of talk-outs in a second-grade classroom. Baseline—before experimental conditions. Praise plus a favorite activity—systematic praise and permission to engage in a favorite classroom activity contingent on not talking out. Straws plus surprise—systematic praise plus token reinforcement (straws) backed by the promise of a surprise at the end of the week. B₂—withdrawal of reinforcement. Praise—systematic praise and attention for handraising and ignoring of talking out. (From: Hall, R. V., Fox, R., Willard, D., Goldsmith, L., Emerson, M., Owen, M., Davis, F., & Porcia, E. [1971]. The teacher as observer and experimenter in the modification of disputing and talking-out behaviors. *Journal of Applied Behavior Analysis*, **4**, 141–149. Copyright 1971 The Society for the Experimental Analysis of Behavior, Inc. Reproduced by permission.)

ignored, assumed to be present but disregarded, or recognized and handled in a relatively cumbersome (and controversial) fashion. In contrast, time series analysis depends upon the serial dependency in the data, adjusts to the specific dependency relationships among data points, and provides separate analyses for level and slope changes in light of special characteristics of the data. Another important feature of the analysis is that it does not depend upon stable baselines. Evaluation of single-case designs through visual inspection is facilitated when there is no slope in baseline or even a slope in the direction opposite to that predicted by the intervention effects. In contrast, time series analysis can be readily applied even when there is a trend toward improved performance in baseline, as illustrated earlier. The separate analyses of the changes in level and slope provide a reliable criterion in cases where visual inspection may be particularly difficult to invoke. Notwithstanding the desirable features of time series analysis, several issues need to be considered before using the analysis in applied research.

Number of Data Points. Time series analysis depends on a relatively large number of data points to identify the model that best describes the data (Box &

Jenkins, 1970). The nature of the underlying data is revealed through autocorrelations of different lags. In conventional analyses, large sample sizes are important to achieve statistical power. In time series analysis, the large sample (of data points) is necessary to identify the processes within the series itself and to select a model that fits the data.

Precisely what constitutes a large or sufficient number of observations depends on several factors such as the nature of the data, the types of changes across phases, variability within a phase, and other parameters that characterize a given series. However, the number of data points usually advocated is much greater than the number typically available in applied or clinical investigations. For example, various authors have suggested that at least 50 (G. V. Glass et al., 1974), and preferably 100 (Box & Jenkins, 1970), observations are required for estimating autocorrelations. Fewer observations have been used (e.g., data with 10 to 20 observations) in applied research and have detected statistically significant changes (R. R. Jones et al., 1977). Yet applied investigations often employ relatively short phases lasting only a few days to demonstrate intervention effects. In such cases, time series analyses will not be applicable.

Prevalence of Serial Dependency in Single-Case Data. Time series analysis in behavioral research has been advocated because of the concern over serial dependency in the data for a single subject. Intuitively one might expect serial dependency because multiple data points are generated by the same subject over time and because any influence on a particular occasion may spread (i.e., continue) to other occasions as well. Thus data from one occasion to the next are likely to be correlated, and the correlation is likely to attenuate over time as new factors impinge on the subject. In the middle and late 1970s, when time series analyses began to receive attention in single-case research, it seemed as if serial dependency were likely to be the rule rather than the exception (e.g., Hartmann, 1974; Kratochwill et al., 1974; Thoresen & Elashoff, 1974; R. R. Jones et al., 1977). Moreover, empirical evaluation of published single-case data indicated that the prevalence of serial dependency was quite high (e.g., 83% of nonrandomly selected instances) (R. R. Jones et al., 1977). However, in recent years questions have been raised about the prevalence of significant autocorrelation and hence the need for time series, as opposed to conventional, analyses. For example, one evaluation of applied research has suggested that only a minority of studies (less than 30%) shows serial dependency (Kennedy, 1976). The basis for the discrepancy in the prevalence of serial dependency is not readily clear, particularly since R. R. Jones et al. (1977) and Kennedy (1976) selected published investigations from the same journal. In general, whether data from a particular subject are serially dependent should not be assumed but should be tested directly. The difficulty is that computing autocorrelation itself requires multiple data points to detect a statistically significant effect, and a small number of data points may not permit precise evaluation of the processes involved in the data.

General Comments. Time series analysis has been used increasingly within the last several years. The increased availability of publications on the topic (e.g., Gottman, 1981; McCleary & Hay, 1980) and several computer programs (Hartmann et al., 1980; Horne et al., 1982) may be fostering increased use of time series analyses. Nevertheless, use of the analysis has been relatively limited for several reasons. The tests are complex and involve multiple steps that are not easily described in terms familiar to most researchers. For example, serial dependency and autocorrelation, two of the less esoteric notions underlying time series analysis, are not part of the usual training of researchers who conduct group studies in the social sciences. More in-depth examination of time series analysis and its underlying rationale introduces many concepts that depart from conventional statistical techniques and training (see Gottman, 1981). In addition, requirements for conducting time series analysis may not foster widespread adoption within applied behavioral research. The relatively brief phases typically used in single-case experimental designs make the test difficult to apply and perhaps, simply, inappropriate. Recent controversy over whether single-case data as a rule are serially dependent raises questions for some about the need for time series analysis. Nevertheless, time series analyses have been appropriately applied in several demonstrations and provide a valuable addition to statistical analyses of single-case data.

9.6. RANDOMIZATION TESTS

Several different tests useful for single-case experiments are based on the notion of assigning treatments randomly to different occasions (e.g., days or sessions) (Edgington, 1980b, 1984; Levin, Marascuilo, & Hubert, 1978; Wampold & Furlong, 1981b). At least two treatments, or conditions, are required; one of which may be baseline (A) and the other an intervention (B), and therefore these tests are useful for evaluating ATDs (see chapter 8). Prior to the experiment, the total number of occasions that the treatments will be implemented must be specified, along with the number of occasions on which each specific condition will be applied. Once these decisions are made, A and B (or A, B, C . . . *n*) conditions are assigned randomly to each session or day of the experiment, with the restriction that the number of occasions for each meets the prespecified totals. Each day, one of the conditions is administered according to the randomized schedule planned in advance.

The null hypothesis of the randomization test is that the client's response on the dependent measure(s) is not influenced by the condition in effect on that occasion (e.g., baseline or intervention). If the condition makes no difference, performance on any particular day will be a function of factors unrelated to the condition in effect. The random assignment of treatments to occasions in effect randomly assigns responses of the subject to the treatments. The obtained data are assumed to be the same as those that would have been obtained under any other random ordering of the treatments to occasions. Thus the null hypothesis attributes differences between conditions to the chance assignment of one condition rather than the other to particular occasions. To test the null hypothesis, a sampling distribution of the differences between the conditions under every equally likely assignment of the same response measures to occasions of A and B is computed. From this distribution, one can determine the probability of obtaining a difference between treatments as large as the one that was actually obtained.[4]

Data analysis

Consider as an illustration an investigation designed to evaluate the effect of teacher praise on the attentive behavior of a disruptive student. To use the randomization test, the investigator must decide in advance the number of days of the study and the number of days that each of two (or more) conditions will be administered. Assume for present purposes that the investigator wishes to compare the effects of ordinary classroom practices (baseline or A Condition) with a reinforcement program based on praise (intervention or B Condition). To facilitate computations, suppose that the duration of the study is decided in advance to be 8 days and that each condition will be in effect for 4 days. (The statistical test does not require an equal number of days for each condition.) On each of the 8 days, either condition A or condition B is in effect, until each is administered for 4 different days. Each day, observations of teacher and child performance are made, and they provide the data to evaluate the effects of the different conditions.

The prediction is that praise (Condition B) will lead to higher levels of attentive behavior than ordinary classroom practices (Condition A). Stated as a one-tailed (directional) hypothesis, Condition B is expected to lead to higher scores than Condition A. Under the null hypothesis, any difference between means for the two conditions is due solely to chance differences in performance on the occasions to which A and B conditions were randomly assigned. To determine whether the differences are sufficient to reject the null hypothesis, the mean level of performance is computed separately for each condition, and the difference between these means is derived.

Hypothetical data for the example appear in Table 9-2 (upper portion). The mean difference between A and B Conditions is 43.75, also shown in the table (lower portion). Whether this difference is statistically significant is determined by estimating the probability of obtaining scores this discrepant in the predicted direction when conditions have been assigned randomly to occasions.

TABLE 9-2. Percentage of Intervals of Attentive Behavior
Across Days and Treatments (Hypothetical Data)

				DAYS			
A	B	A	A	B	A	B	B
20	50	15	10	60	25	65	70

COMPARING TREATMENT MEANS

A	B
20	50
15	60
10	65
25	70

$\Sigma A = 70$ $\Sigma B = 245$
$\overline{X}_A = 17.50$ $\overline{X}_B = 61.25$

$\overline{X}_B > \overline{X}_A = 43.75$

The random assignment of conditions to occasions makes several combinations of the obtained data equally probable. Actually, 70 different combinations $(8!/4!4!)$ are possible. The question for computing statistical significance is: What proportion of the different combinations (of assigning conditions to occasions) would provide as large a difference between means as 43.75?[5]

A critical region of the sampling distribution is identified to evaluate the statistical significance of the obtained difference. The critical region is based on the level of confidence the investigator selects for the statistical test (e.g., α = .05) and the number of combinations of data possible. At the .05 level of confidence for the present example, the critical level would be $.05 \times 70$ (or the level of confidence times the number of possible combinations). The result would be 3.5. When a critical region is not an integer, selection of the larger whole number is recommended (Conover, 1971). In the present example, the larger whole number would be 4. With this critical region, the four combinations of the obtained data that are the least likely under the null hypothesis must be found. The least likely combination of data of course is one in which the A and B mean difference in the predicted direction is the greatest possible given the obtained scores. For the present example, the critical region consists of the four combinations of the obtained data allocated to A and B conditions that maximize the difference between the two means. The four data permutations that constitute the critical region are obtained by reallocating the obtained data to A and B conditions in such a way that the differences between conditions are the greatest in the predicted direction.

Table 9-3 presents permutations of the obtained data that reflect the four least likely combinations. The table was derived by first reallocating data points

TABLE 9-3. Critical Region for the Obtained Data from the Hypothetical Example

A				TOTAL FOR A OCCASIONS	$\overline{X}_A$	B				TOTAL FOR B OCCASIONS	$\overline{X}_B$	$\overline{X}_A > \overline{X}_B$
20	10	15	25	(70)	17.50	50	60	65	70	(245)	61.25	43.75
20	10	15	50	(95)	23.75	25	60	65	70	(220)	55.00	31.25
50	10	15	25	(100)	25.00	20	60	65	70	(215)	53.75	28.75
60	10	15	20	(105)	26.25	25	50	65	70	(210)	52.50	26.25

Note. All other combinations of the obtained data (allocated to A and B treatments) are not in the critical region using .05 as a level of significance for a one-tailed test.

to conditions that yielded the greatest difference between A and B, then the combination of data points that could show the next greatest difference, and so on. A total of four combinations was selected because this is the number of combinations that reflects the critical region for the .05 level of confidence. Thus the critical region consists of the n set of data combinations in the predicted direction that are the least likely to have occurred by chance (where n = the number of combinations that constitutes the critical region). The question for the randomization test is whether the difference between means obtained in the original data is equal to or greater than one of the mean differences included in the critical region. The obtained mean difference (43.75) equals the most extreme value in the critical region and hence is a statistically significant effect. The actual probability of the difference being this large, given random assignment of conditions to occasions, is $1/70$ or $p = .014$. When the data represent the least probable combination of data (given a one-tailed null hypothesis), the probability equals 1 divided by the total number of possible data combinations.

In the above example, a one-tailed test was performed. For a two-tailed test, the critical region is at both ends (tails) of the distribution. The number of data combinations that constitute the critical region is unchanged for a given level of confidence. However, the number of combinations is divided among the two tails. Because of the division of the critical region into two tails, the probability level of an obtained mean difference is doubled. Thus, if the above example utilized a two-tailed test, the probability level of the obtained difference would be $2/70$ or $p = .028$.

Considerations and limitations

Special Features. An advantage of randomization tests is that they do not rely on some of the assumptions of conventional tests such as random sampling of subjects from a population or normality of the population distribution. Also, serial dependency is not a problem that affects application of the tests. Dependency may exist in the data. Yet the test is based on the null hypothesis that there would be identical responses across occasions if the conditions were presented in a different order. Every order of presenting treatments should lead to an identical pattern of data (assuming the null hypothesis). Serial dependency does not affect the estimation of the sampling distribution of the statistic from which the inference of significance is drawn.

Computational Difficulties. An important issue regarding the use of randomization tests is the computation of the critical region. For a given confidence level, the investigator must compute the number of different ways in which the obtained scores could result from random assignment of conditions to occasions. When the number of occasions for assigning treatments exceeds 10 or 15, even obtaining the possible arrangements of the data by computer becomes monumental (Conover, 1971; Edgington, 1969). Thus, for most applications of randomization tests in single-case research, computation of the statistic in the manner described above may be prohibitive.

Fortunately convenient approximations of the randomization test are available that permit use of the test without the cumbersome computation of the critical region. The approximations depend on the same conditions as the randomization test does, namely, the random assignment of treatments to occasions. The approximations include the familiar t and F tests for two or more conditions, respectively. The t and F tests are identical in computation to conventional t and F, discussed earlier. Yet there is one important difference in the test itself. Serial dependency makes conventional t and F tests inappropriate. The use of t or F as an approximation of randomization tests avoids the problem of serial dependency. Because the treatments are assigned to occasions in a random order across all occasions, t and F provide a close approximation to the randomization distribution (Box & Tiao, 1965; Moses, 1952). Serial dependency does not interfere with this approximation.

For example, in the earlier example (Table 9-2), a t test for independent groups could be applied to approximate the randomization distribution where degrees of freedom is based on the number of A and B occasions ($df = n_1 + n_2 - 2$). The data yield a $t(6) = 8.17, p < .001$), which is less than the probability obtained with the exact analysis from the randomization test ($p = .014$). In cases in which the critical region is not easily computed, t and F can provide useful approximations if the conditions are randomly assigned to occasions in the design.

An alternative to the use of the t test is to approximate the randomization distribution with the Mann-Whitney U Test. To employ this test, the A and B data points are ranked from 1 to n (the number of treatment occasions) without reference to the treatment conditions from which each value is derived. The null hypothesis of no difference between treatments may be rejected if the ranks associated with one treatment tend to be larger than the values of the other treatment. The distribution from which this determination is made is available in published tables (Conover, 1971) and need not be computed for each set of data unless A plus B occasions are relatively large (e.g., over 20). The Mann-Whitney U is a convenient test that may be used in place of t and has been described in other sources (see Conover, 1971; Kirk, 1968).

Practical Restrictions. A few practical considerations influence the utility of randomization tests (Kazdin, 1980a; see also chapter 8). First, the use of the tests as described here requires that the subject's performance change rapidly (or reverse) across conditions. Thus, when conditions are changed from one day to the next (from A to B or B to A), performance must respond quickly to reflect treatment effects. Although rapid shifts in performance are often found when conditions are withdrawn or altered in applied research, this is not always the case. Without consistently rapid reversals in performance, differences between A and B conditions may not be detected. In situations where performance does not reverse, where there is a carryover effect from one condition to the next, or where attempting to reverse behavior is undesirable for clinical or ethical reasons, use of the randomization test may be limited.

A second and related issue involves the fact that it may not be feasible to allow different conditions such as baseline (A) and treatment (B) or multiple treatments (C, D, etc.) to vary on a daily basis. Such conditions cannot be implemented and shifted rapidly in applied settings to meet the requirements of the statistic. For example, a randomization test might be used to compare baseline (A) and token economy (B) conditions among patients on a psychiatric ward. Because of random assignment of conditions to days, the AB conditions will be alternated frequently to meet the requirements of the design. Yet to alternate conditions on a daily basis would be extremely difficult in most settings. One cannot easily implement an intervention such as a token economy for 1 or 2 days, remove it on the next, implement it again for 1 or 2 days, and so on, as dictated by the design.

There is a solution that overcomes this practical obstacle. Rather than alternation of conditions on a daily basis, a fixed block of time (e.g., 3 days or 1 week) could serve as the unit for alternating treatment. Whenever A is implemented, it would be in place for 3 consecutive days or a week; when B is assigned, the time period would be the same. The mean (or total) score for each period (rather than for each day) serves as the unit for computing the randomization test. The AB conditions are still assigned in a random order, but a given condition stays in effect whenever it is assigned for a period longer than one day. Thus the different conditions need not be shifted daily. Moreover, because of random assignment, a given condition is likely to be assigned for two or more consecutive occasions (periods). This would increase the length of the period in which a particular condition is in effect (e.g., 6 days if two consecutive 3-day periods of a particular condition are assigned). Thus the problem of rapidly shifting treatments would be partially ameliorated. If fixed blocks of several days rather than single days constitute the occasions, the mean score for a block as a whole is the datum used to compute the test. Because a block of days of a condition counts as only one occasion, several blocks will be required to achieve a relatively large number of occasions. A small number of occasions may restrict the possibility of obtaining statistically significant effects when treatments differ in their effects. Thus, when fixed blocks of several days are used to define the occasion, the number of days of the investigation will be longer than if individual days are used as the occasion. The practicality of extending the duration of time that defines an occasion needs to be weighed against the feasibility of extending the overall duration of the project.

In general, randomization tests provide a useful set of statistical techniques for single-case research. The availability of convenient (and familiar) approximations to the randomization distribution makes the tests more readily accessible to most users than such tests as time series analysis. The major problems delimiting use of the tests pertain to the need to assign conditions to occasions on a random basis and to show that treatment effects can be reversed rapidly as the conditions are changed.

9.7. THE R_n TEST OF RANKS

A test of ranks, referred to as R_n, has been proposed for evaluating data obtained in multiple baseline designs (Revusky, 1976; Wolery & Billingsley, 1982). The test requires that data be collected across several different baselines (e.g., different individuals, behaviors, or situations). Whether the intervention produces a statistically reliable effect is determined by evaluating the performance of each of the baselines at the point when the intervention is introduced. For example, in a multiple baseline design across individuals, the statistical comparison is completed by ranking scores of each subject at the point when the intervention is introduced for any one of the subjects. Each individual is considered a subexperiment. When Condition B is introduced for a subject, the performance of all subjects (including those for whom treatment is withheld) is ranked. The sum of the ranks across all subexperiments each time the treatment is introduced constitutes the statistic R_n.

An essential feature of the test is that the intervention is applied to different baselines in a random order. Thus the rationale underlying R_n follows that of randomization tests as outlined earlier. Because the baseline (e.g., person or behavior) that receives the intervention is determined randomly, the combination of ranks at the point of intervention for all subjects will be randomly distributed if the intervention has no effect. On the other hand, if the behavior of the client who receives the intervention changes at the point of intervention, compared with persons who have yet to receive the intervention, this should be reflected in the ranks. If each subject in turn shows a change when the intervention is introduced, this would be reflected in the sum of the ranks (or R_n) across all subjects, and it suggests that the ranks are not the likely result of random factors. R_n requires several different baselines or subexperiments to evaluate whether change at the point of treatment is reliable. At the .05 level of confidence the minimum requirement for detecting a statistically significant effect is four baselines (i.e., persons, behaviors, or situations).

Data analysis

Application of the R_n can be illustrated in a hypothetical example in which an intervention is applied to increase the amount of time that five aggressive children engage in appropriate and cooperative play during recess at school. To fulfill the requirements of the multiple baseline design, data are gathered for the target behaviors. For present purposes, assume that the data consist of the percentage of intervals (e.g., 30 sec) observed during recess in which the child engages in appropriate play. Treatment is introduced to different children at different points in time. The child who receives treatment first, second, and so on is always determined randomly.

Table 9-4 provides hypothetical data on the percentage of intervals of appropriate play across 10 days. As is evident in the table, baseline is in effect for everyone for 5 days. On the sixth day, one child is *randomly* selected to receive the intervention (B), whereas all other children continue under baseline (A) conditions. On successive days, a different child is exposed to the intervention. The ranking procedure is applied to each subexperiment at the point when the intervention is introduced. On each occasion that the intervention is introduced (which includes Days 6–10 in the example), the children are ranked. The lowest rank is given to the child who has the highest score (if a high score is in the desired direction).[6] In the example, on Days 6–10, the child with the highest amount of appropriate play at each point of intervention receives the rank of 1, the next highest the rank of 2, and so on. When the intervention is introduced to the first child, all children are ranked. When the intervention is introduced on subsequent occasions, all children except those who previously received the intervention are ranked. Even though all subjects are ranked when the intervention is introduced, not all ranks are used. R_n consists of the sum of the ranks for those subjects who receive the intervention at the point that the intervention is introduced. If treatment is ineffective, the ranks of these persons should be randomly distributed, i.e., include numbers ranging from 1 to the n number of baselines. If treatment is effective, the point of intervention should result in low ranks for each subject at that point (if low numbers are assigned to the most extreme score in the predicted direction of change).

TABLE 9-4. Percentage of Intervals of Appropriate Play
for Five Children Studied in a Multiple Baseline Design (Hypothetical Data)

					DAYS					
	1	2	3	4	5	6	7	8	9	10
1	45	30	35	50	40	30a	70b			
2	60	75	80	60	50	70a	50a	65a	80b	
3	20	20	25	10	30	80b				
4	55	60	40	45	50	40a	75a	90b		
5	30	25	20	30	20	30a	30a	40a	35a	50b

(Row label: Children)

Ranks = 1 2 1 1 1 $\Sigma R = 6$

Note. Days 1 through 5 served as baseline (a) days for all subjects and are unmarked.
a = control or baseline, b = experimental or intervention point for a child.

As is evident in Table 9-4, hypothetical data show that the child who receives the intervention at a given point in time, with the exception of Subject 1, receives the lowest rank (i.e., 1 or 1st place) for performance on that occasion. Summing the ranks for all children exposed to the intervention yields $R_n = 6$. The significance of the ranks for designs employing different numbers of subjects (or baselines) can be determined by examining Table 9-5. The table provides a one-tailed test for R_n. (A two-tailed test, of course, can be computed by doubling the probability level for the tabled columns.) To return to the above example, $R_n = 6$ for 5 subjects (one-tailed test) is equal to the tabled value required for the .05 level (see arrow). Thus the data in the hypothetical example permit rejection of the null hypothesis of no treatment effect.

Considerations and limitations

Rapidity of Behavior Change. In the above example, the rankings were assigned to the different baselines (children) at the point when the intervention was introduced (i.e., on the first day). However, it is quite possible, and indeed likely, that intervention effects would not be evident on the first day that the intervention was applied. With some interventions, slow and gradual improvements may be expected, or performance may even become slightly worse before becoming better. The statistic can still be used without necessarily applying the ranks on the first day of the intervention for each baseline.

The intervention can be evaluated on the basis of *mean* performance for a given person (behavior or situation) across several days rather than on the basis of a change in level (at the point of intervention) on the first day that the intervention is introduced. For example, the intervention could be introduced for one person and withheld from others for several days or a week. The

TABLE 9-5. Maximum values of R_n significant
at the indicated one-tailed probability levels when the
experimental scores tend to be smaller than the control scores.

NO. OF	SIGNIFICANCE LEVEL				
SUBJECTS	0.05	0.025	0.02	0.01	0.005
4	4				
5	6	5	5	5	
6	8	7	7	7	6
7	11	10	10	9	8
8	14	13	13	12	11
9	18	17	16	15	14
10	22	21	20	19	18
11	27	25	24	23	22
12	32	30	29	27	26

Note. Table provides significance for a one-tailed test. The number of subjects in the table also can be used to denote the number of responses or situations across which baseline data are gathered, depending on the variation of the multiple baseline design. (From Revusky, S. H. [1967]. Some statistical treatments compatible with individual organism methodology. *Journal of Experimental Analysis of Behavior*, **10**, 319–330. Copyright 1976 Society for the Experimental Analysis of Behvior, Inc. Reproduced by permission.)

rankings could be made on the basis of the mean performance across the entire week while the intervention was in effect. Mean performance of the target child would be compared with the mean of the other persons, and ranks would be assigned on the basis of each person's mean for that time period. Using means across days is likely to provide a more stable estimate of actual performance, to allow the intervention to operate on behavior, and consequently to reflect intervention effects more readily than evaluation based on the first day that the intervention is applied. Also, by using averages, the statistic takes into account the usual manner in which multiple baseline designs are conducted where the intervention is continued for several days for one person (baseline) before being introduced to the next person.[7]

If ranks are to be based on several days rather than a single day, additional considerations become important. *First*, the duration employed to evaluate treatment changes within subjects should be specified in advance. If intervention effects are expected to take a certain period of time, the precise number of days (or a conservative estimate) should be specified. The mean for that period is then used when the ranks are assigned. *Second*, the duration for introducing the treatment and for computing mean performance should be constant across all subjects. These two features ensure that randomness will not be influenced by *post hoc* treatment of the data and capitalization on chance fluctuations in performance.

Differences in Responses Across Baselines. If the scores across the different baselines vary markedly from each other in absolute magnitude, it may be difficult to reflect change using R_n. The scores may vary so much that when the intervention is introduced to one subject, and change occurs, the amount of change does not bring the person's score higher (or lower) than the level of another person who has continued in baseline conditions. The intervention may have led to change, but this is not reflected in the rankings because of discrepancies in the magnitude of scores across subjects.

For example, in Table 9-4, compare the hypothetical performance of Child 2 and Child 5. The performance of Child 2 was higher during baseline than was the performance of Child 5 when treatment was introduced. Had treatment been introduced to Child 5 before Child 2, the rank assigned to Child 5 would not have been as low as it was in the example. This would have been an artifact of the differences in absolute levels of performance of the subjects rather than of the ineffectiveness of the intervention. In general, the ranking procedure, as described thus far, does not take into account the differences in baseline magnitudes.

A simple data transformation can be used to ameliorate the problem of different response magnitudes. The transformation corrects for the different initial levels of baseline responding (Revusky, 1967). The formula for the transformation is

$$\frac{B_i - \overline{A}_i}{\overline{A}_i}$$

Where B_i = performance level for Subject i when the experimental intervention is introduced, and $\overline{A}_i$ = mean performance across all baseline days for the same subject.

Use of the transformation is the same as examination of the change in percentage of responding from baseline to treatment. The raw scores for each subject (i.e., for each baseline) are transformed when the intervention is introduced to any one subject. The ranks are computed on the basis of the transformed scores. In general, the transformation might be used routinely because of its simplicity and the likelihood that responses would have different magnitudes that could obscure the effects of treatment. Where response levels are widely discrepant during baseline, the transformation will be especially useful.

9.8. THE SPLIT-MIDDLE TECHNIQUE

The split-middle technique provides a method of describing the rate of behavior change over time for a single individual or group (White, 1971, 1972, 1974). The technique is designed to reveal a linear trend in the data, to characterize present performance, and to predict future performance. By describing the rate of behavior change, one can estimate the likelihood that the client's behavior will attain a particular goal. The technique permits examination of the trend or slope within phases and comparison of slopes across phases. Rate of behavior (frequency/time) has been advocated as the most useful measure for this method. The advantage of rate for purposes of plotting trends is that no upper limit exists. Theoretically at least there is no ceiling effect that can limit the slope of the trend. Yet the method can be applied to other performance measures than rate that are often used in applied research such as intervals, discrete categorization, and duration. Special charting paper has been advocated for the use of the split-middle techniques that allows graphing of performance in semilog units.[8] The special charting paper increases the linearity of the data, may enhance predictive validity, and is easily employed by practitioners (White, 1972, 1974). However, the split-middle technique can be used with ordinary graph paper with arithmetic (equal interval) units rather than log units on the ordinate.

The split-middle technique has been proposed primarily to *describe* the process of change within and across phases rather than to be used as an inferential statistical technique. The descriptive purposes are achieved by plotting trends within baseline and intervention phases to characterize client progress. Statistical significance can be examined once the trend lines have been determined.

Data description

The split-middle technique involves multiple steps. The technique begins with graphically plotting the data. From the data within a given phase, a trend, or *celeration line*, is constructed to characterize the rate of performance over time. (The term *celeration* derives from the notions of acceleration and deceleration if the trend is ascending or descending, respectively.) The celeration line predicts the direction and the rate of change.

To illustrate computation of the celeration line, consider hypothetical data plotted in Figure 9-4. (The example will utilize *rate* of performance and *semilog* units to illustrate recommended use of the method.) The data in the upper panel are from one phase of an A-B-A-B (or other) design plotted on a semilog chart. The manner in which the celeration line is computed will be conveyed with data from only one phase, although in practice celeration lines would be computed and plotted separately for each phase.

The first step for computing a celeration line in a phase is to divide the phase in half by drawing a vertical line at the *median number of sessions* (or days). The second step is to divide each of these halves in half again. (When there is an uneven number of days, the vertical line is drawn through the data point that is the median day rather than between two data points.) The dividing lines should always result in an equal number of points on each side

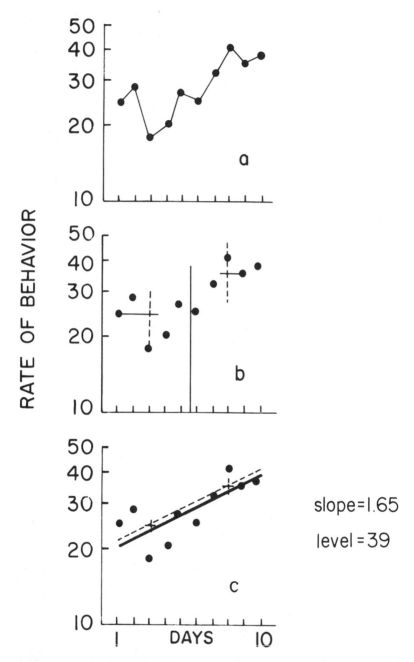

slope=1.65

level=39

FIGURE 9-4. Hypothetical data during one phase of an A-B-A-B design (*top panel—a*), with steps to determine the median data points in each half of the phase (*middle panel—b*), and with the original data (dashed) and adjusted (solid) celeration line (*bottom panel—c*).

of the division. The next step is to determine the *median rate of performance* for the first and second halves of the phase. This median refers to the data points that form the dependent measure rather than to the number of sessions.

Two potentially confusing points should be resolved. *First*, although the sessions are divided into quarters, only the first division (halves) is employed at this stage. *Second*, the median data value within each half of the sessions is selected. These medians are based on the ordinate (dependent variable values) rather than the abscissa (number of days). To obtain the data point that is the median within each half, one merely counts from the bottom (ordinate) up toward the top data point for each half. The data point that constitutes the median value within each half is selected. A horizontal line is drawn through the median at each half of the phase until the line intersects the vertical line dividing each half.

Figure 9-4b shows the above three steps, namely, a division of the data into quarters and the selection of median values within each half. Within each half of the data, a vertical and horizontal line intersect. The next step is finding the slope, which entails drawing a line connecting the points of intersection between the two halves.

The final step is to determine whether the line that results "splits" all of the data, in other words, is the *split-middle* line or slope. The split-middle slope is that line that is situated so that 50% of the data fall on or above the line and 50% fall on or below the line. The line is adjusted to divide the data in this fashion. In practice the line is moved up or down to the point at which all of the data are divided. The adjusted line remains parallel to the original line.

Figure 9-4c shows the original line (dotted) and the line (solid) after it has been adjusted to achieve the split-middle slope. Note that the original line did not divide the data so that an equal number of points fell above and below the line. The adjustment achieves this "middle" slope by altering the level of the line (and not the slope). (In some cases, the original line may not have to be adjusted.)

The celeration line reflects the rate of behavior change, which can also be expressed numerically. White (1974) has used the weekly rate of change as the basis of calculating rate, although any time period that might be more meaningful for a given situation can be employed. To calculate the rate of change, a point of the celeration line (Day_x) that passes through a given value on the ordinate is determined. The data value on the ordinate for the celeration line 7 days later (i.e., Day_{x+7}) is obtained. To compute the rate of change, the numerically larger value (either Day_x or Day_{x+7}) is divided by the smaller value.

The procedure can be applied to the data in Figure 9-4c. At Day 1, the celeration line is at 20. Seven days later, the line is at approximately 33. Applying the above computations, the ratio for the rate of change is 1.65.

Because the celeration line is accelerating, this indicates that the average rate of responding for a given week is 1.65 times greater than it was for the prior week. The ratio merely expresses the slope of the line.

The level of the slope can be expressed by noting the level of the celeration line on the last day of the phase. In the above example, the level is approximately 39. When separate phases are evaluated (e.g., baseline and intervention), the levels of the celeration lines refer to the last day of the first phase and the first day of the second phase, as will be discussed below.

For each phase in the experimental design, separate celeration lines are drawn. The slope of each line is expressed numerically. The change across phases is evaluated by comparing the levels and slopes. Consider hypothetical data for A and B phases, each with its separate celeration line, in Figure 9-5. To estimate the change in level, a comparison is made between the last data point in baseline (approximately 22) and the first data point during the intervention (approximately 28). The larger value is divided by the smaller value, yielding a ratio of 1.27. The ratio merely expresses how much higher (or lower) the intersection of the different celeration lines is. Similarly, for a change in slope, the larger slope is divided by the smaller slope, yielding a value in the example of 1.52. The change in level and slope summarizes the differences in performance across phases.

Statistical analysis

It should be reiterated that the split-middle procedure has been advocated as a technique to describe the process of change in an individual's behavior rather than as a tool to assess statistical significance. However, statistical significance of change across phases can be evaluated once the celeration lines have been calculated.

To determine whether there is a statistically significant change in behavior across phases, a simple statistical test has been proposed (White, 1972). Again, consider change across A and B phases in an A-B-A-B design. The null hypothesis upon which the test is based is that there is no change in performance across A and B phases. If this hypothesis is true, then the celeration line of the baseline phase should be a valid estimate of the celeration line of the intervention phase. Assuming the intervention had no effect, the split-middle slope of baseline should be the split-middle slope of the intervention phase, as well. Thus 50% of the data in the intervention or B phase should fall on or above and 50% of the data should fall on or below the slope of baseline when that slope is *projected* into the intervention phase.

To complete the statistical test, the slope of the baseline phase is extended or projected through the intervention phase. Consider the example of hypothetical data in Figure 9-5, which shows the celeration line computed and

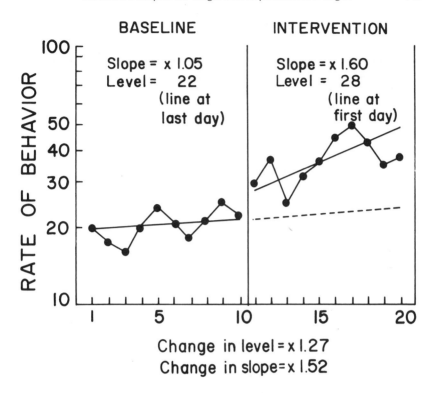

FIGURE 9-5. Hypothetical data across baseline (A) and intervention (B) phases, with separate celeration lines for each phase (solid lines). The dashed line represents an extension of the celeration line for the baseline phase.

extended from baseline into the intervention phase. For purposes of the statistical test, it is assumed that the probability of a data point during the intervention phase falling above the projected celeration line of baseline is 50% (i.e., $p = .5$), given the null hypothesis of no change across phases. A binomial test can be used to determine if the number of data points that are above the projected slope in the intervention phase is of a sufficiently low probability to reject the null hypothesis.[9]

Using this procedure for the data in Figure 9-5, 10 of 10 data points during the intervention phase fall above the projected slope of baseline. Applying the binomial test to determine the probability of obtaining all 10 data points above the slope, $p = \binom{10}{10}\frac{1}{2}^{10}$ yields a $p < .001$. Thus the null hypothesis can be rejected; the data in the intervention phase are significantly different from the data of the baseline phase. The results do not convey whether the level and/or slope account for the differences but only that the data overall depart from one phase to another.

318   Single-case Experimental Design

Considerations and limitations

Utility of the Test. The primary purpose of the split-middle technique is to describe the data in a summary fashion and to predict the outcome given the rate of change. The utility of the test is that it provides a computationally simple technique for characterizing data and for examining if trends change across phases. In the usual case of data presentation in single-case research, summary statistics are often restricted to describing mean changes across phases (see Kazdin, 1982b). The split-middle technique can provide additional descriptive information on the level, slope, and changes in these characteristics over time (see Wolery & Billingsley, 1982).

Since a major purpose of the technique is to predict behavior rather than to determine statistical significance of change, it is appropriate to examine the extent to which this purpose is adequately achieved. White (1974) presented data based upon "several thousand" analyses of classroom performance. The analyses determined the accuracy of predicting behavior using the split-middle procedure at different points in the future. As might be expected, the extent to which the predictions approximated the actual data depended upon the number of data points upon which the prediction was based and upon the amount of time into the future that was predicted. For example, on the basis of 7 days of data, performance one week into the future would be successfully predicted (with a narrow margin of error) 64% of the time; for performance 3 weeks into the future, predictions were successful 50% of the time. With 11 days of data, predictions one week into the future were successful 89% of the time; for performance 3 weeks into the future, predictions were successful 81% of the time.

The predictive uses of the split-middle technique have been accorded important applied significance. If the data suggest that behavior is not changing at a sufficient rate to obtain a particular goal, the intervention can be altered. Thus the technique may provide useful information that leads the investigator to change the intervention as needed.

Statistical Inferences. Several different tests have been proposed to assess change based on information obtained from plotting slope and level (see White, 1972; Wolery & Billingsley, 1982). Most of these tests also rely on the binomial as illustrated above. As E. S. Edgington (personal communication, August, 1974) has noted, the binomial may not be valid when applied to data that show a trend during baseline. Consider the following circumstances in which the binomial might lead to misinterpretation. A random set of numbers could be assigned randomly as data points to baseline and intervention phases. On the basis of chance alone, baseline occasionally would show an accelerating or decelerating slope. If the data points in the A phase show a slope, it is unlikely that the data points in the B phase will show the same slope. The randomness of the process of assigning data points to phases would make identical trends possible but very unlikely. Hence if there is an initial trend in baseline, it is quite possible that data in the intervention phase on the basis of chance alone would fall above or below the projected slope of baseline. The binomial test might show a statistically significant effect even though the numbers were assigned randomly and no intervention was implemented. Thus problems may exist in drawing inferences using the binomial test when trend is evident in baseline (or the condition from which a projected celeration line is made).

The split-middle technique has been infrequently reported in published investigations as either a descriptive or an inferential procedure. Thus important questions about the statistical techniques and the problems they may introduce remain to be elaborated. The conditions in which the binomial test represents the probability of the distribution of data points across phases, given the null hypothesis, are not well explored. Nevertheless, as a descriptive tool, the split-middle technique provides important information about level and slope changes that is usually not reported.

9.9. EVALUATION OF STATISTICAL TESTS: GENERAL ISSUES

Single-case designs provide a wide array of options for the applied researcher. Statistical techniques available for such designs are numerous. Selected tests were reviewed to convey the breadth of options available. Additional variations of these analyses, as well as different tests, have also been described (e.g., Edgington, 1982; Tryon, 1982).

Some of the analyses discussed have wider applicability than others. Single-case designs generally involve a comparison of two or more phases. This one characteristic raises the possibility of time series, split-middle, randomization, and t tests. The options were illustrated and discussed in the context of A-B-A-B and multiple baseline designs, but they can also be applied to other designs such as the changing-criterion designs, and alternating or simultaneous treatment designs.[10] Despite the flexibility of various tests, several considerations and sources of caution warrant mention.

First, statistical evaluation of single-case (or any other) data only addresses the issue of whether the change is statistically significant over the course of separate conditions. When statistical significance is obtained, this does not of course provide any necessary clues about the basis for a change in behavior. Conclusions about the basis for the change derive from the experimental design rather than from the mere demonstration of statistical significance. Thus statistical evaluation of an A-B design does not elevate the sophistication of the comparison. Drawing conclusions between the effect of an intervention and behavior assumes an adequate design independent of the techniques to evaluate the data.

Second, the analyses outlined above only addresses the statistical significance and not the clinical significance of the changes. Although rules of science have depended upon levels of confidence as a criterion to decide veridical effects, no leap is warranted from levels of confidence to the applied value of the finding. Clinical significance, as noted earlier, refers to the importance of the change and entails different criteria from those invoked for statistical analyses.

Clinical significance is usually viewed as a more stringent criterion than statistical significance because many statistically reliable effects can be obtained without clear or detectable impact on everyday client functioning. It is generally true that, with clinically significant effects, behavior change is especially marked and hence typically statistically significant. There are also cases, however, where clinically significant effects might be evident where statistical tests might not be applicable and or where statistical significance is not clear. For example, for clinical cases where complete amelioration of the problem is achieved in one trial (e.g., Creer, Chai, & Hoffman, 1977), statistical significance would be difficult if not impossible to demonstrate with conventional techniques. The main point is that statistical and clinical significance need to be kept distinct in applied research. A statistically significant difference obtained in applied single-case research may lead the investigator to conclude that the intervention was effective. In this context, *effective* refers to effective in producing a statistically reliable change and not necessarily effective in ameliorating the clinical problem to which the intervention was applied.

Finally, the statistical techniques mentioned above invoke special conditions that may limit their use in many applied investigations. For example, a randomization test of means and R_n require assigning conditions randomly (to occasions or baselines). Yet it is easy to consider many situations in hospitals, classrooms, or institutional settings where this requirement could not be invoked. Different sorts of problems are raised with other statistical tests. For example, protracted baseline phases are difficult to justify but could be essential in order to apply such tests of time series analyses.

An important characteristic of single-case designs is that they are quite flexible. Design changes are made in part as a function of the client's responses to alternative interventions. This is unlike between-group studies, where designs are usually worked out well in advance and subjects are run in a predetermined fashion. There are important implications for the applicability of statistical tests to these different design practices. The statistical analyses reviewed earlier often entail conditions that must be planned in advance of the study. Insofar as these conditions restrict the flexibility of the investigator, their application in any given case may present problems. Experimental design considerations already constrain clinical applications in some instances because of temporary suspensions of treatment (reversal phases) or delays in introducing treatment (multiple baseline designs). Statistical analyses need to be considered carefully in advance because they may place additional restrictions on the manner in which treatment is implemented.

Statistical analyses should not be viewed as practical obstacles for the investigator. The tests can assist and overcome many problems of evaluation. For example, when ideal conditions for data evaluation through visual inspection are not obtained, descriptive and inferential statistics may greatly facilitate interpretation of outcome. A prime example would be where there is initial trend in baseline. An investigator ordinarily might hope and wait for an asymptote to be reached to facilitate subsequent evaluation of intervention effects. Yet alternative statistical analyses such as time series analyses and split-middle techniques can be quite helpful because they examine intervention effects in light of prior trends in the data. Thus statistical techniques can also make important practical contributions to applied research.

9.10. CONCLUSIONS

The present chapter has discussed specific statistical tests for single-case experimental designs and considerations dictated by their use. The availability of multiple statistics provides the investigator with diverse options for the single-case. A few salient considerations underlying all of the tests warrant reiteration. To begin with, the appropriateness of utilizing statistical criteria for the evaluation of applied behavioral interventions remains a major source of controversy. Statistical analysis is seen by many proponents of single-case research as a violation of the rationale for conducting research with the individual subject. Thus whether statistical tests should be used to draw inferences from single-case research remains an issue.

On this issue, it is important to distinguish experimental designs (e.g., single-case and between-group designs), methods of data evaluation (e.g., visual inspection and statistical analyses), and types of research (e.g., basic or applied). There are no necessary connections between particular types of research, designs, and analyses. Thus use of statistical analyses does not necessarily conflict with single-case designs or their purposes. When research attempts to develop a technology of behavior change and to achieve clinically important effects, statistical analyses will definitely be of limited value. Small effects that pass beyond a threshold of traditional levels of confidence may not address the priorities of applied research. Yet there are several uses of statistics, detailed earlier, that may contribute to the goals of applied research.

Another issue important to mention is that the use of statistical tests may have implications for the manner in which a particular intervention needs to be implemented. For example, the random assignment of treatment to occasions or subjects may compete with clinical priorities. Exigencies of clinical settings may delimit the applicability of diverse procedures upon which various statistical tests depend. Yet in many situations, there is flexibility in deciding the research design. Awareness of statistical tests on the part of the investigator may lead to different arrangements of the intervention that do not impact on clinical care. In some cases, the investigator may have other options for data evaluation in addition to visual inspection.

Statistical analyses for single-case research have been used relatively infrequently. Their use is likely to increase, albeit slowly, for different reasons. Concerns over the interjudge reliability of visual inspection and increased dissemination of statistical analyses for single-case designs and the computer programs for their execution are two influences pointing in the direction of increased utilization. Interventions are applied in increasingly diverse settings, and experimental control over factors that minimize variability is more difficult to obtain. Statistical analyses may be helpful in evaluating interventions where data requirements for visual inspection are not readily obtained. The present chapter illustrated several options for statistical analyses and the problems attendant upon their use.

NOTES

1. As the lag increases, the correlation becomes somewhat less stable, in part, because of the decrease in the number of pairs of observations upon which the coefficient can be based (Holtzman, 1963).

2. Although the statistical significance of autocorrelations can be approximated by testing them as correlations in the usual manner, Anderson (1942) has provided tables for the exact test. (See also Anderson, 1971, and Ezekiel & Fox, 1959.)

3. Baer (1977a) has articulately stated the similarities and differences in the rationales underlying statistical analysis and visual inspection. Both methods of data evaluation attempt to avoid Type I and Type II error. Type I error refers to concluding that the intervention produced a veridical effect when in fact the results are attributed to chance. Type II errors refers to concluding that the intervention did not produce a veridical effect when in fact it did. Typically, researchers give a higher priority to avoiding a Type I error. In statistical analyses, the probability of committing a Type I error is specified (by the level of confidence of the statistical test or α). With visual inspection, the probability of a Type I error is not known. Hence, to avoid chance effects, the investigator searches for highly consistent effects that can be readily seen. By minimizing the probability of a Type I error, researchers increase the probability of making a Type II error. Investigators who rely on visual inspection are more likely to commit Type II errors than investigators who rely on statistical analyses. Thus reliance on visual inspection

will tend to overlook and discount many reliable but weak effects. From the standpoint of developing an effective applied technology of behavior change, Baer (1977a) has argued persuasively that minimizing Type I errors leads to identification of a few variables whose effects are consistent and potent across a wide range of conditions. Thus visual inspection may be suited for the special goals of applied research. For other research purposes (e.g., testing of alternative theories), weak but reliable effects may be important to detect, and the priorities of erring in one direction rather than another might change.

4. The randomization test discussed and illustrated here is one of many available tests (see Edgington, 1969, 1984). The specific one selected, which compares means from different conditions, is likely to be of special interest in single-case experiments where performance is compared across phases.

5. The example selected here is devised for computational simplicity. It is unlikely that an investigator would be interested in only eight occasions for evaluating two different phases (baseline and intervention). In addition, it is also unlikely that the nonoverlapping distributions of the magnitude included in the example would be subjected to a statistical test.

6. As a general guideline, ranks are assigned so that the lowest number is given to the baseline that shows the highest level of performance in the desired direction. An easy rule of thumb is to assign "first place" (a rank of 1) to the highest or lowest score that represents the "best" performance in terms of the dependent measure. Thus 1 might be assigned to the highest performance of social skills or the lowest performance of self-abusive behavior. Second, third, and subsequent ranks are assigned accordingly for lower scores in the therapeutic direction.

7. In addition to the use of R_n to evaluate changes in means, a recent extension has illustrated evaluation of changes in trends combining R_n and split-middle techniques (see Wolery & Billingsley, 1982).

8. The semilog units refer to the fact that the scale on the ordinate is logarithmic but the scale on the abscissa is not. The effect of this arrangement is to ensure that there is no zero origin on the graph and that low and high rates of performance can be readily represented. The chart can be used for behaviors with extremely high or low rates. Rates of behavior can vary from .0006944 per minute (i.e., one every 24 hours) to 1000 per minute. (The semilog chart paper has been developed by Behavior Research Company, Kansas City, KS.) Adoption of the charting procedure has not been widespread in applied research. Hence it is useful to note that the split-middle technique can be used with ordinary graph paper.

9. The binomial applied to the split-middle slope test would be the probability of attaining x data points above the projected slope:

$$f(x) = {n \atop x} \; p^x q^{n-x} \text{ (or simply } {n \atop x} \; p^n),$$

Where n = the number of total data points in Phase B

$x =$ the number of data points above (or below) the projected slope

$p = q = .5$ by definition of the split-middle slope

p and $q =$ the probability of data points appearing above or below the slope given the null hypothesis

10. Other design options may raise special issues for statistical tests. For example, in a changing criterion design, the intervention may be introduced in such a way that only gradual and small changes in behavior are sought. Obviously, one might not wish to test for changes in level in such instances, because abrupt changes at the point of introducing the intervention might not be expected. In an alternating- or simultaneous-treatment design of special interest, it is not the change from one phase to another but rather whether separate interventions implemented in the same phase differ significantly. Analyses discussed previously can be adopted to these circumstances (e.g., see Edgington, 1982; Kratochwill & Levin, 1980).

CHAPTER 10

Beyond the Individual: Replication Procedures

10.1 INTRODUCTION

Replication is at the heart of any science. In all sciences, replication serves at least two purposes: first, to establish the reliability of previous findings; and, second, to determine the generality of these findings under differing conditions. These goals, of course, are intrinsically interrelated. Each time that certain results are replicated under different conditions, this not only establishes generality of findings, but also increases confidence in the reliability of these findings. The emphasis of this chapter, however, is on replication procedures for establishing generality of findings.

In chapter 2 the difficulties of establishing generality of findings in applied research were reviewed and discussed. The problem in generalizing from a heterogeneous group to an individual limits generality of findings from this approach. The problem in generalizing from one individual to other individuals who may differ in many ways limits generality of findings from a single-case. One answer to this problem is the replication of single-case experiments. Through this procedure, the applied researcher can maintain his or her focus on the individual, but establish generality of findings for those who differ from the individual in the original experiment. Sidman (1960) has outlined two procedures for replicating single case experiments in basic research: direct replication and systematic replication. In applied research a third type of replication, which we term *clinical replication*, is assuming increasing importance.

The purpose of this chapter is to outline the procedures and goals of replication strategies in applied research. Examples of each type of replication series will be presented and criticized. Guidelines for the proper use of these procedures in future series will be suggested from current examples judged to be successful in establishing generality of findings. Finally, the feasibility of large-scale replication series will be discussed in light of the practical limitations inherent in applied research.

10.2 DIRECT REPLICATION

Direct replication of single-case experiments have often appeared in professional journals. As noted above, these series are capable of determining both reliability of findings and generality of findings across clients. In most cases, however, the very important issue of generality of findings has not been discussed. Indeed, it seems that most investigators employing single-case methodology, as well as editors of journals who judge the adequacy of such endeavors, have been concerned primarily with reliability of findings as a goal in replication series rather than generality of findings. That is, most investigators have been concerned with demonstrating that certain results can or cannot be replicated in subsequent experiments rather than with systematically observing the replications themselves to determine generality of findings. However, since any attempt to establish reliability of a finding by replicating the experiment on additional cases also provides information on generality, many applied researchers have conducted direct replication series yielding valuable information on client generality. Examples of several of these series will be presented below.

Definition of direct replication

For our purposes, we agree basically with Sidman's (1960) definition of direct replication as "... replication of a given experiment by the same investigator" (p. 73). Sidman divided direct replication into two different procedures: repetition of the experiment on the same subject and repetition on different subjects. While repetition on the same subject increases confidence in the reliability of findings and is used occasionally in applied research (see chapter 5), generality of findings across clients can be ascertained only by replication on different subjects. More specifically, direct replication in applied research refers to administration of a given procedure by the same investigator or group of investigators in a specific setting (e.g., hospital, clinic, or classroom) on a series of clients homogeneous for a particular behavior disorder (e.g., agoraphobia, compulsive hand washing). While it is recognized that, in applied research, clients will always be more heterogeneous on background variables such as age, sex, or presence of additional maladaptive behaviors than in basic research, the conservative approach is to match clients in a replication series as closely as possible on these additional variables. Interpretation of mixed results, where some clients benefit from the procedure and some do not, can then be attributed to as few differences as possible, thereby providing a clearer direction for further experimentation. This point will be discussed more fully below.

Direct replication as we define it can begin to answer questions about generality of findings across clients but cannot address questions concerning generality of findings across therapists or settings. Furthermore, to the extent that clients are homogeneous on a given behavior disorder (such as agoraphobia), a direct replication series cannot answer questions on the results of a given procedure on related behavior disorders such as claustrophobia, although successful results should certainly lead to further replication on related behavior disorders. A close examination of several direct replication series will serve to illustrate the information available concerning generality of findings across clients.

Example one: Two successful replications

The first example concerns one successful experiment and two successful replications of a therapeutic procedure. This early clinical series examined the effects of social reinforcement (praise) on severe agoraphobic behavior in three patients (Agras et al., 1968). This series was also one of the first evaluations of direct-exposure-based treatments for phobia that have become the treatment of choice today (Mavissakalian & Barlow, 1981b). This procedure has also come to be known as reinforced practice (Leitenberg, 1976) and self-observation therapy (Emmelkamp, 1982). The procedure was straightforward.

All patients were hospitalized. Severity of agoraphobic behavior was measured by observing the distance the patients were able to walk on a course from the hospital to a downtown area. Landmarks were identified at 25-yard intervals for over one mile. The patients were asked two or more times a day to walk as far as they could on the course without feeling "undue tension." Their report of distance walked was surreptitiously checked from time to time by an observer to determine reliability, precise feedback of progress in terms of increases in distance was provided, and this progress was socially reinforced with praise and approval during treatment phases and ignored during withdrawal phases. In the first patient, increases in time spent away from the center were praised first, but as this resulted in the patient simply standing outside the front door of the hospital for longer periods, the target behavior was changed to distance. Because baseline procedures were abbreviated, this design is best characterized as a B-A-B design (see chapter 5). The comparison, then, is between treatment (praise) and no treatment (no praise).

For purposes of generality across clients, it is important to note that the patients in this experiment were rather heterogeneous, as is typically the case in applied research. Although each patient was severely agoraphobic, all had numerous associated fears and obsessions. The extent and severity of agoraphobic fears differed. One subject was a 36-year-old male with a 15-year agoraphobic history. He was incapacitated to the extent that he could manage a 5-minute drive to work in a rural area only with great difficulty. A second subject was a 23-year-old female with only a one-year agoraphobic history. This patient, however, could not leave her home unaccompanied. The third subject, a 36-year-old female, also could not leave her home unaccompanied, but had a 16-year agoraphobic history. In fact, this patient had to be sedated and brought to the hospital in an ambulance. In addition, these 3 patients presented different background variables such as personality characteristics and cultural variations (one patient was European).

The results from one of the cases (the male) are presented in Figure 10-1. Reinforcement produced a marked increase in distance walked, and withdrawal of reinforcement resulted in a deterioration in performance. Reintroduction of reinforcement in the final phase produced a further increase in distance walked. These results were replicated on the remaining 2 patients.

At least three conclusions can be drawn from these data. The first conclusion is that the treatment was effective in modifying agoraphobic behavior. The second conclusion is that within the limits of these data, the results are reliable and not due to idiosyncracies present in the first experiment, since two replications of the first experiment were successful. The third conclusion, however, is of most interest here. The procedure was clearly effective with 3 patients of different ages, sex, duration of agoraphobic behavior, and cultural backgrounds. For purposes of generality of findings, this series of experiments would be strengthened by a third replication (a total of 4 subjects). But the consistency of the results across 3 quite different patients enables one to draw initially favorable conclusions on the general effectiveness of this procedure across the population of agoraphobic clients through the process of logical generalization (Edgington, 1967).

On the other hand, if one client had failed to improve or improved only slightly such that the result was clinically unimportant, an immediate search would have had to be made for procedural or other variables responsible for the lack of generality across clients. Given the flexibility of this experimental design, alterations in procedure (e.g., adding additional reinforcers, changing the criterion for reinforcement) could be made in an attempt to achieve clinically important results. If mixed results such as these were observed, further replication would be necessary to determine which procedures were most efficacious for given clients (see section 2.2, chapter 2).

In this series, however, these steps were not necessary due to the uniformly successful outcomes, and some preliminary statements about client generality were made. The next step in this series, then, would be an attempt to replicate the results systematically, that is, across different situations and therapists. It

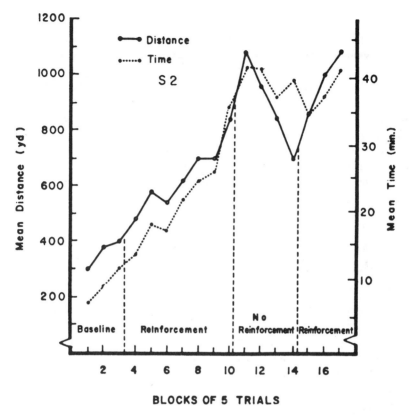

FIGURE 10-1. The effects of reinforcement and nonreinforcement upon the performance of an agoraphobic patient (Subject 2). (Figure 2, p. 425, from: Agras, W. S., Leitenberg, H., and Barlow, D. H. [1968]. Social reinforcement in the modification of agoraphobia. *Archives of General Psychiatry*, **19**, 423–427. Copyright 1968 by American Medical Association. Reproduced by permission.)

is evident that the preliminary series, which was carried out in Burlington, Vermont, does not address questions on effectiveness of techniques in different settings or with different therapists. It is entirely possible that characteristics of the therapist or the particular structure of the course that the agoraphobic walked facilitated the favorable results. Thus these variables must be systematically varied to determine generality of findings across all important clinical domains. In fact, this step was taken many times. Using procedures that were operationally quite similar to those described above, but carrying different labels, Marks (1972) successfully treated a variety of severe agoraphobics in an urban European setting (London) using, of course, different therapists, and Emmelkamp (1974, 1982) treated a long series of Dutch agoraphobics.

In fact, further experimentation over a period of 10 years revealed that while this intervention was repeatedly successful with thousands of cases, reinforcement, feedback, and other techniques served primarily to motivate practice with or exposure to feared objects or situations and that this was the primary therapeutic ingredient (see Mavissakalian and Barlow, 1981b, for a review). One strong cue was the rising baseline in Figure 10-1 where agoraphobics' behavior was improving with practice or exposure alone. Ideally, of course, reinforcement should not have been introduced until the baseline stabilized (see section 3, chapter 3). When this was tested properly in subsequent single-case experimentation, the power of pure exposure, even in the absence of external motivating variables such as praise, was demonstrated (Leitenberg, Agras, Edwards, Thomson, & Wincze, 1970). But the purpose of these illustrations is to examine the process of establishing generality of findings through replication and it is to this topic that we now return.

Example two: Four successful replications
with design alterations during replications

A second rather early example of a direct replication series will be presented because the behavior is clinically important (compulsive rituals), and the issue of client generality within a direct replication series is highlighted because 5 patients participated in the study (Mills, Agras, Barlow, & Mills, 1973). In this experiment, what was a new treatment at the time—response prevention—was tested. The basic strategy in this experiment and its replications was an A-B-A design: baseline, response prevention, baseline. During replications, however, the design was expanded somewhat to include controls for instructional and placebo effects. For example, two of the replications were carried out in an A-B-BC-B-A design, where A was baseline, B was a placebo treatment, and C was response prevention.

The addition of new control phases during subsequent replication is not an uncommon strategy in single-case design research because each replication is actually a separate experiment that stands alone. When testing a given treatment, however, new variables interacting within the treatment complex that might be responsible for improvement may be identified and "teased out" in later replications. It was noted in chapter 2 that such flexibility of single-case designs allows one to alter experimental procedures *within* a case. Within the context of replication, if a procedure is effective in the first experiment, one has the flexibility to add further, more stringent controls during replication to ascertain more specifically the mechanism of action of a successful treatment. But, to remain a direct replication series within our definition, the major purpose of the series should be to test the effectiveness of a given treatment on a well-defined problem—in this case compulsive rituals—administered by the same therapeutic team in the same setting. Thus the treatment, if successful, must remain the same, and the comparison is between treatment and no treatment or treatment and placebo control.

The first 4 subjects in this experiment were severe compulsive hand washers. The fifth subject presented with a different ritual. All patients were hospitalized on a research unit. All hand washers encountered articles or situations throughout the experiment that produced hand washing. Response prevention consisted of removing the handles from the wash basin wherein all hand washing occurred. The placebo phase consisted of saline injections and oral placebo medication with instructions suggesting improvement in the rituals, but no response prevention. Once again, the design was either A-B-A, with A representing baseline and B representing response prevention, or A-B-BC-B-A, where A was baseline, B was placebo, and C was response prevention. Both self-report measures (number of urges to wash hands) and an objective measure (occasions when the patient approached the sink, recorded by a washing pen—see chapter 4) were administered.

As in the previous series, the patients were relatively heterogeneous. The first subject was a 31-year-old woman with a 2-year history of compulsive hand washing. Previous to the experiment, she had received over one year of both inpatient and outpatient treatment including chemotherapy, individual psychotherapy, and desensitization. She performed her ritual 10 to 20 times a day, each ritual consisting of eight individual washings and rinsings with alternating hot and cold water. The associated fear was contamination of herself and others through contact with chemicals and dirt. These rituals prevented her from carrying out simple household duties or caring for her child.

The second subject was a 32-year-old woman with a 5-year history of hand washing. Frequency of hand washing ranged from 30 to 60 times per day, with an average of 39 during baseline. Unlike with the previous subject, these rituals had strong religious overtones concerning salvation, although fear of contamination from dirt was also present. Prior treatments included two series of electric shock treatment, which proved ineffective.

A third subject was a 25-year-old woman who had a 3-year history of the hand-washing compulsion. Situations that produced the hand washing in this case were associated with illness and death. If an ambulance passed near her home, she engaged in cleansing rituals. Hand washings averaged 30 per day, and the subject was essentially isolated in her home before treatment.

The fourth subject was a 20-year-old male with a history of hand washing for 1½ years. He had been hospitalized for the previous year and was hand washing at the rate of 20 to 30 times per day. The fifth subject, whose rituals differed considerably from the first 4 subjects, will be described below.

Representative results from one case are presented below. Hand washing remained high during baseline and placebo phases and dropped markedly after response prevention. Subjective reports of urges to wash declined slightly during response prevention and continued into follow-up. This decline continued beyond the data presented in Figure 10-2 until urges were minimal. These results were essentially replicated in the remaining three hand washers.

Before discussion of issues relative to replication, experimental design considerations in this series deserve comment. The dramatic success of response prevention in this series is obvious, but the continued reduction of hand washing after response prevention was removed presents some problems in interpretation. Since hand washing did not recover, it is difficult to attribute its reduction to response prevention using the basic A-B-A with-

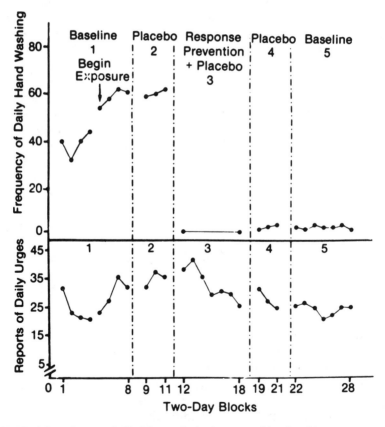

FIGURE 10-2. In the upper half of the graph, the frequency of hand washing across treatment phases is represented. Each point represents the average of 2 days. In the lower portion of the graph, total urges reported by the patient are represented. (Figure 3, p. 527, from: Mills, H. L., Agras, W. S., Barlow, D. H., and Mills, J. R. [1973]. Compulsive rituals treated by response prevention: An experimental analysis. *Archives of General Psychiatry*, **28**, 524–529. Copyright 1973 by American Medical Association. Reproduced by permission.)

drawal design. From the perspective of this design, it is possible that some correlated event occurred concurrent with response prevention that was actually responsible for the gains. Fortunately, the aforementioned flexibility in adding new control phases to replication experiments afforded an experimental analysis from a different perspective. In all patients, hand washing was reasonably stable by history and through both baseline and placebo phases. Hand washing showed a marked reduction *only* when response prevention was introduced. In these cases, baseline and placebo phases were administered for differing amounts of time. In fact, then, this becomes a multiple baseline across subjects (see chapter 7), allowing isolation of response prevention as the active treatment.

Again, this series demonstrates that response prevention works, and replications ensure that this finding is reliable. In addition, the clinical significance of the result is easily observable by inspection, since rituals were entirely elminated in all 4 patients. More importantly, however, the fact that this clinical result was consistently present across 4 patients lends considerable confidence to the notion that this procedure would be effective with other patients, again through the process of logical generalization. It is common sense that confidence in generality of findings across clients increases with each replication, but it is our rule of thumb that a point of diminishing returns is reached after one successful experiment and three successful replications for a total of 4 subjects. At this point, it seems efficient to publish the results so that systematic replication may begin in other settings.

An alternative strategy would be to administer the procedure in the same setting to clients with behavior disorders demonstrating marked differences from those of the first series. Some behavior disorders such as simple phobias lend themselves to this method of replication since a given treatment (e.g., *in vitro* exposure) should theoretically work on many different varieties of simple phobia. Within a disorder such as compulsive rituals, this is also feasible because several different types of rituals are encountered in the clinic (Mavissakalian & Barlow, 1981a; Rachman & Hodgson, 1980). The question that can be answered in the original setting then is: Will the procedure work on other behavior disorders that are topographically different but presumably maintained by similar psychological processes? In other words, would rituals quite different from hand washing respond to the same procedure? The fifth case in this series was the beginning of a replication along these lines.

The fifth subject was a 15-year-old boy who performed a complex set of rituals when retiring at night and another set of rituals when arising in the morning. The night rituals included checking and rechecking the pillow placement and folding and refolding pajamas. The morning rituals were concerned mostly with dressing. This type of ritual has come to be known as *checking* as opposed to previous *washing* rituals. The rituals were extremely time consuming and disruptive to the family's routine. After a baseline phase in which rituals remained relatively stable, the night rituals were prevented, but the morning rituals were allowed to continue. Here again, response prevention dramatically eliminated nighttime rituals. Morning rituals gradually decreased to zero during prevention of night rituals.

The experiment further suggests that response prevention can be effective in the treatment of ritualistic behavior. The implications of this replication, however, are somewhat different from the previous three replications, where the behavior in question was topographically similar. Although the treatment was administered by the same therapists in the same setting, this case does *not* represent a direct replication because the behavior was topographically different. To consider this case as part of a direct replication series, one would have to accept, on an *a priori* basis, the theoretical notion that all compulsive rituals are maintained by similar psychological processes and therefore will respond to the same treatment. Although classification of these under one name (compulsive rituals) implies this, in fact there is some evidence that these rituals are somewhat different and may react differently to response prevention treatments (Rachman & Hodgson, 1980). As such, it was probably inappropriate to include the fifth case in the present series because the clear implication is that response prevention is applicable to all rituals, but only one case was presented where rituals differed.

From the perspective of sound replication procedures, the proper tactic would be to include this case in a second series containing different rituals. This second series would then be the first step in a systematic replication series, in that generality of findings across different behaviors would be established in addition to generality of findings across clients. In fact, response prevention and exposure, combined occasionally with medication, has become the treatment of choice for obsessive-compulsive disorders, based on an extended systematic and clinical replication series that began in the early 1970s (Rachman & Hodgson, 1980; Steketee & Foa, in press; Steketee, Foa, & Grayson, 1982). This series, relying on individual experimental analyses and close examination of individual data from group studies, has also begun to identify patient characteristics that predict failure (e.g., Foa, 1979; Foa et al., 1983), a critical function of any replication series (see section 10.4).

Example three: Mixed results in three replications

The goal of this experiment was an experimental analysis of a new procedure for increasing heterosexual arousal in homosexuals desiring this goal (Herman et al., 1974b). A chance finding in our laboratories suggested that exposure to an explicitly heterosexual film increased heterosexual arousal in separate measurement sessions (see section 2.3, chapter 2). Subsequently, this was tested in an A-B-C-B design, where A was baseline, B was exposure to heterosexual films (the treatment), and C was a control procedure in which the subject was also exposed to erotic films, but the content was homosexual. The measures included changes in penile circumference to homosexual and heterosexual slides (recorded in sessions separate from the treatment sessions) as well as reports of behavior outside the laboratory setting. The purpose of the experiment was to analyze the effect on heterosexual arousal of exposure to films with heterosexual content over and above the effects of simply viewing erotic films, a condition obtaining in the control procedure. Thus the comparison was between treatment and placebo control.

Again, the patients were relatively heterogeneous. The first patient was a 24-year-old male with an 11-year history of homosexuality. During the year preceding treatment, homosexual encounters averaged one to three per day, usually in public restrooms. Also, during this period, the patient had been mugged once, had been arrested twice, and had attempted suicide. The second patient was a 27-year-old homosexual pedophile with a 10-year history of sexual behavior with young boys. The third patient was an 18-year-old male who had not had homosexual relations for several years but complained of a high frequency of homosexual urges and fantasies. The fourth patient, a 38-year-old male, reported a 26-year history of homosexual contacts. Homosexual behavior had increased during the previous 4 years, despite the fact that he had recently married. None of the patients reported previous heterosexual experience with the exception of the fourth subject, who had sexual intercourse with his wife approximately twice a week. Intercourse was successful if he employed homosexual fantasies to produce arousal, but he was unable to ejaculate during intercourse. All patients were seen daily, with the exception of the fourth patient, who was seen approximately three times per week.

Representative results from one case, the first patient, are presented in Figure 10-3. Heterosexual arousal, as measured in separate measurement sessions, increased during exposure to the female (heterosexual) film, dropped considerably when the homosexual film was shown, and rose once again when the female film was reintroduced. The results in this case represent clear and clinically important changes in heterosexual arousal, and the experimental analysis isolated the viewing of the heterosexual film as the procedure responsible for increases. Changes in arousal in the laboratory were accompanied by report of increased heterosexual fantasies and behavior. These results were replicated on Subjects 2 and 3, where similar increases in heterosexual arousal and reports of heterosexual behavior were noted. But the results from the fourth case differed somewhat, thereby posing difficulties in interpretation in this direct replication series (Figure 10-4).

In this case, heterosexual arousal increased somewhat during the first treatment phase, but the increase was quite modest. Withdrawing treatment resulted in a slight drop in heterosexual arousal, which increased once again

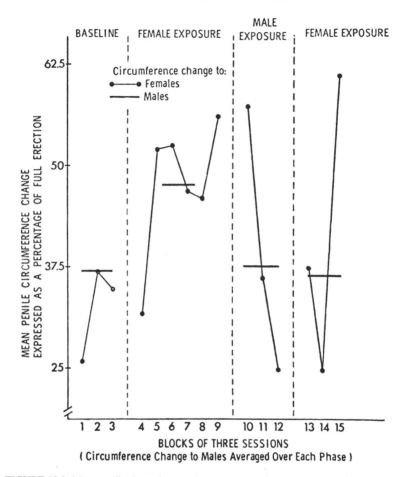

FIGURE 10-3. Mean penile circumference change, expressed as a percentage of full erection, to nude female (averaged over blocks of three sessions) and nude male (averaged over each phase) slides. (Figure 1, p. 338, from: Herman, S. H., Barlow, D. H., and Agras, W. S. [1974]. An experimental analysis of exposure to "explicit" heterosexual stimuli as an effective variable in changing arousal patterns of homosexuals. *Behaviour Research and Therapy*, **12**, 335–346. Copyright 1974 by Pergamon. Reproduced by permission.)

when the heterosexual film was reinstated. This last increase, however, does not become clear until the last point in the phase, which represents only one session. Subsequently, the patient was unable to continue treatment due to prior commitments precluding an extension of this phase, which would have confirmed (or disconfirmed) the increase represented by that one point. Reports of sexual fantasies and behavior were consistent with the modest increases in heterosexual arousal. While some increase in heterosexual fantasies was noted, the patient continued to employ homosexual fantasies occa-

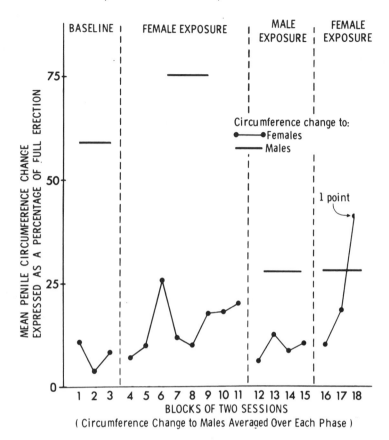

FIGURE 10-4. Mean penile circumference change, expressed as a percentage of full erection, to nude female (averaged over blocks of two sessions) and nude male (averaged over each phase) slides. (Figure 4, p. 342 from: Herman, S. H., Barlow, D. H., and Agras, W. S. [1974]. An experimental analysis of exposure to "explicit" heterosexual stimuli as an effective variable in changing arousal patterns of homosexuals. *Behaviour Research and Therapy*, **12**, 335–346. Copyright 1974 by Pergamon. Reproduced by permission.)

sionally during sexual intercourse with his wife and was still unable to ejaculate.

Again, conclusions in three general areas can be drawn from these data. *First*, exposure to explicit heterosexual films can be an effective variable for increasing heterosexual arousal, as demonstrated by the experimental analysis of the first patient. *Second*, to the extent that the results were replicated directly on three patients, the data are reliable and are not due to idiosyncracies in the first case. It does *not* follow, however, that generality of findings across patients has been firmly established. Although the results were clear and clinically significant for the first 3 patients, results from the fourth patient cannot be considered clinically useful due to the weakness of the effect. In this case, a clear distinction arises between the establishment of functional relationships and the establishment of clinically important generality of findings across clients. As in the first 3 patients, a functional relationship between treatment and heterosexual arousal was demonstrated in the fourth patient. This finding increases our confidence in the reliability of the result. Unlike the first 3 patients, however, the finding was not clinically useful. The conclusion, then, is that this procedure has only limited generality across clients, and the task remains to pinpoint differences between this patient and the remaining patients to ascertain possible causes for the limitations on client generality.

The authors (Herman et al., 1974b) noted that the fourth patient differed in at least two ways from the remaining three. One difference falls under the heading of *background variables* and the other is procedural. *First*, the patient was married and therefore was required to engage in heterosexual intercourse before heterosexual arousal or interest was generated. In fact, he reported this to be quite aversive, which may have hampered the development of heterosexual interest during treatment. The remaining patients had experienced no significant heterosexual behavior prior to treatment. *Second*, this patient was seen less frequently than other patients. At most he was seen three times a week, rather than daily. At times, this dropped to once a week and even once every 3 weeks during periods when other commitments interfered with treatment. It is possible that this factor retarded development of heterosexual interest. To the extent that this was a procedural problem, rather than a variable that the patient brought with him to the experiment, it would have been possible to alter the procedure prior to the beginning of the experiment or even during the experiment (i.e., require daily attendance). If this alteration had been undertaken and similar results (the weak effect) had ensued, it might have limited the search for causes of the weak effect to just the background variables, such as the ongoing aversive heterosexual behavior. Of course, this procedural variable was not thought to be important when the experiment was designed. In fact, failures to replicate are always occurring in direct replication series. Another good example was presented in the study by Ollendick et al. (1981) in chapter 8 (Figures 8-3 and 8-4). In this comparison of two treatments in an ATD, one treatment was more effective than another for the first subject, but just the opposite was true for the second subject. Because the investigators were close to the data, they speculated on one seemingly obvious reason for this discrepancy. Thus, pending a subsequent test of their hypothesis, they have already taken the first step on the road to tracking down intersubject variability and establishing guidelines for generality of findings. The investigators themselves are always in the best position to identify, and subsequently test, putative sources of lack of generality of findings.

The issue of interpreting mixed results and looking for causes of failure illustrates an important principle in replication series. We noted above that subjects in a direct replication series should be as homogeneous as possible. If subjects in a series are not homogeneous, the investigator is gambling (Sidman, 1960). If the procedure is effective across heterogeneous subjects, he or she has won the gamble. If the results are mixed, he or she has lost. More specifically, if one subject differs in three or four definable ways from previous subjects, but the data are similar to previous subjects, then the experimenter has won the gamble by demonstrating that a procedure has client generality *despite* these differences. If the results differ in any significant manner, however, as in the example above, the experimenter cannot know which of the three, four, or more variables was responsible for the differences. The task remains, then, to explore systematically the effects of these variables and track down causes of intersubject variability.

In basic research with animals, one seldom sees this type of gamble in a direct replication series, because most variables are controlled and subjects are highly homogeneous. In applied research, however, clients always bring to treatment a variety of historical experiences, personality variables, and other background variables such as age and sex. To the extent that a given treatment works on 3, 4, or 5 clients, the applied researcher has already won a gamble even in a direct replication series, because a failure could be attributed to any one of the variables that differentiate one subject from another. In any event, we recommend the conservative approach whenever possible, in that subjects in a direct replication series should be homogeneous for aspects of the target behavior as well as background variables. The issue of gambling arises again when one starts a systematic replication series because the researcher must decide on the number of ways he or she wishes the systematic replication series to differ from the original direct series.

Example four: Mixed results in nine replications

Although all subjects demonstrated some improvement in the study described above, the data are more variable in a direct replication series. Such is the case in the following study, where attempts to modify delusional speech in 10 paranoid schizophrenics produced mixed results (Wincze et al., 1972). In this procedure the effects of feedback and token reinforcement on delusional speech were evaluated. Feedback consisted of reading sentences with a high probability of eliciting a particular patient's delusional behavior. If the patient responded delusionally, he or she would be informed that the response was incorrect and given the correct response. For instance, one patient thought he was Jesus Christ. If he answered affirmatively when asked this question, he would be told that he was not Jesus Christ, who lived 2,000 years ago, but rather Mr. M., who was 40 years old. If he answered correctly, he would be so informed. During token reinforcement phases, the patient received tokens redeemable for food and recreational activities, contingent on nondelusional speech in the sessions. Sessions consisted of 15 questions each day. Tokens were also administered to some patients for nondelusional talk on the ward in addition to the contingencies within sessions; but, for our purposes, we will discuss only the effects of feedback and token reinforcement on delusional talk within sessions.

All patients were chronic paranoid schizophrenics who had been hospitalized at least 2 years (the range covered from 2 to 35 years). Six males and four females participated, with an age range from 25 to 67. Level of education ranged from eighth grade through college. Thus these patients were, again, heterogeneous on many background variables.

The experimental design for the first 5 patients consisted of baseline procedures followed by feedback and then token reinforcement. In some cases, token reinforcement on the ward, in addition to tokens within sessions, was introduced toward the end of the experiment. Additional baseline phases were introduced whenever feedback or reinforcement produced marked decreases in delusional talk. For Subjects 6 through 10, the first feedback and token reinforcement in-session phases were withdrawn, to examine the effects of token reinforcement when it was presented first in the treatment sequence.

All data were presented individually in the experiment so that any functional relations between treatments and delusional speech were apparent. Individual data from the first patient are presented in Figure 10-5 to illustrate the manner of presentation. In this particular case, the baseline phase following the first feedback phase was omitted because no improvement was noted during feedback. Results from all patients are summarized in Table 10-1.

In 5 out of 10 cases, feedback alone produced at least a 20% decrease in delusional speech within sessions. In two cases, this decrease in delusional speech was clinically impressive both in magnitude and in the consistent trend in behavior throughout the phase (Subjects 2 and 8). In the remaining 3 patients, the magnitude of the decrease and/or the behavior trend across the feedback phase was relatively weak. For instance, Table 10-1 indicates that the last two data points in the feedback phase for Subject 9 were considerably lower than the last two data points in the preceding baseline phase (a drop of 49.8%). But the extreme variability in data across the feedback phase indicates that this was a weak effect. A withdrawal of feedback and return to baseline procedures was not associated with a clear reversal in delusional speech (at least a 20% increase) in any of the 5 patients who improved, although the finding is particularly important for those 2 patients who demonstrated improvement of clinical proportions. Thus it was not demonstrated that feedback was the variable responsible for improvement within treatment sessions.

If the marked improvement of Subjects 2 and 8 had been replicated on additional patients, one would be tempted to undertake a further experimen-

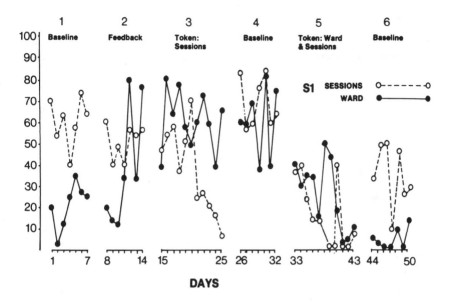

FIGURE 10-5. Percentage delusional talk of Subject 1 during therapist sessions and on ward for each experimental day. (Figure 1, p. 254, from: Wincze, J. P., Leitenberg, H., and Agras, W. S. [1972]. The effects of token reinforcement and feedback on the delusional verbal behavior of chronic paranoid schizophrenics. *Journal of Applied Behavior Analysis*, **5**, 247–262. Copyright 1972 by Society for Experimental Analysis of Behavior. Reproduced by permission.)

tal analysis to determine which variables were responsible for the improvement. The lack of replication, however, suggests that this would not be a fruitful line of inquiry.

The results from token reinforcement were quite different. This procedure was administered to 9 patients. Six (Subjects 1, 2, 4, 5, 8 and 9) improved—an improvement that was confirmed by a return of delusional speech when token reinforcement was removed. Subject 7 also improved, but delusional speech did not reappear when token reinforcement was removed. In all of these patients, the decrease was substantial both in percentage of delusional speech and in trends across the token phase.

Several conclusions can be drawn from these data. In terms of reduction of delusional speech within sessions, the experimental analysis demonstrated that token reinforcement was effective, and replication indicated that the finding had some reliability. Generality of findings across clients, however, is limited. Two patients did not improve during administration of token reinforcement. As Sidman (1960) noted, the failure to replicate on all subjects does *not* detract from the successes in the remaining subjects. Token reinforcement is clearly responsible for improvement in those subjects to the

TABLE 10-1. Mean Percentage Delusional Talk of Each S Based on Last Two Data Points of Each Phase in Therapist Sessions and on the Ward

SUBJECTS | PHASE SEQUENCES

SUBJECTS	BASELINE	FEEDBACK	BASELINE	TOKEN: SESSIONS	BASELINE	TOKEN: WARD AND SESSIONS	BONUS	BASELINE
S1 Sessions	68.1	59.8	—	11.6	61.4	1.6	—	28.2
S1 Ward	26.2	50.4	—	52.9	56.7	7.4	—	11.3
S2 Sessions	83.0	1.6	13.3	—	—	—	—	—
S2 Ward	16.6	5.9	0.0	—	—	—	—	—
S3 Sessions	91.3	73.0	—	3.3	91.3	11.6	5.0	64.7
S3 Ward	27.0	9.9	—	36.3	5.0	21.6	4.6	4.0
S4 Sessions	76.4	66.4	68.1	21.6	61.4	—	29.9	61.4
S4 Ward	27.0	2.6	24.2	4.4	13.3	—	0.0	3.2
S5 Sessions	86.3	51.5	64.7	24.9	59.8	18.3	21.6	38.2
S5 Ward	48.3	79.2	70.6	61.9	51.7	45.1	4.6	29.2

SUBJECTS	BASELINE	TOKEN: SESSIONS	BASELINE	FEEDBACK	BASELINE	TOKEN: WARD AND SESSIONS	BONUS	BASELINE
S6 Sessions	79.7	64.7	76.4	68.1	—	66.4	78.0	83.0
S6 Ward	58.2	79.5	50.7	56.6	—	78.8	69.6	25.7
S7 Sessions	89.6	59.8	69.7	48.1	63.1	48.1	36.5	71.4
S7 Ward	23.0	12.5	19.1	9.1	18.8	14.0	37.4	20.9
S8 Sessions	86.3	18.3	49.8	8.3	0.0	—	—	—
S8 Ward	6.9	3.3	0.0	0.0	0.0	—	—	—
S9 Sessions	79.7	13.3	54.8	5.0	20.0	1.7	—	51.5
S9 Ward	13.4	8.9	44.9	16.3	34.8	3.4	—	14.0
S10 Sessions	83.0	66.4	73.0	64.7	—	66.4	—	—
S10 Ward	16.6	33.1	8.2	11.3	—	58.2	—	—

Note. Table 2, p. 258, from: Wincze, J. P., Leitenberg, H., and Agras, W. S. (1972). The effects of token reinforcement and feedback on the delusional verbal behavior of chronic paranoid schizophrenics. *Journal of Applied Behavior Analysis*, **5**, 247–262. Copyright 1972 by Society for Experimental Analysis of Behavior. Reproduced by permission.

extent that the experimental design was sound (internally valid). However, applied researchers cannot stop here, satisfied that the procedure seems to work well enough on most cases, since the practicing clinician would be at a loss to predict which cases would improve with this procedure. In fact, because the authors (Wincze et al., 1972) noted that these two cases actually deteriorated on the ward during this treatment, the search for accurate predictions of success becomes all the more important to the clinician. Thus a careful search for differences that might be important in these cases should ensue, leading to a more intensive functional investigation and experimental manipulation of those factors that contribute to success or failure.

In view of the additional fact that all subjects in this series demonstrated little generalization of improvement from session to ward behavior, analysis of this treatment is in a very preliminary state and, as Wincze et al. (1972) pointed out, ". . . much work needs to be done in order to predict when a given type of behavioral intervention is likely to succeed in a given case" (p. 262).

Finally, it seems important to make a methodological point on the size of this series. While the nine replications in this series yielded a wealth of data, a more efficient approach might have been to stop after four or five replications and conduct a functional analysis of failures encountered. In the unlikely event that failures did not occur in the initial replication series, the results would be strong enough to generate systematic replication in other research settings, where failures would almost certainly appear, leading to a search for critical differences at this point. If failures did appear in this shorter series, the investigators could immediately begin to determine factors responsible for variant data rather than continue direct replications that would only have a decreasing yield of information as subjects accumulated. Perhaps for this reason, one encounters few direct replication series with an N of seven or more. One notable exception is a multiple-baseline-across-subjects experiment on seven anorexics, where, unfortunately for both experimental and clinical reasons, all patients improved substantially (Pertschuk, Edwards, & Pomerleau, 1978).

Example five: Simultaneous replication

Finally, a method of conducting simultaneous replications has been suggested by J. A. Kelly (Kelly, 1980; Kelly, Laughlin, Clairborne, & Patterson, 1979). This procedure is very useful when one is intervening with a coexisting group. Examples would be group therapy for any of a number of problems such as phobia and assertiveness, or interventions in a classroom or on a hospital ward. In this procedure, any number of subjects in the group can be treated simultaneously in a particular experimental design, but individual data would be plotted separately. Figure 10-6 illustrates this strategy with hypothetical data originally presented by J. A. Kelly (1980). In this hypothetical strategy, the experimental design was a multiple baseline across behaviors for six subjects. Three different aspects of social skills were repeatedly assessed by role playing. Intervention then proceeded for all six subjects on the first social skill, followed by the second social skill, and so on. In this hypothetical example, of course, all subjects did very well, with particular aspects of social skills improving only when treated. Naturally, this strategy need not be limited to a multiple-baseline-across-behaviors design. Almost any single-subject design, such as an alternating treatments design or a standard withdrawal design, could be simultaneously replicated.

From the point of view of replication, this is a very economical and conservative way to proceed. It is economical because it is less time consuming to treat six clients in a group than it is to treat six clients individually. But one still has the advantage of observing individual data repeatedly measured from six different subjects. Naturally, this is only possible where opportunities for group therapy exist. Furthermore, the procedure is conservative because fewer variables are different from client to client. The gamble taken by the investigator in a replication series with increasing heterogeneity or diversity of subjects or settings was mentioned above. To repeat, if a replication fails, the more differences there are in subjects, settings, timing of the intervention, and so forth, the harder it is to track down the cause of the failure for replication during subsequent experimentation. If all subjects are treated simultaneously in the same group, at the same time, then one can be relatively sure that the intervention procedures, as well as setting and temporal factors, are identical. If there is a failure to replicate, then the investigator should look elsewhere for possible causes, most likely in background variables or personality differences in the subjects themselves.

Of course, treating clients in group therapy has its own special kind of setting. If one were interested in the generality of these findings to individual treatment settings, the first step in a systematic replication series would be to test the procedure in subjects treated individually. Also, when groups of individuals are treated simultaneously, one cannot stop the series at just any time to begin examining for causes of failures if they occur. However, this is not really a problem as long as the groups remain reasonably small (e.g., three to six), such that the investigator would be unlikely to accumulate a large number of failures before having an opportunity to begin the search for causes. Other examples of simultaneous replication can be found in an experiment by E. B. Fisher (1979) mentioned in chapter 8.

Guidelines for direct replication

Based on prevailing practice and accumulated knowledge on direct replication, we would suggest the following guidelines in conducting a direct replication series in applied research:

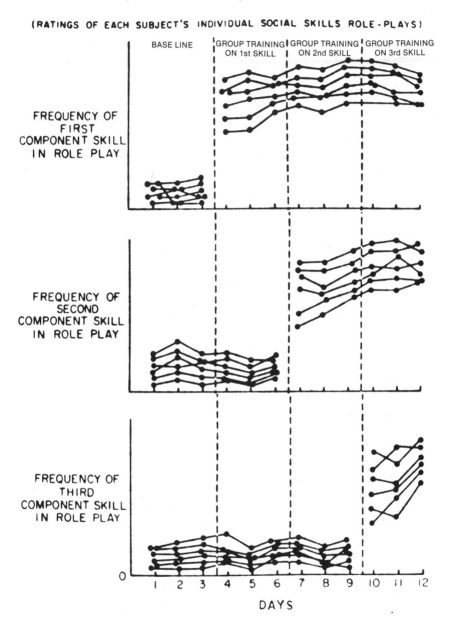

FIGURE 10-6. Graphed hypothetical data of simultaneous replications design. (Figure 2, p. 306 from: Kelly, J. A., Laughlin, C., Claiborne, M., & Patterson, J. [1979]. A group procedure for teaching job interviewing skills to formerly hospitalized psychiatric patients. *Behavior Therapy*, **10**, 299–310. Copyright 1979 by Association for Advancement of Behavior Therapy. Reproduced by permission.)

1. Therapists and settings should remain constant across replications.
2. The behavior disorder in question should be topographically similar across clients, such as a specific phobia.
3. Client background variables should be as closely matched as possible, although the ideal goal of identical clients can never be attained in applied research.
4. The procedure employed (treatment) should be uniform across clients, until failures ensue. If failures are encountered during replication, attempts should be made to determine the cause of this intersubject variability through improvised and fast-changing experimental designs (see section 2.3, chapter 2). If the search is successful, the necessary alteration in treatment should be tested on additional clients who share the characteristics or behavior of the first client who required the alteration. If the search for sources of variability is not successful, differences in that particular client from other successful clients should be noted for future research.
5. One successful experiment and three successful replications are usually sufficient to generate systematic replication of topographically different behaviors in the same setting or of the same behavior in different settings. This guideline is not as firm as those preceding, because results from a study containing one unusual or significant case may be worth publishing, or an investigator may wish to continue direct replication if experimentally successful but clinically "weak" results are obtained. Generally, though, after one experiment and three successful replications, it is time to go on to systematic replication.

    On the other hand, if direct replication produces mixed success and failure, then investigators must decide when to stop the series and begin to analyze reasons for failure in what is essentially a new series, because the procedure or treatment presumably will change. If one success is followed by two or three failures, then neither the reliability of the procedure nor the generality of the finding across clients has been established, and it is probably time to find out why. If two or three successes are mixed in with one or two failures, then the reliability of the procedure would be established to some extent, but the investigator must decide when to begin investigating reasons for lack of client generality. In any case, it does not appear to be sound experimental strategy to continue a direct replication series indefinitely, when both successes and failures are occurring.
6. Broad client generality cannot be established from one experiment and three replications. Although a practitioner can observe the extent to which an individual client who responded to treatment in a direct replication series is similar to his or her client and can proceed accordingly with the treatment, chances are the practitioner may have a client with a topographically similar behavior disorder who is different in some clinically

important way from those in the series. Fortunately, as clinical and systematic replication ensues with other therapists in other settings, many more clients with different background variables are treated, and confidence in generality of findings across clients, which was established in a preliminary manner in the first series, is increased with each new replication.

10.3 SYSTEMATIC REPLICATION

Sidman (1960) noted that where direct replication helps to establish generality of findings among members of a species, ". . . systematic replication can accomplish this and at the same time extend its generality over a wide range of situations" (p. 111). In applied research, we have noted that direct replication can begin to establish generality of findings across clients but cannot answer questions concerning applicability of a given procedure or functional relationship in different therapeutic settings or by different therapists. Another limitation of the initial direct replication series is an inability to determine the effectiveness of a procedure proven effective with one type of behavior disorder on a related but topographically different behavior disorder.

Definition of systematic replication

We can define systematic replication in applied research as any attempt to replicate findings from a direct replication series, varying settings, behavior change agents, behavior disorders, or any combination thereof. It would appear that any successful systematic replication series in which one or more of the above-mentioned factors is varied also provides further information on generality of findings across clients because new clients are usually included in these efforts.

Example: Differential attention series

There are now many examples of mature, important, systematic replication series in applied research. Extant series on time-out procedures (see J. M. Johnston & Pennypacker, 1980), exposure-based treatments for phobia (see Mavissakalian & Barlow, 1981) and social skills training with a variety of populations (e.g., Bornstein, Bellack & Hersen, 1980; Hersen & Bellack, 1976; Turner, Hersen, & Bellack, 1978; Wells, Hersen, Bellack, & Himmelhoch, 1979), among others, have established broad generality for what are now common therapeutic interventions. But one of the most extensive and advanced systematic replication series has been in progress since the early 1960s. The purpose of this series has been to determine the generality of the effectiveness of a single intervention technique, often termed *differential attention*. Differential attention consists of attending to a client contingent on the emission of a well-defined desired behavior. Usually such attention takes the form of positive interaction with the client consisting of praise, smiling, and so on. Absence of the desired behavior results in withdrawal of attention, hence "differential" attention. This series, consisting of over 100 articles, has provided practitioners with a great deal of specific information on the effectiveness of this procedure in various settings with different behavior disorders and behavior change agents. Preliminary success in this area has generated a host of books advocating use of the technique in various settings, particularly with children in the home or classroom, most often in combination with other procedures such as other types of reinforcing or mildly punishing consequences including time-out (e.g., Forehand & McMahon, 1981; Patterson, 1982; Ross, 1981; Sulzer-Azaroff & Mayer, 1977). What is perhaps more important is that articles in this series have noted certain occasions when the procedure fails, leading to a clearer delineation of the generality of this technique in all relevant domains in the applied area. A brief review of findings from this series in the various important domains of applied research will illustrate the process of systematic replication.

Differential attention: Adult psychotic behaviors

One of the first reports on differential attention appeared in 1959 (Ayllon & Michael). This report contained several examples of the application of differential attention to institutionalized patients in a state hospital. The therapists in all cases were psychiatric nurses or aides. The purpose of this early demonstration was to illustrate to personnel in the hospital the possible clinical benefits of differential attention. Thus differential attention was applied to most cases in an A-B design, with no attempt to demonstrate experimentally its controlling effects. In several cases, however, an experimental analysis was performed. One patient was extremely aggressive and required a great deal of restraint. One behavior incompatible with aggression was sitting or lying on the floor. Four-day baseline procedures revealed a relatively low rate of being on the floor. Social reinforcement by nurses increased the behavior, resulting in decreased aggression. Subsequent withdrawal of social reinforcement produced decreases in the behavior and increases in aggression. Unfortunately, ward personnel could not tolerate this, and the patient was restrained once again, aborting a return to social reinforcement. The resultant A-B-A design was sufficient, however, to demonstrate the effects of social reinforcement in this setting for this class of behavior.

This early experiment suggested that differential attention could be effective when applied by nurses or aides as therapists. These successes sparked replication by these investigators in additional cases. Other psychotic behavior in adult psychiatric wards modified by differential attention or a combination of differential attention and other procedures included faulty eating behavior (Ayllon & Haughton, 1964) and towel hoarding (Ayllon, 1963). These early studies were the beginning of the systematic replication series, in that topographically different behavior responded to differential attention.

Another problem behavior in adult psychiatric wards considered more central to psychiatric psychopathology is psychotic verbal behavior such as delusions or hallucinations. An early example of the application of differential attention to delusions was reported by Rickard, Dignam, and Horner (1960), who attended (smiled, nodded, etc.) to a 60-year-old male during periods of nondelusional speech and withdrew attention (minimal attention) during delusional speech. Therapists were psychologists. Initially, nondelusional speech increased to almost maximal levels (9 minutes out of a 10-minute session) during periods of attention and decreased during the minimal attention condition. Later, even minimal attention was sufficient to maintain nondelusional speech. A 2-year follow-up (Rickard & Dinoff, 1962) revealed maintenance of these gains and reports of generalization to hospital settings. Unfortunately, only one patient was included in this experiment, precluding any preliminary conclusion on generality of findings across other patients.

Ayllon and Haughton (1964) followed this up with a series of 3 adult patients in a psychiatric ward who demonstrated bothersome delusional or psychosomatic verbal behavior. In all three cases, differential attention was effective in controlling the behavior, as demonstrated by an A-B-C-B design, where A was baseline, B was social attention, and C was withdrawal of attention. Here, as in other reports by Ayllon and his associates, therapists were nurses or aides. This early experiment was a good direct replication series in its own right but, more importantly, served to systematically replicate findings from the single-case reported by Rickard, Dignam, and Horner (1960). In Ayllon and Haughton's experiment, therapists were nurses or aides, rather than psychologists, and the setting was, of course, a different psychiatric ward. Despite these factors, differential attention again produced control over deviant behavior in adults on a psychiatric ward. This independent, systematic replication provides a further degree of confidence in the effectiveness of the technique with psychotic behavior and in its generality across therapists and settings.

After these early attempts to control psychotic behavior of adults on psychiatric wards through differential attention, Ayllon and his associates moved on to stronger reinforcers and developed the token economy (Ayllon & Azrin, 1968), abandoning for the most part their work on the exclusive use of differential attention. The impact of this early work was not lost on clinical investigators, however, and the importance of differential attention on adult wards of hospitals was once again demonstrated in a very clever experiment by Gelfand, Gelfand, and Dobson (1967). These investigators observed six psychotic patients on an inpatient psychiatric ward, to determine sources of social attention contingent on disruptive or psychotic behavior. At the same time, they noted who was most successful in ignoring behaviors among the groups on the ward (i.e., other patients, nurses' aides, or nurses). Results indicated that other patients reinforced these behaviors least and ignored them the most effectively, followed by nurses' aides and nurses. Thus the personnel most responsible for implementing therapeutic programs, the nurses, were providing the greatest amount of social reinforcement contingent on undesirable behavior. This study does not, of course, demonstrate the controlling effects of differential attention. But, growing out of earlier experimental demonstrations of the effectiveness of this procedure, this study highlighted the potential importance of this factor in maintaining undesirable behavior on inpatient psychiatric units and led to further replication efforts on other wards.

After the appearance of these early studies analyzing the effects of differential attention, most investigators working in these settings moved on to more comprehensive, multifaceted treatment programs incorporating a variety of treatment components in addition to differential attention (e.g., Liberman, Neuchterlein, & Wallace, 1982; Monti, Corriveau, & Curran, 1982; Paul & Lentz, 1977). For example, the well-known and very successful program devised and described by Paul and Lentz (1977) included a comprehensive point system, or token economy, as well as other structured training procedures.

The exciting therapeutic program devised by Liberman, Wallace, and their colleagues (Wallace et al., in press) emphasized a very detailed and meticulous approach to training in social and life skills necessary for functioning outside of the institutional setting. Some of these skills include recreational planning, food preparation, locating and moving into an apartment, money management, job interviews, anger and stress control, long-term planning, and dealing with friendship or dating situations. While a token economy or point system is not part of this program, differential attention in terms of praise for completion of assignments and so forth is woven throughout the various modules or treatment components. Largely as a result of this integration, few, if any, studies analyzing the effects of differential attention in isolation with this population have appeared recently.

Comment on replication procedures

It is safe to say that the impact of this work on adult wards has been substantial, and differential attention to psychotic behavior is now a common therapeutic procedure on many wards. More importantly, it has been thoroughly integrated into comprehensive psychosocial treatment programs for these populations (e.g., Paul & Lentz, 1977; Wallace et al., in press). In retrospect, however, there are many methodological faults with this series, leading to large gaps in our knowledge, which could have been avoided had replication been more systematic.

While differential attention was successfully administered on psychiatric wards in several different parts of the country across the range of therapists or ward personnel typically employed in these settings and across a variety of psychotic behaviors, from motor behavior through inappropriate speech, only a few studies contained experimental analyses. On the other hand, many of the reports would come under the category of case studies (A-B designs with measurement). Certainly, this preliminary series on institutionalized patients would be much improved had each class of behavior (e.g., verbal behavior, withdrawn behavior, inappropriate behavior, aggressive or other motor behaviors) been subjected to a direct replication series with three or four patients and then systematically replicated in other settings with other therapists.

This procedure most likely would have produced some failures. Reasons for these failures could then have been explored, providing considerably more information to clinicians and ward personnel on the limitations of differential attention. As it stands, Ayllon and Michael (1959) reported a failure but did not describe the patient in any detail or the circumstances surrounding the failure. This type of reporting leads to undue confidence in a procedure among naive clinicians; when failures do occur, disappointment is followed by a tendency to eliminate the procedure entirely from therapeutic programs. In this specific case, however, what has happened is that differential attention has been incorporated into more comprehensive programs without adequate analysis of its contribution. With some cases or in some settings it may be either important or superfluous. In other cases it may even be detrimental (see Herbert et al., 1973).

This early series also illustrated a second use of the single-case study (A-B). In chapter 1 we noted that case studies can suggest initially that a new technique is clinically effective, which can lead to more rigorous experimental demonstration and direct replication. In a systematic replication series the single-case study makes another appearance. Many reports are published that include only one case, but replicate an earlier direct replication series in either an experimental or an A-B form. Usually the reports are from different settings and contain a slight twist, such as a new form of the behavior disorder or a slight modification of the procedure. While these reports are less desirable from the larger viewpoint of a systematic replication series, the fact is that they are published. When a sufficient number accumulate, these reports can provide considerable information on generality of findings. We will return to this point later.

Differential attention: Other adult behaviors

The early success of differential attention and positive reinforcement procedures in general with institutionalized patients led to application of this procedure to other adult behavior disorders in other settings.

Most of these examples were published as single-case reports. Some of these single-cases contain a functional analysis of differential attention; others are A-B designs wth measurements. For instance, Brookshire (1970) eliminated crying in a 47-year-old male suffering from multiple sclerosis by attending to incompatible verbal behavior. Other single-case examples include Brady and Lind's (1961) modification of hysterical blindness through differential attention to a visual task in a hospital setting. A hospital setting was also utilized to test the effectiveness of differential attention on a conversion reaction, specifically astasia-abasia, or stumbling and falling while walking (Agras, Leitenberg, Barlow, & Thomson, 1969). Praise combined with ignoring stumbling resulted in improvement in this case. In another setting, these procedures also proved effective on a similar case (Hersen, Gullick, Matherne, & Harbert, 1972). Psychogenic vomiting was treated in a hospital setting by Alford, Blanchard, and Buckley (1972) who ignored vomiting and withdrew social contact immediately after vomiting. Therapists in this case were nurses. The authors cite success of this procedure on vomiting in a child (Wolf, Birnbrauer, Williams, & Lawler, 1965) as a rationale for attempting it with an adult. More recently, Redd has extended this work by demonstrating the usefulness of differential attention in controlling retching and vomiting in cancer patients undergoing chemotherapy (e.g., Redd, 1980). Specifically, nurses seem able to manage the well-known conditioned nausea response using differential attention.

Various other case studies along these lines were published. Many of the studies describe slight modification of the procedure or some variation in the behavior disorder. As in the treatment of psychotic patients, differential attention also was combined with other treatment variables such as other forms of positive reinforcement or punishment in many research reports, making it difficult to specify the exclusive effects of differential attention.

From an historical viewpoint, one of the more interesting studies on differential attention was reported by Truax (1966), who reanalyzed tape recordings of Carl Rogers' therapy sessions. He discovered that Rogers responded differently (i.e., positively) to five classes of verbal behavior over a number of therapy sessions, and four of these classes increased in frequency. This is reminiscent of the verbal conditioning studies (e.g., Greenspoon, 1955) and suggests, in a non-experimental A-B fashion, that differential attention is operative in a variety of different psychotherapeutic approaches. But, once again, few if any studies examining the effects of differential attention in isolation with non-psychotic adult populations have occurred in recent years.

The reasons for this seem to be very similar to those described above in series on institutionalized psychotic patients. That is, differential attention has been "co-opted" into larger treatment packages without further analysis of its effects. One good example is marital therapy. In a large, early series Goldstein (1971) used differential attention procedures with 10 women who were experiencing marital difficulties. Specifically, these women were instructed on attending to desired behaviors emitted by their husbands and ignoring undesirable behaviors. Using a time series analysis, statistically significant changes occurred in eight out of ten cases. To the extent that these changes were clinically as well as statistically significant, these uncontrolled case studies suggested that differential attention was effective in this context. Since that time, marital therapies based broadly on social learning principles have become well developed and are widely used for the treatment of marital distress (Jacobson & Margolin, 1979; Liberman et al., 1980; O'Leary & Turkewitz, 1981). Most of these programs contain a variety of interventions, including comunications training, problem solving, and instructions on altering various dyadic patterns of behavior. Embedded within these approaches, however, is a strong differential attention component. For example, when leading marital therapists describe their actual approaches in great detail (e.g., L. F. Wood & Jacobson, 1984), these treatments include training in expressions of appreciation and praise contingent on desirable partner behavior. Often this is most prominent in the early stages of therapy. For example, during "caring days" husbands and wives are taught to express appreciation for positive qualities or behaviors of their spouses. Ways in which spouses would like their partners to express appreciation are carefully explored in the therapy session. These types of expressions, most often including positive verbal feedback of some sort or another, are then integrated into the couples' daily lives. Unfortunately, this treatment component has never been evaluated systematically, and thus, once again, we are not sure of the specific conditions in which it succeeds or fails.

Comment on replication procedures

Thus the deficits and faults in this area are similar to those encountered in the series with psychotic adults described above. Evidence exists that differential attention *can* be effective in a number of settings (e.g., inpatient, outpatient, or home) when applied by different therapists (e.g., doctors, nurses, or wives) on a number of different behavioral problems. The difficulty here is with the dearth of experimental analyses and direct replication in each new setting or with each new problem. Nevertheless, clinical investigators have for the most part not followed the type of detailed technique-building approach described in chapter 2 that would ensure that treatment programs, such as marital therapy, be as powerful as they might be.

Differential attention: Children's behavior disorders

In fact, differential attention procedures applied to adults, whether psychotic or nonpsychotic, comprise only a small part of the work reported in this area. The greatest number of experimental inquiries on the effectiveness of differential attention have been conducted with children, and this series represents what is probably the most comprehensive systematic replication series to data. One of the earliest studies on the application of differential attention to behavior problems of a child was reported by C. D. Williams (1959), who instructed parents to withdraw attention from nighly temper tantrums. When an aunt unwittingly attended to tantrum behavior, tantrums increased and were extinquished once again by withdrawal of attention.

Table 10-2 presents summaries of replication efforts in this series since that time. Studies reported in this table used differential attention as the sole or, at least, a very major treatment component. Studies where differential attention was a minor part of a treatment package, such as parent training, were for the most part omitted. It is certainly possible that a few additional studies were inadvertently excluded. In the table, it is important to note the variety of clients, problem behaviors, therapists, and settings described in the studies, because generality of findings in all relevant domains is entirely dependent on the diversity of settings, clients, and the rest employed in such studies. One should also note that the bulk of this work occurred in the late 1960s and early 1970s, with a decrease in published research since that time. Unlike the examples above, this is due to the fact that many of the goals of this systematic replication series were completed. We will discuss this issue further.

Most replication efforts through 1965 presented an experimental analysis of results from a single-case (see Table 10-2). A good example of the early studies was presented by Allen et al. (1964), who reported that differential attention was responsible for increased social interaction with peers in a socially isolated preschool girl. The setting for the demonstration was a classroom, and the behavior change agent, of course, was the teacher. While most of the early studies contained only one case, the experimental demonstration of the effectiveness of differential attention in different settings with different therapists began to provide information on generality of findings across all-important domains. These replications increased confidence in this procedure as a generally effective clinical tool. In addition to isolate behavior, the successful treatment of such problems as regressed crawling (Harris, Johnston, Kelley, & Wolf, 1964), crying (Hart, Allen, Buell, Harris, & Wolf, 1964), and various behavior problems associated with the autistic syndrome (e.g., Davison, 1965) also suggested that this procedure was applicable to a wide variety of behavior problems in children while at the same time providing additional information on generality of findings across therapists and settings.

Although studies of successful application of differential attention to a single-case demonstrated that this procedure is applicable in a wide range of situations, a more important development in the series was the appearance of direct replication efforts containing three or more cases within the systematic replication series. Although reports of single-cases are uniformly successful, or they would not have been published, exceptions to these reports of success can and do appear in series of cases, and these exceptions or failures begin to define the limits of the applicability of differential attention.

For this reason, it is particularly impressive that many series of three or more cases reported consistent success across many different clients, with such behavior disorders as inappropriate social behavior in disturbed hospitalized children (e.g., Laws, Brown, Epstein, & Hocking, 1971), disruptive behavior in the elementary classroom (e.g., Cormier, 1969; R. V. Hall et al., 1971; R. V. Hall, Lund, & Jackson, 1968) or high school classroom (e.g., Schutte & Hopkins, 1970), chronic thumb-sucking (Skiba, Pettigrew, & Alden), disruptive behavior in the home (Veenstra, 1971; Wahler, Winkel, Peterson, & Morrison, 1965), and disruptive behavior in brain-injured children (R. V. Hall & Broden, 1967). These improvements occurred in many different settings such as elementary and high school classrooms, hospitals, homes, kindergartens, and various preschools. Therapists included professionals, teachers, aides, parents, and nurses (see Table 10-2).

The consistency of their success was impressive, but as these series of cases accumulated, the inevitable but extremely valuable reports of failures began to appear. Almost from the beginning, investigators noted that differential attention was not effective with self-injurious behavior in children. For instance, Tate and Baroff (1966) noted that in the length of time necessary for differential attention to work, severe injury would result. In place of differential attention, a strong aversive stimulus—electric shock—proved effective in suppressing this behavior. Later, Corte, Wolf, and Locke (1971) found that differential attention was totally ineffective on mild self-injurious behavior in retarded children but, again, electric shock proved effective. Because there are no reports of success in the literature using differential attention for self-injurious behavior, it is unlikely that these cases would have been published at all if differential attention had not proven effective on other behavior disorders. Thus this is an example of a systematic replication series setting the stage for reports of limitations of a procedure.

More subtle limitations of the procedure are reported in series of cases wherein the technique worked in some cases, but not in others. In an early series, Wahler et al. (1965) trained mothers of young, oppositional children in

TABLE 10-2. Summary of Studies on Differential Attention with Children

AUTHORS	CLIENT(s)	N	BEHAVIOR	SETTING	THERAPIST	EXPERIMENTAL ANALYSIS
C. D. Williams (1959)	18-mo.-old female	1	Tantrums	Home	Parents	No
E. H. Zimmerman & Zimmerman (1962)	11-yr.-old males	2	Unproductive classroom behavior	Residential treatment center	Teachers	No
Allen, Hart, Buell, Harris, & Wolf (1964)	4-yr.-old female	1	Isolate behavior	Lab. preschool	Teacher	Yes
Harris, Johnston, Kelley, & Wolf (1964)	3-yr.-old female	1	Crawling behavior	University nursery school	Teacher	Yes
Hart, Allen, Buell, Harris, & Wolf (1964)	4-yr.-old males	2	Crying	Preschool	Teachers	Yes
Davison (1965)	10-yr.-old males	2	Autistic behavior	Private day-care center	Undergraduates	No
Wahler, Winkel, Peterson, & Morrison (1965)	4- to 6-yr.-old males	3	Oppositional behavior	Lab. playroom	Mother	Yes
Allen & Harris (1966)	5-yr.-old female	1	Scratching behavior	Lab. preschool and home	Mother	No
Hawkins, Peterson, Schweid, & Bijou (1966)	4-yr.-old male	1	Tantrums and oppositional behavior	Home	Mother	Yes
Holmes (1966)	9-yr.-old male	1	Underachievement in school and disruptive behavior	Classroom	Teacher	No
M. K. Johnston, Kelley, Harris, & Wolf (1966)	3-yr.-old male	1	Physical activity	Preschool	Teacher	Yes
Allen, Henke, Harris, Baer, & Reynolds (1967)	4½-yr.-old male	1	Short attention span	Lab. preschool	Teachers	Yes
Etzel & Gerwitz (1967)	6- and 20-wk.-old infant	2	Crying	Lab.	Professional[a]	Yes
R. V. Hall & Broden (1967)	5- and 6-yr.-old males and 9-yr.-old female with CNS dysfunction	3	Behavior considered by staff to be interfering with their developmental progress	Experimental educational unit	Parents and teachers	Yes

356

TABLE 10-2. Summary of Studies on Differential Attention with Children *(Continued)*

AUTHORS	CLIENT(s)	N	BEHAVIOR	SETTING	THERAPIST	EXPERIMENTAL ANALYSIS
Sloane, Johnston, & Bijou (1967)	4-yr.-old male	1	Extreme aggression, temper tantrums, and excessive fantasy play	Remedial nursery school	Teachers	Yes
Buell, Stoddard, Harris, & Baer (1968)	3-yr.-old female	1	Lack of cooperative play and participation in preschool program	Preschool program	Teacher	Yes
Carlson, Arnold, Becker, & Madsen (1968)	8-yr.-old female	1	Tantrums	Classroom	Teacher	No
Ellis (1968)	4- and 5-yr.-old males	5	Aggressive behavior	Lab. school	Teacher and helper	Yes
B. V. Hall, Lund, & Jackson (1968)	Elementary school pupils	6	Disruptive and dawdling study behavior	Poverty area classroom	Teachers	Yes
R. V. Hall, Panyan, Rabon, & Broden (1968)	3 classrooms (1st, 6th, 7th grades)	24	Study behavior	Classroom	Teachers	Yes
Hart, Reynolds, Baer, Brawley, & Harris (1968)	5-yr.-old female	1	Uncooperative play	Preschool	Teacher	Yes
Madsen, Becker, & Thomas (1968)	Elementary school pupils	3	Classroom disruption	Classroom	Teachers	Yes
N. J. Reynolds & Risley (1968)	4-yr.-old female	1	Low frequency of talking	Preschool	Teacher	Yes
D. R. Thomas, Becker, & Armstrong (1968)	6- to 11-yr.-old males and females	10	Disruptive behavior	Classroom	Teacher	Yes
D. R. Thomas, Nielson, Kuypers, & Becker (1968)	6-yr.-old male	1	Disruptive behavior	Classroom	Teacher	Yes
Wahler and Pollio (1968)	8-yr.-old male	1	Excessive dependency and lack of aggressive behavior	University clinic	Parents and therapist	Yes
Ward & Baker (1968)	1st-grade children	4	Disruptive behavior	Classroom	Teacher	Yes

357

TABLE 10-2. Summary of Studies on Differential Attention with Children *(Continued)*

AUTHORS	CLIENT(s)	N	BEHAVIOR	SETTING	THERAPIST	EXPERIMENTAL ANALYSIS
Zeilberger, Sampen, & Sloan (1968)	4½-yr.-old male	1	Disobedience and aggressive behavior	Home	Mother	Yes
Brawley, Harris, Allen, Fleming, & Peterson (1969)	7-yr.-old male	1	Autistic behavior	Hospital day-care unit	Professional[a]	Yes
Cormier (1969)	6th- and 8th-grade classes	18	Disruptive behavior and lack of motivation	Classroom	Teachers	Yes
McCallister, Stachowiak, Baer, & Conderman (1969)	High school English class	25	Inappropriate talking and turning around	Classroom	Teacher	Yes
O'Leary, Becker, Evans, & Saudargas (1969)	2nd graders	7	Disruptive classroom behavior	Classroom	Teacher	Yes
Wahler (1969a)	Elementary school age males	2	Oppositional behavior	Home	Parents	Yes
Wahler (1969b)	5- and 8-yr.-old males	2	Oppositional and disruptive behavior	Home and classroom	Parent and teacher	Yes
Broden, Bruce, Mitchell, Carter, & Hall (1970)	2nd-grade males	2	Disruptive behavior	Poverty area classroom	Teacher	Yes
Broden, Hall, Dunlap, & Clark (1970)	7th- and 8th-grade males and females	13	Disruptive classroom behavior	Special education class	Teacher	Yes
J. C. Conger (1970)	9-yr.-old male	1	Encopresis	Home	Mother	Yes
Goodlet, Goodlet, & Dredge (1970)	5- and 7-yr.-old males	2	Disruptive behavior	University lab. classroom	Teacher	Yes
Schutte & Hopkins (1970)	4- to 6-yr.-old females	5	Instruction following	Classroom	Teacher	Yes
Smeets (1970)	18-yr.-old male	1	Rumination and regurgitation	Hospital room	Teacher	Yes
Wahler, Sperling, Thomas, Teeter, & Luper (1970)	4- and 9-yr.-old males	2	"Beginning" stuttering and mildly deviant behavior	Hearing and speech center	Parents	Yes

TABLE 10-2. Summary of Studies on Differential Attention with Children *(Continued)*

AUTHORS	CLIENT(s)	N	BEHAVIOR	SETTING	THERAPIST	EXPERIMENTAL ANALYSIS
J. Wright, Clayton, & Edger (1970)	Severely retarded children	15	Negative behaviors	State residential institution	Ward technicians	No
Buys (1971)	9 problem and 9 control elementary school pupils	18	Deviant classroom behavior	Classroom	Teacher	Yes
Corte, Wolf, & Locke (1971)	Profoundly retarded adolescents	4	Self-injurious behavior	Hospital training lab.	Professional[a]	No
R. V. Hall et al. (1971)	Individual pupils and classroom groups from 1st-grade— junior high school	Ex.# & N 1. 1 2. 1 3. 1 4. 1 5. 30 6. 27	Disruptive and talking-out behavior	White, middle-class and black poverty classroom	Teacher	Yes
Laws, Brown, Epstein, & Hocking (1971)	Severely disturbed 8- and 9-yr.-old males	3	Behavior that interferes with speech and language	State hospital	Speech therapist	Yes
Nordquist (1971)	5½-yr.-old male	1	Enuresis and oppositional behavior	Home	Parents	Yes
Skiba, Pettigrew, & Alden (1971)	8-yr.-old females	3	Thumbsucking	Classroom	Teacher	Yes
J. D. Thomas & Adams (1971)	Well-behaved and remedial primary school pupils	16	Task-related behavior and lowering sound levels	Classroom	Teacher	Yes
Veenstra (1971)	5- to 14-yr.-old siblings	4	Disruptive behavior	Home	Mother	Yes
Vukelich & Hake (1971)	18-yr.-old severely retarded female	1	Choking and grabbing	State hospital	Ward staff	Yes
Yawkey (1971)	7-yr.-old female 7-yr.-old male	2	Poor attending behavior	Classroom	Teacher	Yes

359

TABLE 10-2. Summary of Studies on Differential Attention with Children *(Continued)*

AUTHORS	CLIENT(s)	N	BEHAVIOR	SETTING	THERAPIST	EXPERIMENTAL ANALYSIS
Barnes, Wooton, & Wood (1972)	3- and 4-yr.-old males and females	24	Immature play	Mental health center	Public health nurse	Yes
R. V. Hall et al. (1972)	4- and 8-yr.-old males and 5- and 10-yr.-old females	4	Whining and failure to wear orthodontic device	Home	Parents	Yes
Hasazi & Hasazi (1972)	8-yr.-old male	1	Digit reversal	Classroom	Teacher	Yes
Herbert & Baer (1972)	5-yr.-old male and female	2	Inappropriate behavior in home	Home	Mother	Yes
Kirby & Shields (1972)	13-yr.-old male	1	Nonattending and poor arithmetic	Classroom	Teacher	Yes
Sajwaj, Twardosz, & Burke (1972)	7-yr.-old retarded male	1	Excessive conversation with teacher	Remedial preschool	Teacher	Yes
Twardosz & Sajwaj (1972)	4-yr.-old hyperactive retarded male	1	Sitting	Remedial preschool	Teacher	Yes
Cossairt, Hall, & Hopkins (1973)	3rd- and 4th-grade males and females	12	Low attending and instruction-following behavior	Elementary schools	Teachers	Yes
Herbert et al. (1973)	5- and 6-yr.-old females, 5-, 7- and 8-yr.-old males	6	Deviant	Preschool classroom and observation lab.	Mothers	Yes
Pinkston, Reese, LeBlanc, & Baer (1973)	3½-yr.-old male	1	Aggressive behaviors with peers and low peer interaction	Preschool classroom	Teacher	Yes
Budd, Green, & Baer (1976)	3-yr.-old female	1	Noncompliance with instructions and considerable demands for attention	University lab. room	Mother	Yes
Munford & Liberman (1978)	13-yr.-old male	1	Operant coughing	1. Hospital 2. Home	1. Hospital staff 2. Parents	Yes
Varni, Russo, & Cataldo (1978)	11-yr.-old male	1	Delusional speech	Psychiatric hospital	Graduate student	Yes

Professional usually refers to Ph.D., Psychologist, or Psychiatrist.

differential attention procedures. The setting was an experimental preschool. In two out of three cases the mothers were quite successful in modifying oppositional behavior in their children, and an experimental analysis isolated differential attention as the important ingredient. In a third child, however, this procedure was not effective, and an additional punishment (time-out) procedure was necessary. The authors did not offer any explanation for this discrepancy, and there were no obvious differences in the cases that could account for the failure based on descriptions in the article. The authors did not seem concerned with the discrepancy, probably because it was an early effort on the replication series, and the goal was to control the oppositional behavior, which was accomplished when time-out was added. This study was important, however, for it contained the first hint that differential attention might not be effective with some cases of oppositional behavior.

In a later series, after differential attention was well established as an effective procedure, further failures to replicate did elicit concern from the investigator (Wahler, 1968, 1969a). Wahler trained parents of children with severe oppositional behavior in differential attention procedures. Results indicated that differential attention was ineffective across five children, but the addition of time-out again produced the desired changes. Replication in two more cases of oppositional behavior confirmed that differential attention was only effective when combined with a time-out procedure.

In the best tradition of science, Wahler (1969a) did not gloss over the failure of differential attention, although his treatment "package" was ultimately successful. Contemplating reasons for the failure, Wahler hypothesized that in cases of severe oppositional behavior, parental reinforcement value may be extremely low; that is, attention from parents is not as reinforcing. After treatment using the combination of time-out and differential attention, oppositional behavior was under control, even though time-out was no longer used. Employing a test of parental reinforcement values, Wahler demonstrated that the treatment package increased the reinforcing value of parental attention, allowing the gain to be maintained. This was the first clear suggestion that therapist variables are important in the application of differential attention, and that with oppositional children particularly, differential attention alone may be ineffective due to the low reinforcing value of parental attention.

Although differential attention occasionally has been found ineffective in other settings, such as the classroom (O'Leary et al., 1969), other investigators actually observed deleterious effects under certain conditions (e.g., Herbert et al., 1973; Sajwaj & Hedges, 1971). For example, Herbert et al. (1973) trained mothers in the use of differential attention in two separate geographical locations (Kansas and Mississippi). Although preschools were the settings in both locations, the design and function of the preschools were quite dissimilar. Clients were children with a variety of disruptive and deviant behaviors, including hyperactivity, oppositional behavior, and other inappropriate social behaviors. These young children presented different background variables, from familial retardation through childhood autism and Down's syndrome, and they came from differing socioeconomic backgrounds. The one similarity among the six cases (two from Mississippi, four from Kansas) was that differential attention from parents was not only ineffective but detrimental in many cases, in that deviant behavior increased, and dangerous and surprising side effects appeared. Deleterious effects of this procedure were confirmed in extensions of A-B-A designs, where behavior worsened under differential attention and improved when the procedure was withdrawn.

These results were, of course, surprising to the authors, and discovery of similar results in two settings through personal communication prompted the combining of the data into a single publication. In this particular report the investigators were unable to pinpoint reasons for these failures. As the authors note, ". . . the results were not peculiar to a particular setting, certain parent-child activities, observation code or recording system, experimenter or parent training procedure. Subject characteristics also were not predictive of the results obtained" (Herbert et al., 1973, p. 26). But in one case where time-out was added, disruptive behavior declined. In fact, Sajwaj and Dillon (1977) analyzed a large portion of their systematic replication series and found a ratio of 87 individual successes to only 27 individual failures. In many of the cases that failed, the addition of another procedure, such as time-out, quickly converted the failure to a success. More recent studies have continued to find that adding time-out corrects differential attention failures (Roberts, Hatzenbuehler, & Bean, 1981).

As noted above, the number of articles analyzing the effects of differential attention with children has dropped off markedly in recent years, as is evident in Table 10-2. Most likely this is due to widespread confidence in its general applicability. But another reason is that the field has moved on. As was the case with various adult behaviors, differential attention has been fully incorporated into a package treatment, usually referred to as *parent training* (e.g., Forehand & McMahon, 1981). This package consists of additional components to differential attention, such as time-out and training in the discrimination of certain instructions or commands. Since this package has been well worked out, the field is now more concerned with results from a clinical replication analysis of the treatment package than with continued systematic replications of the differential attention procedure attempting to determine what conditions predict failure. Yet, in 1979 Wahler, Berland, and Coe referred to these occasional failures of differential attention as one of the anomalies of operant interventions.

Comment on replication

In our view, data on failures are a sign of the maturity of a systematic replication series. Only when a procedure is proven successful through many replications, do negative results assume this importance. But these failures do not detract from the successful replications. The effectiveness of differential attention has been established repeatedly. These data do, however, indicate that there are conditions that even today are not fully understood that limit generality of effectiveness and that practitioners must proceed with caution (Wahler et al., 1979).

In conclusion, this advanced systematic replication series on differential attention has generated a great deal of confidence among practitioners. The evidence indicates that it can be effective with adults and children with a variety of behavioral problems in most any setting. The clinically oriented books and monographs widely advocating its use, most often in combination with other procedures as part of a treatment package (Forehand & McMahon, 1981; Jacobson & Margolin, 1979; Patterson, 1982; Paul & Lentz, 1977), have made this procedure available to numerous professionals concerned with behavior change, as well as to the consuming public. In fact, most editors of appropriate journals probably would not consider accepting another article on differential attention unless it illustrated a clear exception to the effectiveness of this procedure, as did the Herbert et al. (1973) report.

However, the process of establishing generality of findings across all relevant domains is a slow one indeed, and it will probably be years before we know all we should about this treatment or other treatments currently undergoing systematic replication. As we pointed out in the context of adult psychotic behavior, investigators probably proceeded too quickly to incorporating differential attention into various package treatments without fully understanding the limits of its effects. Even with the very informative and complete systematic replication series on childhood problems, we do not yet know what predicts failure from differential attention. In fact, there are many promising hypotheses to account for these failures (Paris & Cairns, 1972; Sajwaj & Dillon, 1977; Wahler, 1969a; Warren & Cairns, 1972). But these have not yet been explored in the applied setting. Until the time that the process of systematic replication reveals the precise limitations of a procedure, clinicians and other behavior change agents should proceed with caution, but also with hope and confidence that this powerful process will ultimately establish the conditions under which a given treatment is effective or ineffective.

Guidelines for systematic replication

The formulation of guidelines for conducting systematic replication is more difficult than for direct replication due to the variety of experimental efforts that comprise a systematic replication series. However, in the interest of providing some structure to future systematic replication, we will attempt to provide an outline of the general procedures necessary for sound systematic replication in applied research. These procedures or guidelines fall into four categories.

1. Earlier we defined systematic replication in applied research as any attempt to replicate findings from a direct replication series, varying settings, behavior change agents, behavior disorders, or some combination thereof. Ideally, then, the systematic replication should begin with sound direct replication where the reliability of a procedure is established and the beginnings of client generality are ascertained. If results in the initial experiment and three or more replications are uniformly successful, then the important work of testing the effectiveness of the procedure in other settings with other therapists and so on can begin. If a series begins with a report of a single case (as it often does), then the first order of business is to initiate a direct replication series on this procedure, so that the search for exceptions can begin.

2. Investigators evaluating systematic replication should clearly note the differences among their clients, therapists, or settings from those in the original experiment. In a conservative systematic replication, one, or possibly two, variables differ from the original direct replication. If more than one or two variables differ, this indicates that the investigator is "gambling" somewhat (Sidman, 1960). That is, if the experiment succeeds, the series will take a large step forward in establishing generality of findings. If the experiment fails, the investigator cannot know which of the differing variables or combination of variables was responsible for the change and must go back and retrace his or her steps. Whether scientists take the gamble depends on the setting and their own inclinations; there is no guideline one could suggest here without also limiting the creativity of the scientific process. But it *is* important to be fully aware of previous efforts in the series and to list the number of ways in which the current experiment differs from past efforts, so that other investigators and clinicians can hypothesize along with the experimenter on which differences were important in the event of failure. In fact, most good scientists do this (e.g., Herbert et al., 1973).

3. Systematic replication is essentially a search for exceptions. If no exceptions are found as replications proceed, then wide generality of findings is established. However, the purpose of systematic replication is to define the conditions under which a technique will succeed or fail, and this means a search for exceptions or failures. Thus any experimental tactics that hinder the finding and reporting of exceptions are of less value than an experimental design that highlights failure. Of those experimental procedures

typically found in a systematic replication series (e.g., see Table 10-2), two fall into this category: the experimental analysis containing only one case and the group study.

As noted above, the report of a single-case, particularly when accompanied by an experimental analysis, can be a valuable addition to a series in that it describes another setting, behavior disorder, or other item where the procedure was successful. Reports of single-cases also may lead to direct and systematic replication, as in the differential attention series. Unfortunately, however, failures in a single-case are seldom published in journals. Among the numerous successful reports of single-case studies contained in the differential attention series, very few reported a failure, although it is our guess that differential attention has failed on many occasions, and these failures simply have not been reported.

The group study suffers from the same limitation because failures are lost in the group average. Again, group studies can play an important role in systematic replication in that demonstration that a technique is successful with a given group, as opposed to individuals in the group, may serve an important function (see section 2.9). In the differential attention series, several investigators thought it important to demonstrate that the procedure could be effective in a classroom as a whole (e.g., Ward & Baker, 1968). These data contributed to generality of findings across several domains. The fact remains, however, that failures will not be detected (unless the whole experiment fails, in which case it would not be published), thus leading us no closer to the goal of defining the conditions in which a successful technique fails. In clinical replication, or *field testing*, described below, one has more flexibility in examining results from large groups of treated clients as long as it is possible to pinpoint individuals who succeed or fail.

4. Finally, the question arises: When is a systematic replication series over? For direct replication series, it was possible to make some tentative recommendations on a number of subjects, given experimental findings. With systematic replication, no such recommendations are possible. In applied research, we would have to agree with Sidman's (1960) conclusion concerning basic research that a series is never over, because scientists will always attempt to find exceptions to a given principle, as well they should. It may be safe to say that a series is over when no exception to a proven therapeutic principle can be found, but, as Sidman pointed out, this is entirely dependent on the complexity of the problem and the inductive reasoning of clinical researchers who will have to judge in the light of new and emerging knowledge which conditions could provide exceptions to old principles. Of course, series will eventually begin to "fade away," as with the differential attention series, when wide generality of applicability has been established.

Fortunately, practitioners do not have to wait for the end of a series to apply interim findings to their clients. In these series, knowledge is cumulative. A clinician may apply a procedure from an advanced series, such as differential attention, with more confidence than procedures from less advanced series (Barlow, 1974). However, it is still possible through inspection of these data to utilize those new procedures with a degree of confidence dependent on the degree to which the experimental clients, therapists, and settings are similar to those facing the clinician. At the very least, this is a good beginning to the often discouraging and sometimes painful process of clinical trial and error.

10.4 CLINICAL REPLICATION

A somewhat different type of replication process occurs only in applied research. We have termed this process *clinical replication* (Hersen & Barlow, 1976). Clinical replication is an advanced replication procedure in which a treatment package containing two or more distinct procedures is applied to a succession of clients with multiple behaviors or emotional problems that cluster together; in other words, the usual and customary types of multifaceted problems that present to practitioners such as conduct problems in children, depression, schizophrenia, or autism.

Direct replication was defined as the administration of a given procedure by the same investigator or group of investigators in a specific setting (e.g., hospital, clinic, classroom) on a series of clients homogeneous for a particular behavior disorder such as agoraphobia or compulsive hand washing. As this definition implies, one component of a treatment procedure is applied to one well-defined problem in succeeding clients. Similarly, systematic replication examines the effectiveness of this functional relationship across multiple settings, therapists, and (related) behaviors. Most often, direct and systematic replications are testing only one component of what eventually becomes a treatment package, as in the examples above.

In constructing an effective treatment package, however, it is very important that one develop and test treatments for one problem at a time, with the eventual goal of combining successful treatments for all coexisting problems. This is the technique-building strategy suggested by Bergin and Strupp (1972). For example, one of the direct replication series described above tested the effects of a specified treatment on delusional speech, which, of course, is often one component of schizophrenia (Wincze et al., 1972). If this series were consistently successful, the applied researcher might begin to test treatments for coexisting problems in these patients, such as social isolation or thought disorders, if these were present. When successful procedures had been developed for all coexisting problems, the next step would be to establish generality of findings by replicating this treatment package on additional patients who present a similar combination of problems. This would be clinical replication (e.g., Wallace, 1982). The insertion of differential attention, time-out, and other well-tested procedures into a "parenting" package is a good example of technique building resulting in a treatment ready for clinical replication.

Another name for clinical replication, then, could be field testing, because this is where clinicians and practitioners take newly developed treatments or newly modified treatments and apply them to the common, everyday problems encountered in their practice. While this process can be carried out by either full-time clinical investigators or scientist-practitioners (Barlow et al., 1983), establishing the widest possible client and setting generality would require substantial participation by full-time practitioners. The job of these practitioners, then, would be to apply these treatments to large numbers of their clients while observing and recording successes and failures and analyzing through experimental strategies, where possible, the reasons for this individual variation. But even if practitioners are not inclined to analyze causes for failures in the application of a particular treatment package, full descriptions of these failures will be extremely important for those investigators who are in a position to carry on this search (Barlow et al., 1983).

Thus, while all facets of single-case experimental research are much closer to the procedures in clinical or applied practice than to other types of research methodology (see below), clinical replication in its most elementary form becomes almost identical with the activities of practitioners.

Definition of clinical replication

We would define clinical replication as the administration by the same investigator or practitioner of a treatment package containing two or more distinct treatment procedures. These procedures would be administered in a specific setting to a series of clients presenting similar combinations of multiple behavioral and emotional problems. Obviously, this type of replication process is advanced in that it should be the end result of a systematic, technique-building applied research effort, which should take years.

Of course, there are many clinical replication series in the literature describing the application of comprehensive treatments that did not benefit from careful technique-building strategies. One good example is the Masters and Johnson series describing the treatment of sexual dysfunction. Because of this weakness, this treatment approach, which enjoys wide application, is now coming under increasing attack as one that does not have wide generality of effectiveness (Zilbergeld & Evans, 1980). And, since no technique-building strategy preceded the introduction of this treatment, we have no idea why.

Example: Clinical replication with autistic children

One of the best examples of a clinical replication series is the work of Lovaas and his colleagues with autistic children (e.g., Lovaas, Berberich, Perloff, & Schaeffer, 1966; Lovaas, Schaeffer, & Simmons, 1965; Lovaas & Simmons, 1969). The diagnosis of autism fulfills the requirements of clinical replication in that it subsumes a number of behavioral or emotional problems and is a major clinical entity. Lovaas, Koegel, Simmons, and Long (1973) listed eight distinct problems that may contribute to the autistic syndrome: (1) apparent sensory deficit, (2) severe affect isolation, (3) self-stimulating behavior, (4) mutism, (5) echolalic speech, (6) deficits in receptive speech, (7) deficits in social and self-help behaviors, and (8) self-injurious behavior. Step-by-step, they developed and tested treatments for each of these behaviors, such as self-destructive behavior (Lovaas & Simmons, 1969), language acquisition (e.g., Lovaas et al., 1966), and social and self-help skills (Lovaas, Freitas, Nelson, & Whalen, 1967). These procedures were tested in separate direct replication series on the initial group of children. The treatment package constructed from these direct replication series was administered to subsequent children presenting a sufficient number of these behaviors to be labeled autistic.

Lovaas et al. (1973) presented the results and follow-up data from the initial clinical replication series for 13 children. Results were presented in terms of response of the group as a whole, as well as of individual improvement across the variety of behavioral and emotional problems. While these data are complex, they can be summarized as follows. All children demonstrated increases in appropriate behaviors and decreases in inappropriate behaviors. There were marked differences in the amount of improvement. At least one child was returned to a normal school setting, while several children improved very little and required continued institutionalization. In other words, each child improved, but the change was not clinically dramatic for several children.

Because clinical replication is similar to direct replication, it can be analyzed in a similar fashion, and conclusions can be made in two general areas. *First*, the treatment package can be effective for behaviors subsumed under the autistic syndrome. This conclusion is based on (1) the initial experimental analysis of each component of the treatment package in the original direct replication series (e.g., Lovaas & Simmons, 1969) and (2) the withdrawal and reintroduction of this whole package in A-B-A-B fashion in several children (Lovaas et al., 1973). *Second*, replication of this finding across all subjects indicates that the data are reliable and not due to idiosyncracies in one child. It does not follow, however, that generality across children was established. As in example 3 in the section on direct replication (10.2), the results were clear and clinically significant for several children, but the results were also weak and clinically unimportant for several children. Thus the package has only limited generality across clients, and the task remains to pinpoint differences between children who improved and those who did not improve. From these differences, possible causes for limitations on client generality should emerge.

In fact, children in this series were quite heterogeneous. In many respects, this was due to an inherent difficulty in clinical replication—the vagueness and unreliability of many diagnostic categories. As Lovaas et al. (1973) pointed out, ". . . the delineation of 'autism' is one area that will demand considerably more work. It has not been a particularly useful diagnosis. Few people agree on when to apply it" (p. 156). It follows that heterogeneity of clients will most likely be greater than in a direct replication series, where the target behavior is well defined and clients can be matched more closely.

Thus the causes of failure in a series with mixed results are more difficult to ascertain, due to the greater number of differences among individuals. Nevertheless, it is necessary to pinpoint these differences and begin the search for intersubject variability. As Lovaas et al (1973) concluded:

> Finally a major focus of future research should attempt more functional descriptions of autistic children. As we have shown, the children responded in vastly different ways to the treatment we gave them. We paid scant attention to individual differences when we treated the first twenty children. In the future, we will assess such individual differences. (p. 163)

In the meantime, child clinicians would do well to examine closely the exemplary series by Lovaas and his associates to determine logical generalization to children under their care.

Taking cues from this initial clinical replication series, the investigators in this research group have since improved their treatment package, based on a long-term analysis of individual differences, and hypothesized reasons for failure or minimal success. Subsequent experimental analyses have isolated procedures and strategies that seem to improve the training program as a whole (e.g., Koegel & Schreibman, 1982; Schreibman, Koegel, Mills, & Burke, in press). These innovations, with particular emphasis on parent training, combined with new and more valid measures of overall change, have made possible another more advanced clinical replication series currently under way.

Guidelines for clinical replication are similar to those for direct replication when series are relatively small and contain four to six clients. A detailed discussion of series containing 20, 50, or even 100 clients was presented in Barlow et al. (1983).

10.5. ADVANTAGES OF REPLICATION
OF SINGLE-CASE EXPERIMENTS

In view of the reluctance of clinical researchers to carry out the large-scale replication studies required in traditional experimental design (Bergin & Strupp, 1972), one might be puzzled by the seeming enthusiasm with which investigators undertake replication efforts using single-case designs, as evidenced by the differential attention series and other less advanced series. A quick examination of Table 10-2 demonstrates that there is probably little or no savings in time or money when compared to the large-scale collaborative factorial designs initially proposed by Bergin and Strupp (1972). No fewer clients are involved and, in all likelihood, more applied researchers and settings are involved. Why, then, does this replication tactic succeed when Bergin and Strupp concluded that the alternative could not be implemented? In our view, there are four very important but rather simple reasons.

First, the effort is *decentralized*. Rather than in the type of large collaborative factorial study necessary to determine generality of findings at a cost of millions of dollars, the replication efforts are carried out in many settings such that funding, when available, is dispersed. This, of course, is more practical for government or other funding sources, who are not reluctant to award $10,000 to each of 100 investigators but would be quite reluctant to award $100,000 to one group of investigators. Often, of course, these small studies involving three or four subjects are unfunded. Also, rather than administering a large collaborative study from a central location where all scientists or therapists are to carry out prescribed duties, each scientist administers his or her own replication effort based on his or her ideas and views of previous findings (see Barlow et al., 1983). What is lost here is some efficiency, since there is no guarantee that the next obvious step in the replication series will be carried out at the logical time. What is gained is the freedom and creativity of individual scientists to attack the problem in their own ways.

Second, systematic replication will continue because the professional contingencies are favorable to its success. The professional contingencies in this case are publications and the accompanying professional recognition. Initial efforts in a series experimentally demonstrating success of a technique on a single case are publishable. Direct replications are publishable. Systematic replications are publishable each time the procedure is successful in a different setting or with a different behavior disorder or whatever. Finally, after a procedure has been proven effective, failures or exceptions to the success are publishable. It is a well-established principle in psychology that intermittent reinforcement, preferably on a short-variable interval schedule, is more effective in maintaining behavior (in this case the replication series) than the schedule arrangement for a large group study, where years may pass before publishable data are available.

Third, the experimental analysis of the single-case is close to the clinic. As noted in chapter 1, this approach tends to merge the role of scientist and practitioner. Many an important series has started only after the clinician confronted an interesting case. Subsequently, measures were developed, and an experimental analysis of the treatment was performed (Mills et al., 1973). As a result, the data increase one's understanding of the problem, but the client also receives and benefits from treatment. If one plans to treat the patient, it is an easy enough matter to develop measures and perform the necesssary experimental analyses. The recent book mentioned above (Barlow et al., 1983) was designed to explore this potential in our full-time practitioners by demonstrating how they can incorporate these principles into their practices and thereby participate in the research process. This ability to work with ease within the clinical setting, more than any other fact, may ensure the future of meaningful replication efforts.

Finally, as noted above, the results of the series are cumulative, and each new replicative effort has some immediate payoff for the practicing clinician. As this is the ultimate goal of the applied researcher, it is far more satisfactory than participating in a multiyear collaborative study where knowledge or benefit to the clinician is a distant goal.

Nevertheless, the advancement of a systematic replication series is a long and arduous road full of pitfalls and dead ends. In the face of the immediate demands on clinicians and behavior change agents to provide services to society, it is tempting to "grab the glimmer of hope" provided by treatments that prove successful in preliminary reports or case studies. That these hopes have been repeatedly dashed as therapeutic techniques and schools of therapy have come and gone supplies the most convincing evidence that the slow but inexorable process of the scientific method is the only way to meaningful advancement in our knowledge. Although we are a long way from the sophistication of the physical sciences, the single case experimental design with adequate replication may provide us with the methodology necessary to overcome the complex problems of human behavior disorders.

Hiawatha Designs an Experiment

Maurice G. Kendall

(Originally published in *The American Statistician*, Dec. 1959, Vol. 13, No. 5. Reprinted by Permission).

Hiawatha, mighty hunter,
He could shoot ten arrows upwards
Shoot them with such strength and
   swiftness
That the last had left the bowstring
Ere the first to earth descended.
This was commonly regarded
As a feat of skill and cunning.

One or two sarcastic spirits
Pointed out to him, however,
That it might be much more useful
If he sometimes hit the target.
Why not shoot a little straighter
And employ a smaller sample?

Hiawatha, who at college
Majored in applied statistics,
Consequently felt entitled
To instruct his fellow men on
Any subject whatsoever,
Waxed exceedingly indignant
Talked about the law of error,
Talked about truncated normals,
Talked of loss of information,
Talked about his lack of bias,
Pointed out that in the long run
Independent observations
Even though they missed the target
Had an average point of impact
Very near the spot he aimed at
(With the possible exception
Of a set of measure zero).

This, they said, was rather
   doubtful.
Anyway, it didn't matter
What resulted in the long run;
Either he must hit the target
Much more often than at present
Or himself would have to pay for
All the arrows that he wasted.

Hiawatha, in a temper,
Quoted parts of R. A. Fisher
Quoted Yates and quoted Finney
Quoted yards of Oscar Kempthorne
Quoted reams of Cox and Cochran
Quoted Anderson and Bancroft
Practically *in extenso*
Trying to impress upon them
That what actually mattered
Was to estimate the error.

One or two of them admitted
Such a thing might have its uses.
Still, they said, he might do better
If he shot a little straighter.

Hiawatha, to convince them,
Organized a shooting contest
Laid out in the proper manner
By experimental methods
Recommended in the textbooks
(Mainly used for tasting tea, but
Sometimes used in other cases)
Randomized his shooting order

In factorial arrangements
Used the theory of Galois
Fields of ideal polynomials,
Got a nicely balanced layout

And successfully confounded
Second-order interactions.

All the other tribal marksmen
Ignorant, benighted creatures,
Of experimental set-ups
Spent their time of preparation
Putting in a lot of practice
Merely shooting at a target.

Thus it happened in the contest
That their scores were most
    impressive
With one notable exception
This (I hate to have to say it)
Was the score of Hiawatha,
Who, as usual, shot his arrows
Shot them with great strength and
    swiftness
Managing to be unbiased
Not, however, with his salvo
Managing to hit the target.
There, they said to Hiawatha
That is what we all expected.

Hiawatha, nothing daunted,
Called for pen and called for paper
Did analyses of variance
Finally produced the figures
Showing, beyond peradventure,
Everybody else was biased
And the variance components
Did not differ from each other

Or from Hiawatha's
(This last point, one should
    acknowledge

Might have been much more
    convincing
If he hadn't been compelled to
Estimate his own component
From experimental plots in
Which the values all were missing.
Still, they didn't understand it
So they couldn't raise objections.
This is what so often happens
With analyses of variance.)

All the same, his fellow tribesmen
Ignorant, benighted heathens,
Took away his bow and arrows,
Said that though my Hiawatha
Was a brilliant statistician
He was useless as a bowman.
As for variance components,
Several of the more outspoken
Made primeval observations
Hurtful to the finer feelings
Even of a statistician.

In a corner of the forest
Dwells alone my Hiawatha
Permanently cogitating
On the normal law of error,
Wondering in idle moments
Whether an increased precision
Might perhaps be rather better,
Even at the risk of bias,
If thereby one, now and then, could
Register upon the target.

References

Abel, G. G., Blanchard, E. B., Barlow, D. H., & Flanagan, B. (1975, December). *A controlled behavioral treatment of a sadistic rapist.* Paper presented at the meeting of the Association for Advancement of Behavior Therapy, San Francisco.

Agras, W. S. (1975). Behavior modification in the general hospital psychiatric unit. In H. Leitenberg (Ed.), *Handbook of behavior modification* (pp. 547–565). Englewood Cliffs, NJ: Prentice-Hall.

Agras, W. S., Barlow, D. H., Chapin, H. N., Abel, G. G., & Leitenberg, H. (1974). Behavior modification of anorexia nervosa. *Archives of General Psychiatry, 30,* 279–286.

Agras, W. S., Kazdin, A. E., & Wilson, G. T. (1979). *Behavior Thearpy: Toward an applied clinical science.* San Francisco: W. H. Freeman.

Agras, W. S., Leitenberg, H., & Barlow, D. W. (1968). Social reinforcement in the modification of agoraphobia. *Archives of General Psychiatry, 19,* 423–427.

Agras, W. S., Leitenberg, H., Barlow, D. H., Curtis, N. A., Edwards, J. A., & Wright, D. E. (1971). Relaxation in systematic desensitization. *Archives of General Psychiatry, 25,* 511–514.

Agras, W. S., Leitenberg, H., Barlow, D. H., & Thomson, L. E. (1969). Instructions and reinforcement in the modification of neurotic behavior. *American Journal of Psychiatry, 125,* 1435–1439.

Alford, G. S., Blanchard, E. B., & Buckley, M. (1972). Treatment of hysterical vomiting by modification of social contingencies: A case study. *Journal of Behavior Therapy and Experimental Psychiatry, 3,* 209–212.

Alford, G. S., Webster, J. S., & Sanders, S. H. (1980). Covert aversion of two interrelated deviant sexual practices: Obscene phone calling and exhibitionism. A single case analysis. *Behavior Therapy, 11,* 15–25.

Allen, K. E., & Harris, F. R. (1966). Elimination of a child's excessive scratching by training the mother in reinforcement procedures. *Behaviour Research and Therapy, 4,* 79–84.

Allen, K. E., Hart, B. M., Buell, J. S., Harris, F. R., & Wolf, M. M. (1964). Effects of social reinforcement on isolate behavior of a nursery school child. *Child Development, 35,* 511–518.

Allen, K. E., Henke, L. B., Harris, F. R., Baer, D. M., & Reynolds, N. J. (1967). Control of hyperactivity by social reinforcement of attending behavior. *Journal of Educational Psychology, 58,* 231–237.

Allison, M. G., & Ayllon, T. (1980). Behavioral coaching in the development of skills in football, gymnastics, and tennis. *Journal of Applied Behavior Analysis, 13,* 297–314.

Allport, G. D. (1961). *Pattern and growth in personality.* New York: Holt, Rinehart and Winston.

Allport, G. D. (1962). The general and the unique in psychological science. *Journal of Personality, 30,* 405–422.

Altman, J. (1974). Observational study of behavior: Sampling methods. *Behaviour, 49,* 227–267.

American Psychological Association. (1973). *Ethical principles in the conduct of research with human participants.* Washington, DC: Author.

Anderson, R. L. (1942). Distribution of the serial correlation coefficient. *Annals of Mathematical Statistics, 13,* 1–13.

Anderson, R. L. (1971). *The statistical analysis of time series.* New York: Wiley.

Arrington, R. E. (1939). Time-sampling studies of child behavior. *Psychological Monography, 51* ( ).

Arrington, R. E. (1943). Time sampling in studies of social behavior: A critical review of techniques and results with research suggestions. *Psychological Bulletin, 40,* 81–124.

Ashem, R. (1963). The treatment of a disaster phobia by systematic desensitization. *Behaviour Research and Therapy, 1,* 81–84.

Atiqullah, M. (1967). On the robustness of analysis of variance. *Bulletin of the Institute of Statistical Research and Training, 1,* 77–81.

Ault, M. E., Peterson, R. F., & Bijou, S. W. (1968). *The management of contingencies of reinforcement to enhance study behavior in a small group of young children.* Unpublished manuscript.

Ayllon, T. (1961). Intensive treatment of psychotic behavior by stimulus satiation and food reinforcement. *Behaviour Research and Therapy, 1,* 53–61.

Ayllon, T., & Azrin, N. H. (1965). The measurement and reinforcement of behavior of psychotics. *Journal of the Experimental Analysis of Behavior, 8,* 357–383.

Ayllon, T., & Azrin, N. H. (1968). *The token economy: A motivational system for therapy and rehabilitation.* New York: Appleton-Century-Crofts.

Ayllon, T., & Haughton, E. (1964). Modification of symptomatic verbal behavior of mental patients. *Behaviour Research and Therapy, 2,* 87–91.

Ayllon, T., & Michael, J. (1959). The psychiatrist nurse as a behavioral engineer. *Journal of the Experimental Analysis of Behavior, 2,* 323–334.

Azrin, N. H., Holz, W., Ulrich, R., & Goldiamond, I. (1961). The control of the content of conversation through reinforcement. *Journal of the Experimental Analysis of Behavior, 4,* 25–30.

Baer, D. M. (1971). Behavior modification: You shouldn't. In E. Ramp & B. L. Hopkins (Eds.), *A new direction for education: Behavior analysis.* Lawrence, KS: Lawrence University Press.

Baer, D. M. (1977a). "Perhaps it would be better not to know everything." *Journal of Applied Behavior Analysis, 10,* 167–172.

Baer, D. M. (1977b). Reviewer's comment: Just because it's reliable doesn't mean that you can use it. *Journal of Applied Behavior Analysis, 10,* 117–119.

Baer, D. M., & Guess, D. (1971). Receptive training of adjectival inflections in mental retardates. *Journal of Applied Behavior Analysis, 4,* 129–139.

Baer, D. M., Wolf, M. M., & Risley, T. R. (1968). Some current dimensions of applied behavior analysis. *Journal of Applied Behavior Analysis, 1,* 91–97.

Bailey, J. S., Wolf, M. M., & Phillips, E. L. (1970). Home-based reinforcement and the modification of pre-delinquents' classroom behavior. *Journal of Applied Behavior Analysis, 3,* 223–233.

Bakeman, R. (1978). Untangling streams of behavior: Sequential analysis of observational data. In G. P. Sackett (Ed.), *Observing behavior: Vol. 2. Data collection and analysis methods* (pp. 63–78). Baltimore: University Park Press.

Ban, T. (1969). *Psychopharmacology.* Baltimore: Williams & Wilkins.

Bandura, A. (1969). *Principles of behavior modification.* New York: Holt, Rinehart & Winston.

Barker, R. G., & Wright, H. F. (1955). *Midwest and its children: The psychological ecology of an American town.* New York: Harper & Row.

Barlow, D. H. (1974). The treatment of sexual deviation: Towards a comprehensive behavioral approach. In K. S. Calhoun, H. E. Adams, & K. M. Mitchell (Eds.), Innovative treatment methods in psychopathology. New York: John Wiley & Sons, Inc., 1974.

Barlow, D. H. (1980). Behavior therapy: The next decade. *Behavior Therapy, 11,* 315–328.

Barlow, D. H. (Ed.). (1981). *Behavioral assessment of adult disorders.* New York: Guilford Press.

Barlow, D. H., Agras, W. S., Leitenberg, H., Callahan, E. J., & Moore, R. C. (1972). The contributions of therapeutic instructions to covert sensitization. *Behaviour Research and Therapy, 10*, 411–415.

Barlow, D. H., Becker, R., Leitenberg, H., & Agras, W. S. (1970). A mechanical strain gauge for recording penile circumference change. *Journal of Applied Behavior Analysis, 3*, 73–76.

Barlow, D. H., Blanchard, E. B., Hayes, S. C., & Epstein, L. H. (1977). Single case designs and biofeedback experimentation. *Biofeedback and Self-Regulation, 2*, 211–236.

Barlow, D. H., & Hayes, S. C. (1979). Alternating treatments design: One strategy for comparing the effects of two treatments in a single subject. *Journal of Applied Behavior Analysis, 12*, 199–210.

Barlow, D. H., Hayes, S. C., & Nelson, R. O. (1983). *The scientist-practitioner: Research and accountability in clinical and educational settings.* Elmsford, New York: Pergamon Press.

Barlow, D. H., & Hersen, M. (1973). Single case experimental designs: Uses in applied clinical research. *Archives of General Psychiatry, 29*, 319–325.

Barlow, D. H., Leitenberg, H., & Agras, W. S. (1969). Experimental control of sexual deviation through manipulation of the noxious scene in covert sensitization. *Journal of Abnormal Psychology, 74*, 596–601.

Barlow, D. H., Leitenberg, H., Agras, W. S., & Wincze, J. P. (1969). The transfer gap in systematic desensitization: An analogue study. *Behaviour Research and Therapy, 7*, 191–197.

Barlow, D. H., Mavissakalian, M., & Schofield, L. (1980). Patterns of desynchrony in agoraphobia: A preliminary report. *Behaviour Research and Therapy, 18*, 441–448.

Barmann, B. C., Katz, R. C., O'Brien, F., & Beauchamp, K. L. (1981). Treating irregular enuresis in developmentally disabled persons: A study in the use of overcorrection. *Behavior Modification, 5*, 336–346.

Barnes, K. E., Wooton, M., & Wood, S. (1972). The public health nurse as an effective therapist-behavior modifier of preschool play behavior. *Community Mental Health Journal, 8*, 3–7.

Barrera, R. D., & Sulzer-Azaroff, B. (1983). An alternating treatment comparison or oral and total communication training program with echolalic autistic children. *Journal of Applied Behavior Analysis, 16*, 379–395.

Barrett, R. P., Matson, J. L., Shapiro, E. S., & Ollendick, T. H. (1981). A comparison of punishment and DRO procedures for treating stereotypic behavior of mentally retarded children. *Applied Research in Mental Retardation, 2*, 247–256.

Barrios, B. A., & Hartmann, D. P. (in press). Traditional assessment's contributions to behavioral assessment: Concepts, issues, and methodologies. In. R. O. Nelson & S. C. Hayes (Eds.), *Conceptual foundations of behavioral assessment.* New York: Guilford Press.

Barrios, B. A., Hartmann, D. P., & Shigetomi, C. (1981). Fears and anxieties in children. In E. J. Mash & L. G. Terdal (Eds.), *Behavioral assessment of childhood disorders* (pp. 259–304). New York: Guilford Press.

Barron, F., & Leary, T. (1955). Changes in psychoneurotic patients with and without psychotherapy. *Journal of Consulting Psychology, 19*, 239–245.

Barton, E. S., Guess, D., Garcia, E., & Baer, D. M. (1970). Improvement of retardates' mealtime behaviors by timeout procedures using multiple baseline techniques. *Journal of Applied Behavior Analysis, 3*, 77–84.

Bates, P. (1980). The effectiveness of interpersonal skills training on the social acquisition of moderately and mildly retarded adults. *Journal of Applied Behavior Analysis, 13*, 237–248.

Baum, C. G., Forehand, R. L., & Zegiob, L. E. (1979). A review of observer reactivity in adult-child interactions. *Journal of Behavioral Assessment, 1*, 167–178.

Beck, A. T., Rush, A. J., Shaw, B. J., & Emery, G. (1979). *Cognitive therapy of depression.* New York: Guilford Press.

Beck, A. T., Ward, C. H., Mendelson, M., Mock, J., & Erbaugh, J. (1961). An inventory for measuring depression. *Archives of General Psychiatry, 4*, 561–571.

Beck, S. J. (1953). The science of personality: Nomothetic or idiographic. *Psychological Review*, *60*, 353–359.

Bellack, A. S., & Hersen, M. (1977). The use of self-report inventories in behavior assessment. In J. D. Cone & R. P. Hawkins (Eds.), *Behavior assessment: New direction in clinical psychology* (pp. 52–76). New York: Brunner/Mazel.

Bellack, A. S., Hersen, M., & Himmelhoch, J. M. (1981). Social skills training, pharmacotherapy and psychotherapy for unipolar depression. *American Journal of Psychiatry*, *138*, 1562–1567.

Bellack, A. S., Hersen, M., & Turner, S. M. (1976). Generalization effects of social skills training in chronic schizophrenics: An experimental analysis. *Behaviour Research and Therapy*, *14*, 381–398.

Bellack, L., & Chassan, J. B. (1964). An approach to the evaluation of drug effects during psychotherapy: A double-blind study of a single case. *Journal of Nervous and Mental Disease*, *139*, 20–30.

Bergin, A. E. (1966). Some implications of psychotherapy research for therapeutic practice. *Journal of Abnormal Psychology*, *71*, 235–246.

Bergin, A. E., & Lambert, M. J. (1978). The evaluation of therapeutic outcomes. In S. L. Garfield & A. E. Bergin (Eds.), *Handbook of psychotherapy and behavior change: An empirical analysis* (2nd ed.). (pp. 139–191). New York: Wiley.

Bergin, A. E., & Strupp, H. H. (1970). New directions in psychotherapy research. *Journal of Abnormal Psychology*, *76*, 13–26.

Bergin, A. E., & Strupp, H. H. (1972). *Changing frontiers in the science of psychotherapy.* New York: Aldine.

Berk, R. A. (1979). Generalizability of behavioral obvservations: A classification of interobserver agreement and interobserver reliability. *American Journal of Mental Deficiency*, *83*, 460–472.

Berler, E. S., Gross, A. M., & Drabman, R. S. (1982). Social skills training with children: Proceed with caution. *Journal of Applied Behavior Analysis*, *15*, 41–53.

Bernard, C. (1957). *An introduction to the study of experimental medicine.* New York: Dover.

Bernard, M. E., Kratochwill, T. R., & Keefauver, L. W. (1983). The effects of rational-emotive therapy and self-instructional training on chronic hair pulling. *Cognitive Therapy and Research*, *7*, 273–280.

Bickman, L. (1976). Observational methods. In C. Selltiz, L. S. Wrightsman, & S. W. Cook (Eds.), *Research methods in social relations* (pp. 251–290). New York: Holt, Rinehart and Winston.

Bijou, S. W., Peterson, R. F., & Ault, M. E. (1968). A method to integrate descriptive and experimental field studies at the level of data and empirical concepts. *Journal of Applied Behavior Analysis*, *1*, 175–191.

Bijou, S. W., Peterson, R. F., Harris, F. R., Allen, K. E., & Johnston, M. S. (1969). Methodology for experimental studies of young children in natural settings. *Psychological Record*, *19*, 177–210.

Birkimer, J. C., & Brown, J. H. (1979). Back to basics: Percentage agreement measures are adequate, but there are easier ways. *Journal of Applied Behavior Analysis*, *12*, 535–543.

Birnbrauer, J. S., Peterson, C. P., & Solnick, J. V. (1974). The design and interpretation of studies of single subjects. *American Journal of Mental Deficiency*, *79*, 191–203.

Birney, R. C., & Teevan, R. C. (Eds.). (1965). *Reinforcement.* Princeton, NJ: Van Nostrand.

Bittle, R., & Hake, D. F. (1977). A multielement design model for component analysis and cross-setting assessment of a treatment package. *Behavior Therapy*, *8*, 906–914.

Blanchard, E. B. (1981). Behavioral assessment of psychophysiological disorders. In D. H. Barlow (Ed.), *Behavioral assessment of adult disorders* (pp. 239–269). New York: Guilford Press.

Blough, P. M. (1983). Local contrast in multiple schedules: The effect of stimulus discriminability.

Journal of the Experimental Analysis of Behavior, , 39, 427–437.

Boer, A. P. (1968). Application of a single recording system to the analysis of free-play behavior in autistic children. *Journal of Applied Behavior Analysis, 1,* 335–340.

Boice, R. (1983). Observational skills. *Psychological Bulletin, 93,* 3–29.

Bolger, H. (1965). The case study method. In B. B. Wolman (Ed.), *Handbook of clinical psychology* (pp. 28–39). New York: McGraw-Hill.

Boring, E. G. (1950). *A history of experimental psychology.* New York: Appleton-Century-Crofts.

Bornstein, M. R., Bellack, A. S., & Hersen, M. (1977). Social-skills training for unassertive children: A multiple-baseline analysis. *Journal of Applied Behavior Analysis, 10,* 183–195.

Bornstein, M. R., Bellack, A. S., & Hersen, M. (1980). Social skills training for highly aggressive children: Treatment in an inpatient psychiatric setting. *Behavior Modification, 4,* 173–186.

Bornstein, P. H., Bridgewater, C. A., Hickey, J. S., & Sweeney, T. M. (1980). Characteristics and trends in behavioral assessment: An archival analysis. *Behavioral Assessment, 2,* 125–133.

Bornstein, P. H., Hamilton, S. B., Carmody, T. B., Rychtarik, R. G., & Veraldi, D. M. (1977). Reliability enhancement: Increasing the accuracy of self-report through mediation-based procedures. *Cognitive Therapy and Research, 1,* 85–98.

Bornstein, P. H., & Rychtarik, R. G. (1983). Consumer satisfaction in adult behavior therapy: Procedures, problems, and future perspective.*Behavior Therapy, 14,* 191–208.

Bowdlear, C. M. (1955). *Dynamics of idiopathic epilepsy as studied in one case.* Unpublished doctoral dissertation, Case Western Reserve University, Cleveland, Ohio.

Box, G. E. P., & Jenkins, G. M. (1970). *Time series analysis: Forecasting and control.* San Francisco: Holden-Day.

Box, G. E. P., & Tiao, G. C. (1965). A change in level of non-stationary time series. *Biometrika, 52,* 181–192.

Boykin, R. A., & Nelson, R. O. (1981). The effects of instruction and calculation procedures on observers' accuracy, agreement, and calculation correctness. *Journal of Applied Behavior Analysis, 14,* 479–489.

Bradley, L. A., & Prokop, C. K. (1982). Research methods in contemporary medical psychology. In P. C. Kendall & J. N. Butcher (Eds.), *Handbook of research methods in clinical psychology* (pp. 591–649). New York: Wiley.

Brady, J. P., & Lind, D. L. (1961). Experimental analysis of hysterical blindness. *Archives of General Psychiatry, 4,* 331–339.

Brawley, E. R., Harris, F. R., Allen, K. E., Fleming, R. S., & Peterson, R. F. (1969). Behavior modification of an autistic child. *Behavioral Science, 14,* 87–97.

Breuer, J., & Freud, S. (1957). *Studies on hysteria.* New York: Basic Books.

Breuning, S. E., O'Neill, M. J., & Ferguson, D. G. (1980). Comparison of psychotropic drugs, response cost, and psychotropic drug plus response cost procedures for controlling institutionalized mentally retarded persons. *Applied Research in Mental Retardation, 1,* 253–268.

Brill, A. A. (1909). Selected papers on hysteria and other psychoneuroses: Sigmund Freud. *Nervous and Mental Disease Monograph Series, 4.*

Broden, M., Bruce, C., Mitchell, M. A., Carter, V., & Hall, R. V. (1970). Effects of teacher attention on attending behavior of two boys at adjacent desks. *Journal of Applied Behavior Analysis, 3,* 205–211.

Broden, M., Hall, R. V., Dunlap, A., & Clark, R. (1970). Effects of teacher attention and a token reinforcement system in a junior high school special education class. *Exceptional Children, 36,* 341–349.

Brookshire, R. H. (1970). Control of "involuntary" crying behavior emitted by a multiple sclerosis patient. *Journal of Community Disorders, 1,* 386–390.

Browning, R. M. (1967). A same-subject design for simultaneous comparison of three reinforcement contingencies. *Behaviour Research and Therapy, 5,* 237–243.

Browning, R. M., & Stover, D. O. (1971). *Behavior modification in child treatment: An experimental and clinical approach.* Chicago: Aldine.

Brunswick, E. (1956). *Perception and the representative design of psychological experiments.* Berkeley: University of California Press.

Bryant, L. E., & Budd, K. S. (1982). Self-instructional training to increase independent work performance in preschoolers. *Journal of Applied Behavior Analysis, 15,* 259–271.

Budd, K. S., Green, D. R., & Baer, D. M. (1976). An analysis of multiple misplaced parental social contingencies. *Journal of Applied Behavior Analysis, 9,* 459–470.

Buell, J. S., Stoddard, P., Harris, F. R., & Baer, D. M. (1968). Collateral social development accompanying reinforcement of outdoor play in a preschool child. *Journal of Applied Behavior Analysis, 1,* 167–173.

Burgio, L. D., Whitman, T. L., & Johnson, M. R. (1980). A self-instructional package for increasing attending behavior in educable mentally retarded children. *Journal of Applied Behavior Analysis, 13,* 443–459.

Buys, C. J. (1971). Effects of teacher reinforcement on classroom behaviors and attitudes. *Dissertation Abstracts International, 31,* 4884A1–4885A.

Campbell, D. T. (1969). Reforms as experiments. *American Psychologist, 24,* 409–429.

Campbell, D. T., & Fiske, D. W. (1959). Convergent and discriminant validation by the multitrait-multimethod matrix. *Psychological Bulletin, 56,* 81–105.

Campbell, D. T., & Stanley, J. C. (1963). *Experimental and quasi-experimental designs for research.* In D. T. Campbell and J. C. Stanley, *Handbook of Research on Teaching.* Chicago: Rand McNally.

Campbell, D. T., & Stanley, J. C. (1966). *Experimental and quasi-experimental designs for research.* Chicago: Rand McNally.

Carey, R. G., & Bucher, B. (1983). Positive practice overcorrection: The effects of duration of positive practice on acquisition and response relation. *Journal of Applied Behavior Analysis, 16,* 101–111.

Carlson, C. S., Arnold, C. R., Becker, W. C., & Madsen, C. H. (1968). The elimination of tantrum behavior of a child in an elementary classroom. *Behaviour Research and Therapy, 6,* 117–119.

Carver, R. P. (1974). Two dimensions of tests: Psychometric and edumetric. *American Psychologist, 29,* 512–518.

Catania, A. C. (Ed.), (1968). *Contemporary research in operant behavior.* Glenview, IL: Scott, Foresman.

Chaplin, J. P. (1975). *Dictionary of psychology (Rev. Ed.).* New York: Dell Publishing.

Chaplin, J. P., & Kraweic, T. S. (1960). *Systems and theories of psychology.* New York: Holt, Rinehart and Winston.

Chassan, J. B. (1960). Statistical inference and the single case in clinical design. *Psychiatry, 23,* 173–184.

Chassan, J. B. (1962). Probability processes in psychoanalytic psychiatry. In J. Scher (Ed.), *Theories of the mind* (pp. 598–618). New York: Free Press of Glencoe.

Chassan, J. B. (1967). *Research design in clinical psychology and psychiatry.* New York: Appleton-Century-Crofts.

Chassan, J. B. (1979). *Research design in clinical psychology and psychiatry (2nd ed.)* New York: Irvington.

Ciminero, A. R., Calhoun, K. S., & Adams, H. E. (Eds.), (1977). *Handbook of behavioral assessment.* New York: Wiley.

Coates, T. J., & Thoresen, C. E. (1981). Sleep disturbances in children and adolescents. In E. J. Mash & L. G. Terdal (Eds.), *Behavioral assessment of childhood disorders* (pp. 639–678). New York: Guilford.

Cohen, D. C. (1977). Comparison of self-report and overt-behavioral procedures for assessing acrophobia. *Behavior Therapy, 8*, 17–23.

Cohen, J. (1960). A coefficient of agreement for nominal scales. *Educational and Psychological Measurement, 20*, 37–46.

Cohen, J. (1968). Weighted kappa: Nominal scale agreement with provisions for scale disagreement or partial credit. *Psychological Bulletin, 70*, 313–220.

Cohen, L. H. (1976). Clinicians' utilization of research findings. *JSAS Catalog of Selected Documents in Psychology, 6*, 116.

Cohen, L. H. (1979). The research readership and information source reliance of clinical psychologists. *Professional Psychology, 10*, 780–786.

Coleman, R. A. (1970). Conditioning techniques applicable to elementary school classrooms. *Journal of Applied Behavior Analysis, 3*, 293–297.

Cone, J. D. (1977). The relevance of reliability and validity for behavior assessment. *Behavior Therapy, 8*, 411–426.

Cone, J. D. (1979). Confounded comparisons in triple response mode assessment research. *Behavioral Assessment, 1*, 85–95.

Cone, J. D. (1982). Validity of direct observation assessment procedures. In D. P. Hartmann (Ed.), *Using observers to study behavior: New directions for methodology of social and behavioral science* (pp. 67–79). San Francisco: Jossey-Bass.

Cone, J. D., & Foster, S. L. (1982). Direct observations in clinical psychology. In P. C. Kendall & J. N. Butcher (Eds.), *Handbook of research methods in clinical psychology* (pp. 311–354). New York: Wiley.

Cone, J. D., & Hawkins, R. P. (Eds.). (1977). *Behavior assessment: New directions in clinical psychology.* New York: Brunner/Mazel.

Conger, A. J. (1980). Integration and generalization of kappas for multiple raters. *Psychological Bulletin, 88*, 322–328.

Conger, J. C. (1970). The treatment of encopresis by the management of social consequences. *Behavior Therapy, 1*, 386–390.

Conover, W. J. (1971). *Practical nonparametric statistics.* New York: Wiley.

Conrin, J., Pennypacker, H. S., Johnston, J. M., & Rast, J. (1982). Differential reinforcement of other behaviors to treat chronic rumination of mental retardates. *Journal of Behavior Therapy and Experimental Psychiatry, 13*, 325–329.

Cook, T. D., & Campbell, D. T. (Eds.). (1979). *Quasi-experimentation: Design and analysis issues for field settings.* Chicago: Rand McNally.

Cormier, W. H. (1969). Effects of teacher random and contingent social reinforcement on the classroom behavior of adolescents. *Dissertation Abstracts International, 31*, 1615A–1616A.

Corte, H. E., Wolf, M. M., & Locke, B. J. (1971). A comparison of procedures for eliminating self-injurious behavior of retarded adolescents. *Journal of Applied Behavior Analysis, 4*, 201–215.

Cossairt, A., Hall, R. V., & Hopkins, B. L. (1973). The effects of experimenters' instructions, feedback, and praise on teacher praise and student attending behavior. *Journal of Applied Behavior Analysis, 6*, 89–100.

Creer, T. L., Chai, H., & Hoffman, A. (1977). A single application of an aversive stimulus to eliminate chronic cough. *Journal of Behavior Therapy and Experimental Psychiatry, 8*, 107–109.

Cronbach, L. J. (1970). *Essentials of psychological testing* (3rd ed.). New York: Harper & Row.

Cronbach, L. J. (1971). Test validation. In R. L. Thorndike (Ed.), *Educational measurement* (pp. 443–507). Washington: American Council on Education.

Cronbach, L. J., Gleser, G. C., Nanda, H., & Rajaratnam, N. (1972). *The dependability of behavioral measurements: Theory of generalizability for scores and profiles.* New York: Wiley.

Cuvo, A. J., Leaf, R. B., & Borakove, L. S. (1978). Teaching janitorial skills to the mentally retarded: Acquisition, generalization, and maintenance. *Journal of Applied Behavior Analysis, 11*, 345–355.

Cuvo, A. J., & Riva, M. T. (1980). Generalization and transfer between comprehension and production: A comparison of retarded and nonretarded persons. *Journal of Applied Behavior Analysis, 13,* 215–231.

Dalton, K. (1959). Menstruation and acute psychiatric illness. *British Medical Journal, 1,* 148–149.

Dalton, K. (1960a). Menstruation and accidents. *British Medical Journal, 2,* 1425–1426.

Dalton, K. (1960b). School girls' behavior and menstruation. *British Medical Journal, 2,* 1647–1649.

Dalton, K. (1961). Menstruation and crime. *British Medical Journal, 2,* 1752–1753.

Davidson, P. O., & Costello, C. G. (1969). *N = 1: Experimental studies of single cases.* New York: Van Nostrand Reinhold.

Davis, K. V., Sprague, R. L., & Werry, J. S. (1969). Stereotyped behavior and activity level in severe retardates: The effect of drugs. *American Journal of Mental Deficiency, 73,* 721–727.

Davis, V. J., Poling, A. D., Wysocki, T., & Breuning, S. E. (1981). Effects of Phenytoin withdrawal on matching to sample and workshop performance of mentally retarded persons. *Journal of Nervous and Mental Disease, 169,* 718–725.

Davison, G. C. (1965). The training of undergraduates as social reinforcers for autistic children. In L. P. Ullmann & L. Krasner (Eds.), *Case studies in behavior modification* (pp. 146–148). New York: Holt, Rinehart and Winston.

DeProspero, A., & Cohen, S. (1979). Inconsistent visual analysis of intrasubject data. *Journal of Applied Behavior Analysis, 12,* 573–579.

Doke, L. A. (1976). Assessment of children's behavioral deficits. In M. Hersen & A. S. Bellack (Eds.), *Behavioral assessment* (pp. 493–536). Elmsford, New York: Pergamon Press.

Doke, L. A., & Risley, T. R. (1972). The organization of day-care environments: Required *vs* optional activities. *Journal of Applied Behavior Analysis, 5,* 405–420.

Dollard, J., Doob, L. W., Miller, N. E., Mowrer, O. H., & Sears, R. R. (1939). *Frustration and aggression.* New Haven: Yale University Press.

Domash, M. A., Schnelle, J. F., Stomatt, E. L., Carr, A. F., Larson, L. D., Kirchner, R. E., & Risley, T. R. (1980). Police and prosecution systems: An evaluation of a police criminal case preparation program. *Journal of Applied Behavior Analysis, 13,* 397–406.

Drabman, R. S., Hammer, D., & Rosenbaum, M. S. (1979). Assessing generalization in behavior modification with children: The generalization map. *Behavioral Assessment, 1,* 203–219.

Dukes, W. F. (1965). N = 1. *Psychological Bulletin, 64,* 74–79.

duMas, F. M. (1955). Science and the single case. *Psychological Reports, 1,* 65–75.

Dunlap, G., & Koegel, R. L. (1980). Motivating autistic children through stimulus variation. *Journal of Applied Behavior Analysis, 13,* 619–627.

Dunlap, K. (1932). *Habits: Their making and unmaking.* New York: Liveright.

Dyer, K., Christian, W. P., & Luce, S. C. (1982). The role of response delay in improving the discrimination performance of autistic children. *Journal of Applied Behavior Analysis, 15,* 231–240.

Edelberg, R. (1972). Electrical activity of the skin. In N. S. Greenfield & R. A. Sternbach (Eds.), Handbook of psychophysiology (pp. 367–418). New York: Holt, Rinehart and Winston.

Edgington, E. S. (1966). Statistical inference and nonrandom samples. *Psychological Bulletin, 66,* 485–487.

Edgington, E. S. (1967). Statistical inference from N = 1 experiments. *Journal of Psychology, 65,* 195–199.

Edgington, E. S. (1969). *Statistical inference: The distribution-free approach.* New York: Mc-Graw-Hill.

Edgington, E. S. (1972). N = 1 experiments: Hypothesis testing. *Canadian Psychologist, 13,* 121–135.

Edgington, E. S. (1980a). *Randomization tests.* New York: Marcel Dekker.

Edgington, E. S. (1980b). Validity of randomization tests for one-subject experiments. *Journal of*

Educational Statistics, 5, 235-251.

Edgington, E. S. (1982). Nonparametric tests for single-subject multiple schedule experiments. *Behavioral Assessment, 4*, 83-91.

Edgington, E. S. (1983). Response-guided experimentation. *Contemporary Psychology, 28*, 64-65.

Edgington, E. S. (1984). Statistics and single case analysis. In M. Hersen, R. M. Eisler, & P. M. Monti (Eds.). *Progress in Behavior Modification* (Vol. 16). New York: Academic Press.

Edwards, A. L. (1968). *Experimental design in psychological research* (3rd ed.). New York: Holt, Rinehart and Winston.

Egel, A. L., Richman, G. S., & Koegel, R. L. (1981). Normal peer models and autistic children's learning. *Journal of Applied Behavior Analysis, 14*, 3-12.

Eisler, R. M., & Hersen, M. (August, 1973). *The A-B design: Effects of token economy on behavioral and subjective measures in neurotic depression.* Paper presented at the meeting of the American Psychological Association, Montreal.

Eisler, R. M., Hersen, M., & Agras, W. S. (1973). Effects of videotape and instructional feedback on nonverbal marital interaction: An analog study. *Behavior Therapy, 4*, 551-558.

Eisler, R. M., Miller, P. M., & Hersen, M. (1973). Components of assertive behavior. *Journal of Clinical Psychology, 29*, 295-299.

Elkin, T. E., Hersen, M., Eisler, R. M., & Williams, J. G. (1973). Modification of caloric intake in anorexia nervosa: An experimental analysis. *Psychological Reports, 32*, 75-78.

Ellis, D. P. (1968). The design of a social structure to control aggression. *Dissertation Abstracts, 29*, 672A.

Emmelkamp, P. M. G. (1974). Self-observation versus flooding in the treatment of agoraphobia. *Behaviour Research and Therapy, 12*, 229-237.

Emmelkamp, P. M. G. (1982). *Phobic and obsessive-compulsive disorders: Theory, research and practice.* New York: Plenum.

Emmelkamp, P. M. G., & Kwee, K. G. (1977). Obsessional ruminations: A comparison between thought stopping and prolonged exposure in imagination. *Behaviour Research and Therapy, 15*, 441-444.

Epstein, L. H., Beck, S. J., Figueroa, J., Farkas, G., Kazdin, A. E., Daneman, D., & Becker, D. (1981). The effects of targeting improvements in urine glucose on metabolic control in children with insulin dependent diabetes. *Journal of Applied Behavior Analysis, 14*, 365-375.

Epstein, L. H., & Hersen, M. (1974). Behavioral control of hysterical gagging. *Journal of Clinical Psychology, 30*, 102-104.

Epstein, L. H., Hersen, M., & Hemphill, D. P. (1974). Music feedback in the treatment of tension headache: An experimental case study. *Journal of Behavior Therapy and Experimental Psychiatry, 5*, 59-63.

Etzel, B. C., & Gerwitz, J. L. (1967). Experimental modifications of caretaker-maintained highrate operant crying in a 6- and 20-week-old infant (*Infans tyrannotearus*): Extinction of crying with reinforcement of eye contact and smiling. *Journal of Experimental Child Psychology, 5*, 303-317.

Evans, I. M. (1983). Behavioral assessment. In C. E. Wallace (Ed.), *Handbook of clinical psychology: Vol. 1. Theory, research, and practice* (pp. 391-419). Homewood, IL: Dow Jones-Irwin.

Evans, I. M., & Wilson, F. E. (1983). Behavioral assessment on decision making: A theoretical analysis. In M. Rosenbaum, C. M. Franks, & Y. Jaffe (Eds.), *Perspectives on behavior therapy in the eighties* (Vol. 9, (pp. 35-53). New York: Springer Publishing.

Eyberg, S. M., & Johnson, S. M. (1974). Multiple assessment of behavior modification with families: Effects of contingency contracting and order of treated problems. *Journal of Consulting and Clinical Psychology, 42*, 594-606.

Eysenck, H. J. (1952). The effects of psychotherapy: An evaluation. *Journal of Consulting Psychology, 16*, 319-324.

Eysenck, H. J. (1965). The effects of psychotherapy. *International Journal of Psychiatry, 1,* 97–178.

Ezekiel, M., & Fox, K. A. (1959). *Methods of correlation and regression analysis: Linear and curvilinear.* New York: Wiley.

Fairbank, J. A., & Keane, T. M. (1982). Flooding for combat-related stress disorders: Assessment of anxiety reduction across traumatic memories. *Behavior Therapy, 13,* 499–510.

Fisher, E. B. (1979). Overjustification effects in token economies. *Journal of Applied Behavior Analysis, 12,* 407–415.

Fisher, R. A. (1925). On the mathematical foundations of the theory of statistics. In Cambridge Phil. Society (Ed.), *Theory of statistical estimation* (Proceedings of the Cambridge Philosophical Society) England.

Fjellstedt, N., & Sulzer-Azaroff, B. (1973). Reducing the latency of a child's responding to instructions by means of a token system. *Journal of Applied Behavior Analysis, 6,* 125–130.

Fleiss, J. H. (1975). Measuring agreement between two judges on the presence or absence of a trait. *Biometrics, 31,* 651–659.

Foa, E. B. (1979). Failure in treating obsessive-compulsives. *Behaviour Research and Therapy, 17,* 169–175.

Foa, E. B., Grayson, J. B., Steketee, G. S., Doppelt, H. G., Turner, R. M., & Latimer, P. R. (1983). Success and failure in the behavioral treatment of obsessive compulsives. *Journal of Consulting and Clinical Psychology, 51,* 287–297.

Forehand, R. L. (Ed.). (1983). Mini-series on consumer satisfaction and behavior therapy. *Behavior Theraoy, 14,* 189–246.

Forehand, R. L., & McMahon, R. J. (1981). *Helping the noncompliant child: A clinician's guide to parent training.* New York: Guilford Press.

Frank, J. D. (1961). *Persuasion and healing.* Baltimore: Johns Hopkins University Press.

Freund, K., & Blanchard, R. (1981). Assessment of sexual dysfunction and deviation. In M. Hersen & A. S. Bellack (Eds.), *Behavioral assessment: A practical handbook* (2nd ed., pp. 427–455). Elmsford, New York: Pergamon Press.

Frick, T., & Semmel, M. I. (1978). Observer agreement and reliability of classroom observational measures. *Review of Educational Research, 48,* 157–184.

Feuerstein, M., & Adams, H. E. (1977). Cephalic vasomotor feedback in the modification of migraine headache. *Biofeedback and Self-Regulation, 3,* 241–254.

Garfield, S. L., & Bergin, A. E. (Eds.). (1978). *Handbook of psychotherapy and behavior change: An empirical analysis* (2nd ed.). New York: Wiley.

Geer, J. H. (1965). The development of a scale to measure fear. *Behaviour Research and Therapy, 13,* 45–53.

Gelfand, D. M., Gelfand, S., & Dobson, W. R. (1967). Unprogrammed reinforcement of patients' behavior in a mental hospital. *Behaviour Research and Therapy, 5,* 201–207.

Gelfand, D. M., & Hartmann, D. P. (1975). *Child behavior analysis and therapy.* Elmsford, N.Y.: Pergamon Press.

Gelfand, D. M., & Hartmann, D. P. (1984). *Child behavior: Analysis and therapy* (2nd ed.). Elmsford, New York: Pergamon Press.

Geller, E. S., Winett, R. A., & Everett, P. B. (1982). *Preserving the environment: New strategies for behavior change.* Elmsford, New York: Pergamon Press.

Gentile, J. R., Roden, A. H., & Klein, R. D. (1972). An analysis of variance model for the intrasubject replication design. *Journal of Applied Behavior Analysis, 5,* 193–198.

Glass, G. S., Heninger, G. R., Lansky, M., & Talan, K. (1971). Psychiatric emergency related to the menstrual cycle. *American Journal of Psychiatry, 128,* 705–711.

Glass, G. V., Peckham, P. D., & Sanders, J. R. (1972). Consequences of failure to meet assumptions underlying the fixed effects analyses of variance and covariance. *Review of Educational Research, 42,* 237–288.

Glass, G. V., Willson, V. L., & Gottman, J. M. (1974). *Design and analysis of time-series experiments.* Boulder: Colorado Associated University Press.

Goetz, E. M., & Baer, D. M. (1973). Social control of form diversity and the emergence of new forms in children's blockbuilding. *Journal of Applied Behavior Analysis, 6,* 209–217.

Goldfried, M. R., & D'Zurilla, T. J. (1969). A behavioral-analytic model for assessing competence. In C. D. Spielberger (Ed.), *Current topics in clinical and community psychology* (Vol. 1, pp. 151–196). New York: Academic Press.

Goldfried, M. R., & Linehan, M. M. (1977). Basic issues in behavioral assessment. In A. R. Ciminero, K. S. Calhoun, & H. E. Adams (Eds.), *Handbook of behavioral assessment* (pp. 15–46). New York: Wiley.

Goldstein, M. K. (1971). Behavior rate change in marriages: Training wives to modify husbands' behavior. *Dissertation Abstracts International, 32,* 559A.

Goodlet, G. R., Goodlet, M. M., & Dredge, M. (1970). Modification of disruptive behavior of two young children and follow-up one year later. *Journal of School Psychology, 8,* 60–63.

Goodman, L. A., & Gilman, A. (1975). *The pharmacological basis of therapeutics.* New York: Macmillan.

Gorsuch, R. L. (1983). Three models for analyzing limited time-series (*N* of 1) data. *Behavioral Assessment, 5,* 141–154.

Gottman, J. M. (1973). N-of-one and N-of-two research in psychotherapy. *Psychological Bulletin, 80,* 93–105.

Gottman, J. M. (1979). *Marital interaction: Experimental investigations.* New York: Academic Press.

Gottman, J. M. (1981). *Time-series analysis: A comprehensive introduction for social scientists.* Cambridge: Cambridge University Press.

Gottman, J. M., & Glass, G. V. (1978). Analysis of interrupted time-series experiments. In T. R. Kratochwill (Ed.), *Single-subject research: Strategies for evaluating change* (pp. 197–237). New York: Academic Press.

Gottman, J. M., McFall, R. M., & Barnett, J. T. (1969). Design and analysis of research using time series. *Psychological Bulletin, 72,* 299–306.

Greenfield, N. A., & Sternbach, R. A. (Eds.). (1972). *Handbook of psychophysiology.* New York: Holt, Rinehart and Winston.

Greenspoon, J. (1955). The reinforcing effect of two spoken sounds on the frequency of two responses. *American Journal of Psychology, 68,* 409–416.

Greenwald, A. G. (1976). Within-subjects designs: To use or not to use? *Psychological Bulletin, 1976, 83,* 314–320.

Grinspoon, L., Ewalt, J., & Shader, R. (1967). Long term treatment of chronic schizophrenia. *International Journal of Psychiatry, 4,* 116–128.

Hall, C., Sheldon-Wildgen, J., & Sherman, J. A. (1980). Teaching job interview skills to retarded clients. *Journal of Applied Behavior Analysis, 13,* 433–442.

Hall, R. V., Axelrod, S., Tyler, L., Grief, E., Jones, F. C., & Robertson, R. (1972). Modification of behavior problems in the home with a parent as observer and experimenter. *Journal of Applied Behavior Analysis, 5,* 53–74.

Hall, R. V., & Broden, M. (1967). Behavior changes in brain-injured children through social reinforcement. *Journal of Experimental Child Psychology, 5,* 463–479.

Hall, R. V., Cristler, C., Cranston, S. S., & Tucker, B. (1970). Teachers and parents as researchers using multiple baseline designs. *Journal of Applied Behavior Analysis, 3,* 247–255.

Hall, R. V., Fox, R., Willard, D., Goldsmith, L., Emerson, M., Owen, M., Davis, F., & Porcia, E. (1971). The teacher as observer and experimenter in the modification of disputing and talking-out behaviors. *Journal of Applied Behavior Analysis, 4,* 141–149.

Hall, R. V., Lund, D., & Jackson, D. (1968). Effects of teacher attention on study behavior. *Journal of Applied Behavior Analysis, 1,* 1–12.

Hall, R. V., Panyan, M., Rabon, D., & Broden, M. (1968). Instructing beginning teachers in reinforcement procedures which improve classroom control. *Journal of Applied Behavior*

Analysis, 1, 315–322.

Hallahan, D. P., Lloyd, J. W., Kneedler, R. D., & Marshall, K. J. (1982). A comparison of the effects of self- versus teacher-assessment of on-task behavior. *Behavior Therapy, 13,* 715–723.

Halle, J. W., Baer, D. M., & Spradlin, J. E. (1981). Teachers' generalized use of delay as a stimulus control procedure to increase language use in handicapped children. *Journal of Applied Behavior Analysis, 14,* 389–409.

Harbert, T. L., Barlow, D. H., Hersen, M., & Austin, J. B. (1974). Measurement and modification of incestuous behavior: A case study. *Psychological Reports, 34,* 79–86.

Harris, F. R., Johnston, M. K., Kelley, C. S., & Wolf, M. M. (1964). Effects of positive social reinforcement on regressed crawling of a nursery school child. *Journal of Educational Psychology, 55,* 35–41.

Hart, B. M., Allen, K. E., Buell, J. S., Harris, F. R., & Wolf, M. M. (1964). Effects of social reinforcement on operant crying. *Journal of Experimental Child Psychology, 1,* 145–153.

Hart, B. M., Reynolds, N. J., Baer, D. M., Brawley, E. R., & Harris, F. R. (1968). Effect of contingent social reinforcement on the cooperative play of a preschool child. *Journal of Applied Behavior Analysis, 1,* 73–76.

Hartmann, D. P. (1974). Forcing square pegs into round holes: Some comments on "An analysis-of-variance model for the intrasubject replication design." *Journal of Applied Behavior Analysis, 7,* 635–638.

Hartmann, D. P. (1976). Some restrictions in the application of the Spearman-Brown prophecy formula to observational data. *Educational and Psychological Measurement, 36,* 843–845.

Hartmann, D. P. (1977). Consideration in the choice of interobserver reliability estimates. *Journal of Applied Behavior Analysis, 10,* 103–116.

Hartmann, D. P. (1982). Assessing the dependability of observational data. In D. P. Hartmannn (Ed.), *Using observers to study behavior: New directions for methodology of social and behavioral science* (pp. 51–65). San Francisco: Jossey-Bass.

Hartmann, D. P. (1983). Editorial. *Behavioral Assessment, 5,* 1–3.

Hartmann, D. P., & Gardner, W. (1979). On the not so recent invention of interobserver reliability statitics: A commentary on two articles by Birkimer and Brown. *Journal of Applied Behavior Analysis, 12,* 559–560.

Hartmann, D. P., & Gardner, W. (1981). Considerations in assessing the reliability of observations. In E. E. Filsinger & R. A. Lewis (Eds.), *Assessing marriage* (pp. 184–196). Beverly Hills: Sage.

Hartmann, D. P., Gottman, J. M., Jones, R. R., Gardner, W., Kazdin, A. E., & Vaught, R. S. (1980). Interrupted time-series analysis and its application to behavioral data. *Journal of Applied Behavior Analysis, 13,* 543–559.

Hartmann, D. P., & Hall, R. V. (1976). The changing criterion design. *Journal of Applied Behavior Analysis, 9,* 527–532.

Hartmann, D. P., Roper, B. L., & Bradford, D. C. (1979). Some relationships between behavioral and traditional assessment. *Journal of Behavioral Assessment, 1,* 3–21.

Hartmann, D. P., Roper, B. L., & Gelfand, D. M. (1977). Evaluation of alternative modes of child psychotherapy. In B. Lahen & A. Kazdin (Eds.), *Advances in child clinical psychology* (Vol 1, pp. 1–46). New York: Plenum.

Hartmann, D. P., & Wood, D. D. (1982). Observation methods. In A. S. Bellack, M. Hersen, & A. E. Kazdin (Eds.), *International handbook of behavior modification and therapy* (pp. 109–138). New York: Plenum.

Hasazi, J. E., & Hasazi, S. E. (1972). Effects of teacher attention on digit-reversal behavior in an elementary school child. *Journal of Applied Behavior Analysis, 5,* 157–162.

Hawkins, R. P. (1975). Who decided *that* was the problem? Two stages of responsibility for applied behavior analysis. In W. S. Wood (Ed.), *Issues in evaluating behavior modification* (pp. 95–214). Champaign, IL: Research Press.

Hawkins, R. P. (1979). The functions of assessment: Implications for selection and development of devices for assessing repertoires in clinical, educational, and other settings. *Journal of Applied Behavior Analysis, 12,* 501–516.

Hawkins, R. P. (1982). Developing a behavior code. In D. P. Hartmann (Ed.), *Using observers to study behavior: New directions for methodology of social and behavioral science* (pp. 21–35). San Francisco: Jossey-Bass.

Hawkins, R. P., Axelrod, S., & Hall, R. V. (1976). Teachers as behavior analysts: Precisely monitoring student performance. In J. A. Brigham, R. P. Hawkins, J. Scott, & J. F. McLaughlin (Eds.), *Behavior analysis in education: Self-control and reading* (pp. 274–296). Dubuque, IA: Kendall/Hunt.

Hawkins, R. P., & Dobes, R. W. (1977). Behavioral definitions in applied behavior analysis: Explicit or implicit. In B. C. Etzel, J. M. LeBlanc, & D. M. Baer (Eds.), *New directions in behavioral research: Theory, methods, and applications. In honor of Sidney W. Bijou* (pp. 167–188). Hillsdale, NJ: Erlbaum.

Hawkins, R. P., & Dotson, V. A. (1975). Reliability scores that delude: An Alice in Wonderland trip through the misleading characteristics of interobserver agreement scores in interval recording. In E. Ramp & G. Semb (Eds.), *Behavior analysis: Areas of research and application* (pp. 359–376). Englewood Cliffs, NJ: Prentice-Hall.

Hawkins, R. P., & Fabry, B. D. (1979). Applied behavior analysis and interobserver reliability: A commentary on two articles by Birkimer and Brown. *Journal of Applied Behavior Analysis, 12,* 545–552.

Hawkins, R. P., Peterson, R. F., Schweid, E., & Bijou, S. W. (1966). Behavior therapy in the home: Amelioration of problem parent-child relations with the parent in a therapeutic role. *Journal of Experimental Child Psychology, 4,* 99–107.

Hay, L. R., Nelson, R. O., & Hay, W. M. (1980). Methodological problems in the use of participation observers. *Journal of Applied Behavior Analysis, 13,* 501–504.

Hayes, S. C. (1981). Single case experimental design and empirical clinical practice. *Journal of Consulting and Clinical Psychology, 49,* 193–211.

Haynes, S. N. (1978). *Principles of behavioral assessment.* New York: Gardner Press.

Haynes, S. N., & Wilson, C. C. (1979). *Behavioral assessment.* San Francisco: Jossey-Bass.

Hendrickson, J. M., Strain, P. S., Tremblay, A., & Shores, R. E. (1982). Interactions of behaviorally handicapped children: Functional effects of peer social interactions. *Behavior Modification, 6,* 323–353.

Herbert, E. W., & Baer, D. M. (1972). Training parents as behavior modifiers: Self-recording of contingent attention. *Journal of Applied Behavior Analysis, 5,* 139–149.

Herbert, E. W., Pinkston, E. M., Hayden, M. L., Sajwaj, T. E., Pinkston, S., Cordua, G., & Jackson, C. (1973). Adverse effects of differential parental attention. *Journal of Applied Behavior Analysis, 6,* 15–30.

Herman, S. H., Barlow, D. H., & Agras, W. S. (1974a). An experimental analysis of classical conditioning as a method of increasing heterosexual arousal in homosexuals. *Behavior Therapy, 5,* 33–47.

Herman, S. H., Barlow, D. H., & Agras, W. S. (1974b). An experimental analysis of exposure to "explicit" heterosexual stimuli as an effective variable in changing arousal patterns of homosexuals. *Behaviour Research and Therapy, 12,* 335–345.

Herrnstein, R. J. (1970). On the law of effect. *Journal of the Experimental Analysis of Behavior, 13,* 243–266.

Hersen, M. (1973). Self-assessment of fear. *Behavior Therapy, 4,* 241–257.

Hersen, M. (1978). Do behavior therapists use self-report as major criteria? *Behavioral Analysis and Modification, 2,* 328–334.

Hersen, M. (1981). Complex problems require complex solutions. *Behavior Therapy, 12,* 15–29.

Hersen, M. (1982). Single-case experimental designs. In A. S. Bellack, M. Hersen, & A. E. Kazdin (Eds.), *International handbook of behavior modification and therapy* (pp. 167–201). New York: Plenum.

Hersen, M., & Bellack, A. S. (1976). A multiple-baseline analysis of social-skills training in chronic schizophrenics. *Journal of Applied Behavior Analysis, 9,* 239–245.

Hersen, M., & Bellack, A. S. (Eds.), (1981). *Behavioral assessment: A practical handbook* (2nd ed.). Elmsford, New York: Pergamon Press.

Hersen, M., & Breuning, S. E. (Eds.), (in press). *Pharmacological and behavioral treatment: An integrated approach.* New York: Wiley.

Hersen, M., Eisler, R. M., Alford, G. S., & Agras, W. S. (1973). Effects of token economy on neurotic depression: An experimental analysis. *Behavior Therapy, 4,* 392–397.

Hersen, M., Eisler, R. M., & Miller, P. M. (1973). Development of assertive responses: Clinical, measurement, and research considerations. *Behaviour Research and Therapy, 11,* 505–522.

Hersen, M., Gullick, E. L., Matherne, P. M., & Harbert, T. L. (1972). Instructions and reinforcement in the modification of a conversion reaction. *Psychological Reports, 31,* 719–722.

Hersen, M., Miller, P. M., & Eisler, R. M. (1973). Interactions between alcoholics and their wives: A descriptive analysis of verbal and non-verbal behavior. *Quarterly Journal of Studies on Alcohol, 34,* 516–520.

Hilgard, J. R. (1933). The effect of early and delayed practice on memory and motor performances studies by the method of co-twin control. *Genetic Psychology Monographs, 14,* 493–567.

Hinson, J. M., & Malone, J. C., Jr. (1980). Local contrast and maintained generalization. *Journal of the Experimental Analysis of Behavior, 34,* 263–272.

Hoch, P. H., & Zubin, J. (Eds.). (1964). *The evaluation of psychiatric treatment.* New York: Grune & Stratton.

Hollandsworth, J. G., Glazeski, R. C., & Dressel, M. E. (1978). Use of social skills training in the treatment of extreme anxiety and deficit verbal skills in the job interview setting. *Journal of Applied Behavior Analysis, 11,* 259–269.

Hollenbeck, A. R. (1978). Problems of reliability in observational research. In G. P. Sackett (Ed.), *Observing behavior: Vol. 1. Data collection and analysis methods* (pp. 79–98). Baltimore: University Park Press.

Hollon, S. D., & Bemis, K. M. (1981). Self-report and the assessment of cognitive funcitons. In M. Hersen & A. S. Bellack (Eds.), *Behavioral assessment: A practical handbook* (2nd ed.) (pp. 125–174). Elmsford, New York: Pergamon Press.

Holm, R. A. (1978). Techniques of recording observational data. In G. P. Sackett (Ed.), *Observing behavior: Vol. 2. Data collection and analysis methods* (pp. 99–108). Baltimore: University Park Press.

Holmes, D. S. (1966). The application of learning theory to the treatment of a school behavior problem: A case study. *Psychology in the School, 3,* 355–359.

Holtzman, W. H. (1963). Statistical models for the study of change in the single case. In C. W. Harris (Ed.), *Problems in measuring change* (pp. 199–211). Madison, WI: University of Wisconsin Press.

Honig, W. K. (Ed.), (1966). *Operant behavior: Areas of research and application.* New York: Appleton-Century-Crofts.

Hopkins, B. L., Schutte, R. C., & Garton, K. L. (1971). The effects of access to a playroom on the rate and quality of printing and writing of first- and second-grade students. *Journal of Applied Behavior Analysis, 4,* 77–87.

Horne, G. P., Yang, M. C. K., & Ware, W. B. (1982). Time series analysis for single-subject designs. *Psychological Bulletin, 91,* 178–189.

Horner, R. D., & Baer, D. M. (1978). Multiple-probe technique: A variation of the multiple baseline. *Journal of Applied Behavior Analysis, 11,* 189–196.

House, A. E., House, B. J., & Campbell, M. B. (1981). Measures of interobserver agreement: Calculation formulas and distribution effects. *Journal of Behavioral Assessment, 3,* 37–57.

Hubert, L. J. (1977). Kappa revisited. *Psychological Bulletin, 84,* 289–297.

Hundert, J. (1982). Training teachers in generalized writing of behavior modification programs for multihandicapped deaf children. *Journal of Applied Behavior Analysis, 15,* 111–122.

Hurlbut, B. I., Iwata, B. A., & Green, J. D. (1982). Nonvocal language acquisition in adolescents with severe physical disabilities: Blissymbol versus iconic stimulus formats. *Journal of Applied Behavior Analysis, 15,* 241–258.

Hutt, S. J., & Hutt, C. (1970). *Direct observation and measurement of behavior.* Springfield, IL: Charles C Thomas.

Hyman, R., & Berger, L. (1966). Discussion: In H. J. Eysenck (Ed.), *The effects of psychotherapy* (pp. 81–86). New York: International Science Press.

Inglis, J. (1966). *The scientific study of abnormal behavior.* Chicago: Aldine.

Jacobson, N. S., & Margolin, G. (1979). *Marital therapy: Strategies based on social learning and behavior exchange principles.* New York: Brunner/Mazel.

Jayaratne, S., & Levy, R. L. (1979). *Empirical clinical practice.* New York: Columbia University Press.

Johnson, S. M., & Bolstad, O. D. (1973). Methodological issues in naturalistic observation: Some problems and solutions for field research. In L. A. Homerlynck, L. C. Handy, & E. J. Mash (Eds.), *Behavior change: Methodology, concepts, and practice* (pp. 7–67). Champaign, IL: Research Press.

Johnson, S. M., & Lobitz, G. K. (1974). Parental manipulation of child behavior in home observations. *Journal of Applied Behavior Analysis, 7,* 23–31.

Johnston, J. M. (1972). Punishment of human behavior. *American Psychologist, 27,* 1033–1054.

Johnston, J. M., & Pennypacker, H. S. (1981). *Strategies and tactics of human behavioral research.* Hillsdale, NJ: Erlbaum.

Johnston, M. K., Kelley, C. S., Harris, F. R., & Wolf, M. M. (1966). An application of reinforcement principles to development of motor skills of a young child. *Child Development, 37,* 379–387.

Joint Commission on Mental Illness and Health (1961). *Action for mental health.* New York: Science Editions.

Jones, R. R., Reid, J. B., & Patterson, G. R. (1975). Naturalistic observation in clinical assessment. In P. McReynolds (Ed.), *Advances in psychological assessment* (Vol. 3, pp. 42–95). San Francisco: Jossey-Bass.

Jones, R. R., Vaught, R. S., & Reid, J. B. (1975). Time-series analysis as a substitute for single-subject analysis of variance designs. In G. R. Patterson, I. M. Marks, J. D. Matarazzo, R. A. Myers, G. E. Schwartz, & H. H. Strupp (Eds.), *Behavior change, 1974* (pp. 164–169). Chicago: Aldine.

Jones, R. R., Vaught, R. S., & Weinrott, M. R. (1977). Time-series analysis in operant research. *Journal of Applied Behavior Analysis, 10,* 151–167.

Jones, R. R., Weinrott, M. R., & Vaught, R. S. (1978). Effects of serial dependency on the agreement between visual and statistical inference. *Journal of Applied Behavior Analysis, 11,* 277–283.

Jones, R. T., Kazdin, A. E., & Haney, J. I. (1981a). A follow-up to training emergency skills. *Behavior Therapy, 12,* 716–722.

Jones, R. T., Kazdin, A. E., & Haney, J. I. (1981b). Social validation and training of emergency fire safety skills for potential injury prevention and life saving. *Journal of Applied Behavior Analysis, 14,* 249–250.

Kantorovich, N. V. (1928). An attempt of curing alcoholism by associated reflexes. *Novoye v Refleksologii i Fiziologii Nervnoy Sistemy, 3,* 436. Cited by Razran, G. H. S. (1934). Conditional withdrawal responses with shock as the conditioning stimulus in adult human subjects. *Psychological Bulletin, 31,* 111.

Kaufman, K. F., & O'Leary, K. D. (1972). Reward cost and self-evaluation procedures for disrupting adolescents in a psychiatric hospital school. *Journal of Applied Behavior Analysis, 4,* 77–87.

Kazdin, A. E. (1973a). The effect of response cost and aversive stimulation in suppressing punished and non-punished speech dysfluencies. *Behavior Therapy, 4*, 73–82.

Kazdin, A. E. (1973b). Methodological and assessment considerations in evaluating reinforcement programs in applied settings. *Journal of Applied Behavior Analysis, 6*, 517–531.

Kazdin, A. E. (1977). Assessing the clinical or applied significance of behavior change through social validation. *Behavior Modification, 1*, 427–453.

Kazdin, A. E. (1978). *History of behavior modification: Experimental foundations of contemporary research*. Baltimore: University Park Press.

Kazdin, A. E. (1979). Unobtrusive measures in behavioral assessment. *Journal of Applied Behavior Analysis, 12*, 713–724.

Kazdin, A. E. (1980a). Obstacles in using randomization tests in single-case experimentation. *Journal of Educational Statistics, 5*, 253–260.

Kazdin, A. E. (1980b). *Research design in clinical psychology*. New York: Harper & Row.

Kazdin, A. E. (1981). Drawing valid inferences from case studies. *Journal of Consulting and Clinical Psychology, 49*, 183–192.

Kazdin, A. E. (1982a). Observer effects: Reactivity of direct observation. In D. P. Hartmann (Ed.), *Using observers to study behavior: New directions for methodology of social and behavioral science* (pp. 5–19). San Francisco: Jossey-Bass.

Kazdin, A. E. (1982b). *Single-case research designs: Methods for clinical and applied settings*. New York: Oxford University Press.

Kazdin, A. E. (1982c). Sympton substitution, generalization, and response covariation: Implications for psychotherapy outcome. *Psychological Bulletin, 91*, 349–365.

Kazdin, A. E. (in press). *Behavior modification in applied settings*, (3rd ed.). Homewood, IL: Dorsey Press.

Kazdin, A. E., & Bootzin, R. R. (1972). The token economy: An evaluative review. *Journal of Appleid Behavior Analysis, 5*, 343–372.

Kazdin, A. E., & Geesey, S. (1977). Simultaneous-treatment design comparisons of the effects of earning reinforcers for one's peers versus for oneself. *Behavior Therapy, 8*, 682–693.

Kazdin, A. E., & Hartmann, D. P. (1978). The simultaneous-treatment design. *Behavior Therapy, 5*, 912–923.

Kazdin, A. E., & Kopel, S. A. (1975). On resolving ambiguities of the multiple-baseline design: Problems and recommendations. *Behavior Therapy, 6*, 601–608.

Kelly, D. (1980). *Anxiety and emotions: Physiologial basis and treatment*. Springfield, IL: Charles C Thomas.

Kelly, J. A. (1980). The simultaneous replication design: The use of a multiple baseline to establish experimental control in single group social skills treatment studies. *Journal of Behavior Therapy and Expermental Psychiatry, 11*, 203–207.

Kelly, J. A., Laughlin, C., Claiborne, M., & Patterson, J. T. (1979). A group procedure for teaching job interviewing skills to formerly hospitalized psychiatric patients. *Behavior Thearpy, 10*, 299–310.

Kelly, J. A., Urey, J. R., & Patterson, J. T. (1980). Improving heterosocial conversational skills of male psychiatric patients through a small group training procedure. *Behavior Therapy, 11*, 179–188.

Kelly, M. B. (1977). A review of observational data-collection and reliability procedures reported in the *Journal of Applied Behavior Analysis*. *Journal of Applied Behavior Analysis, 10*, 97–101.

Kendall, P. C., & Butcher, J. N. (1982). *Handbook of research methods in clinical psychology*. New York: Wiley.

Kennedy, R. E. (1976). *The feasibility of time-series analysis of single-case experiments*. Unpublished manuscript.

Kent, R. N., & Foster, S. L. (1977). Direct observational procedures: Methodological issues in naturalistic settings. In A. R. Ciminero, K. S. Calhoun, & H. E. Adams (Eds.), *Handbook of behavioral assessment* (pp. 279–329). New York: Wiley.

Kernberg, O. F. (1973). Summary and conclusions of psychotherapy and psychoanalysis: Final report of the Menninger Foundation's psychotherapy research project. *International Journal of Psychiatry, 11,* 62–77.

Kessel, L., & Hyman, H. T. (1933). The value of psychoanalysis as a therapeutic procedure. *Journal of American Medical Association, 101,* 1612–1615.

Kiesler, D. J. (1966). Some myths of psychotherapy research and the search for a paradigm. *Psychological Bulletin, 65,* 110–136.

Kiesler, D. J. (1971). Experimental designs in psychotherapy research. In A. E. Bergin & S. L. Garfield (Eds.), *Handbook of psychotherapy and behavior change: An empirical analysis* (2nd ed.) (pp. 36–74). New York: Wiley.

Kirby, F. D., & Shields, F. (1972). Modification of arithmetic response rate and attending behavior in a seventh-grade student. *Journal of Applied Behavior Analysis, 5,* 79–84.

Kircher, A. S., Pear, J. J., & Martin, G. L. (1971). Shock as punishment in a picture naming task with retarded children. *Journal of Applied Behavior Analysis, 4,* 227–233.

Kirchner, R. E., Schnelle, J. F., Domash, M. A., Larson, L. D., Carr, A. F., & McNees, M. P. (1980). The applicability of a helicopter patrol procedure to diverse areas: A cost-benefit evaluation. *Journal of Applied Behavior Analysis, 13,* 143–148.

Kirk, R. E. (1968). *Experimental design: Procedures for the behavioral sciences.* Glenmont, CA: Brooks/Cole.

Kistner, J., Hammer, D., Wolfe, D., Rothblum, E., & Drabman, R. S. (1982). Teacher popularity and contrast effects in a classroom token economy. *Journal of Applied Behavior Analysis, 15,* 85–96.

Knapp, T. J. (1983). Behavior analysts' visual appraisal of behavior change in graphic display. *Behavioral Assessment, 5,* 155–164.

Knight, R. P. (1941). Evaluation of the results of psychoanalytic therapy. *American Journal of Psychiatry, 98,* 434–466.

Koegel, R. L., & Schreibman, L. (1982). *How to teach autistic and other severely handicapped children.* Lawrence, KS: H & H Enterprises.

Kraemer, H. C. (1979). One-zero sampling in the study of primate beahvior. *Primates, 20,* 237–244.

Kraemer, H. C. (1981). Coping strategies in psychiatric clinical research. *Journal of Consulting and Clinical Psychology, 49,* 309–319.

Krasner, L. (1971a). Behavior therapy. *Annual Review of Psychology, 22,* 483–532.

Krasner, L. (1971b). The operant approach in behavior therapy. In A. E. Bergin & S. L. Garfield (Eds.), *Handbook of psychotherapy and behavior change: An empirical analysis* (pp. 612–653). New York: Wiley.

Kratochwill, T. R. (1978a). Foundations of time-series research. In T. R. Kratochwill (Ed.), *Single-subject research: Strategies for evaluating change* (pp. 1–101). New York: Academic Press.

Kratochwill, T. R. (Ed.) (1978b). *Single-subject research: Strategies for evaluating change.* New York: Academic Press.

Kratochwill, T. R., Alden, K., Demuth, D., Dawson, D., Panicucci, C., Arntson, P., McMurray, N., Hempstead, J., & Levin, J. R. (1974). A further consideration in the application of an analysis-of-variance model for the intrasubject replication design. *Journal of Applied Behavior Analysis, 7,* 629–633.

Kratochwill, T. R., & Brody, G. H. (1978). Single subject designs: A perspective on the controversy over employing statistical inference and implications for research and training in behavior modification. *Behavior Modification, 2,* 291–307.

Kratochwill, T. R., & Levin, J. R. (1980). On the applicability of various data analysis procedures to the simultaneous and alternating treatment designs in behavior therapy research. *Behavioral Assessment, 2,* 353–360.

Lacey, J. I. (1959). Psychophysiological approaches to the evaluation of psychotherapeutic process and outcome. In E. A. Rubinstein & M. B. Parloff (Eds.), *Research in psychotherapy* (pp. 160–208). Washington, DC: National Publishing Co.

Lang, P. J. (1968). Fear reduction and fear behavior: Problems in treating a construct. In J. M. Shlien (Ed.), *Research in psychotherapy* (Vol. 3, pp. 90–102). Washington, DC: American Psychological Association.

Last, C. G., Barlow, D. H., & O'Brien, G. T. (1983). Comparison of two cognitive strategies in treatment of a patient with generalized anxiety disorder. *Psychological Reports, 53,* 19–26.

Laws, D. R., Brown, R. A., Epstein, J., & Hocking, N. (1971). Reduction of inappropriate social behavior in disturbed children by an untrained paraprofessional therapist. *Behavior Therapy, 2,* 519–533.

Lawson, D. M. (1983). Alcoholism. In M. Hersen (Ed.), *Outpatient behavior therapy: A clinical guide* (pp. 143–172). New York: Grune & Stratton.

Lazarus, A. A. (1963). The results of behavior therapy in 126 cases of severe neurosis. *Behaviour Research and Therapy, 1,* 69–80.

Lazarus, A. A. (1973). Multi-modal behavior therapy: Treating the BASIC ID. *Journal of Nervous and Mental Disease, 156,* 404–411.

Lazarus, A. A., & Davison, G. C. (1971). Clinical innovation in research and practice. In A. E. Bergin & S. L. Garfield (Eds.), *Handbook of psychotherapy and behavior change: An empirical analysis* (pp. 196–213). New York: Wiley.

Leitenberg, H. (August, 1973). Interaction designs. Paper read at American Psychological Association, Montreal.

Leitenberg, H. (1973). The use of single-case methodology in psychotherapy research. *Journal of Abnormal Psychology, 82,* 87–101.

Leitenberg, H. (1976). Behavioral approaches to treatment of neuroses. In H. Leitenberg (Ed.), *Handbook of behavior modification and behavior therapy* (pp., 124–167). Englewood Cliffs, NJ: Prentice-Hall.

Leitenberg, H., Agras, W. S., Edwards, J. A., Thomson, L. E., & Wincze, J. P. (1970). Practice as a psychotherapeutic variable: An experimental analysis within single cases. *Journal of Psychiatric Research, 7,* 215–225.

Leitenberg, H., Agras, W. S., Thomson, L. E., & Wright, D. E. (1968). Feedback in behavior modification: An experimental analysis of two phobic cases. *Journal of Applied Behavior Analysis, 1,* 131–137.

Leonard, S. R., & Hayes, S. C. (1983). Sexual fantasy alternation. *Journal of Behavior Therapy and Experimental Psychiatry, 14,* 241–249.

Levin, J. R., Marascuilo, L. A., & Hubert, L. J. (1978). N = Nonparametric randomization tests. In T. R. Kratochwill (Ed.), *Single-subject research: Strategies for evaluating change* (pp. 167–197). New York: Academic Press.

Levy, R. L., & Olson, D. G. (1979). The single-subject methodology in clinical practice: An overview. *Journal of Social Service Research, 3,* 25–49.

Lewin, K. (1933). Vectors, cognitive processes and Mr. Tolman's criticism. *Journal of General Psychology, 8,* 318–345.

Lewinsohn, P. M., & Libet, J. (1972). Pleasurable events, activity schedules, and depression. *Journal of Abnormal Psychology, 79,* 291–295.

Lewinsohn, P. M., Mischel, W., Chaplin, W., & Barton, R. (1980). Social competence and depression: The roles of illusory self-perceptions. *Journal of Abnormal Psychology, 89,* 203–212.

Liberman, R. P., Davis, J., Moon, W., & Moore, J. (1973). Research design for analyzing drug-environment-behavior interactions. *Journal of Nervous and Mental Disease, 156,* 432–439.

Liberman, R. P., Neuchterlein, K. H., & Wallace, C. J. (1982). Social skills training in the nature of schizophrenia. In Curran, J. P. & Monti, P. M. (Eds.), *Social skills training* (pp. 1–56). New York: Guilford Press.

Liberman, R. P., & Smith, V. (1972). A multiple baseline study of systematic desensitization in a patient with multiple phobias. *Behavior Therapy, 3,* 597–603.

Liberman, R. P., Wheeler, E. G., DeVisser, L. A., Kuehnel, J., & Kuehnel, T. (1980). *Handbook of marital therapy.* New York: Plenum.

Lick, J. R., Sushinsky, L. W., & Malow, R. (1977). Specificity of Fear Survey Schedule items and the prediction of avoidance behavior. *Behavior Modification, 1,* 195–204.

Light, F. J. (1971). Measures of response agreement for qualitative data: Some generalizations and alternatives. *Psychological Bulletin, 76,* 365–377.

Lindsley, O. R. (1962). Operant conditioning techniques in the measurement of psychopharmaco-logical response. In J. H. Nodine & J. H. Moyer (Eds.), *Psychosomatic medicine: The first Hahnemann symposium on psychosomatic medicine* (pp. 373–383). Philadelphia: Lea & Febiger.

Linehan, M. M. (1980). Content validity: Its relevance to behavioral assessment. *Behavioral Assessment, 2,* 147–159.

Lovaas, O. I., Berberich, J. P., Perloff, B. F., & Schaeffer, B. (1966). Acquisition of imitiative speech by schizophrenic children. *Science, 161,* 705–707.

Lovaas, O. I., Freitas, L., Nelson, K., & Whalen, C. (1967). The establishment of imitation and its use for the development of complex behavior in schizophrenic children. *Behaviour Research and Therapy, 5,* 171–181.

Lovaas, O. I., Koegel, R., Simmons, J. Q., & Long, J. D. (1973). Some generalization and follow-up measures on autistic children in behavior therapy. *Journal of Applied Behavior Analysis, 5,* 131–166.

Lovaas, O. I., Schaeffer, B., & Simons, J. Q. (1965). Experimental studies in childhood schizophrenia: Building social behaviors using electric shock. *Journal of Experimental Research in Personality, 1,* 99–109.

Lovaas, O. I., & Simmons, J. Q. (1969). Manipulation of self-destruction in three retarded children. *Journal of Applied Behavior Analysis, 2,* 143–157.

Luborsky, L. (1959). Psychotherapy. In P. R. Farnsworth & Q. McNemar (Ed.), *Annual review of psychology* (pp. 317–344). Palo Alto, CA: Annual Review.

Lyman, R. D., Richard, H. C., & Elder, I. R. (1975). Contingency management of self-report and cleaning behavior. *Journal of Abnormal Child Psychology, 3,* 155–162.

Madsen, C. H., Becker, W. C., & Thomas, D. R. (1968). Rules, praise, and ignoring: Elements of elementary classroom control. *Journal of Applied Behavior Analysis, 1,* 139–150.

Malan, D. H. (1973). Therapeutic factors in analytically oriented brief psychotherapy. In R. H. Gosling (Ed.), *Support, innovation and autonomy* (pp. 187–205). London: Tavistock.

Malone, J. C., Jr. (1976). Local contrast and Pavlovian induction. *Journal of the Experimental Analysis of Behavior, 26,* 425–440.

Mandell, R. M., & Mandell, M. P. (1967). Suicide and the menstrual cycle. *Journal of the American Medical Association, 200,* 792–793.

Mann, R. A. (1972). The behavior-therapeutic use of contingency contracting to control an adult behavior problem: Weight control. *Journal of Applied Behavior Analysis, 5,* 99–109.

Mann, R. A., & Baer, D. M. (1971). The effects of receptive language training on articulation. *Journal of Applied Behavior Analysis, 4,* 291–298.

Mann, R. A., & Moss, G. R. (1973). The therapeutic use of a token economy to manage a young and assaultive inpatient population. *Journal of Nervous and Mental Disease, 157,* 1–9.

Mansell, J. (1982). Repeated direct replication of AB designs (Letter to the Editor). *Journal of Behaviour Therapy and Experimental Psychiatry, 13,* 261–262.

Marks, I. M. (1972). Flooding (implosion) and allied treatments. In W. S. Agras (Ed.), *Behavior modification: Principles and clinical applications* (pp. 151–213). Boston: Little, Brown.

Marks, I. M. (1981). New developments in psychological treatments of phobias. In M. R. Mavissakalian & D. H. Barlow (Eds.), *Phobia: Psychological and pharmacological treatment* (pp. 175–199). New York: Guilford Press.

Marks, I. M., & Gelder, M. G. (1967). Transvestism and fetishism: Clinical and psychological changes during faradic aversion. *British Journal of Psychiatry, 113,* 711–729.

Martin, G., Pallotta-Cornick, A., Johnstone, G., & Celso-Goyos, A. (1980). A supervisory strategy to improve work performance for lower functioning retarded clients in a sheltered workshop. *Journal of Applied Behavior Analysis, 13,* 185–190.

Martin, P. J., & Lindsey, C. J. (1976). Irregular discharge as an unobtrusive measure of . . . something: Some additional thoughts. *Psychological Reports, 38,* 627–630.

Mash, E. J., & Makohoniuk, G. (1975). The effects of prior information and behavioral predictability on observer accuracy. *Child Development, 46,* 513–519.

Mash, E. J., & Terdal, L. G. (Eds.). (1981). *Behavioral assessment of childhood disorders.* New York: Guilford Press.

Matson, J. L. (1981). Assessment and treatment of clinical fears in mentally retarded children. *Journal of Applied Behavior Analysis, 14,* 287–294.

Matson, J. L. (1982). The treatment of behavioral characteristics of depression in the mentally retarded. *Behavior Therapy, 13,* 209–218.

Mavissakalian, M. R., & Barlow, D. H. (1981a). Assessment of obsessive-compulsive disorders. In D. H. Barlow (Ed.), *Behavioral assessment of adult disorders* (pp. 209–239). New York: Guilford Press.

Mavissakalian, M. R., & Barlow, D. H. (1981b). Phobia: An overview. In M. R. Mavissakalian & D. H. Barlow (Eds.), *Phobia: Psychological and pharmacological treatment* (pp. 1–35). New York: Guilford Press.

Mavissakalian, M. R., & Barlow, D. H. (Eds.). (1981c). *Phobia: Psychological and pharmacological treatment.* New York: Guilford Press.

Max, L. W. (1935). Breaking up a homosexual fixation by the conditioned reaction techique: A case study. *Psychological Bulletin, 32,* 734.

May, P. R. A. (1973). Research in psychotherapy and psychoanalysis. *International Journal of Psychiatry, 1,* 78–86.

McCallister, L. W., Stachowiak, J. G., Baer, D. M., & Conderman, L. (1969). The application of operant conditioning techniques in a secondary school classroom. *Journal of Applied Behavior Analysis, 2,* 277–285.

McCleary, R., & Hay, R. A., Jr. (1980). *Applied time series analysis for the social sciences.* Beverly Hills: Sage.

McCullough, J. P., Cornell, J. E., McDaniel, M. H., & Meuller, R. K. (1974). Utilization of the simultaneous treatment design to improve student behavior in a first-grade classroom. *Journal of Consulting and Clinical Psychology, 42,* 288–292.

McFall, R. M. (1970). Effects of self-monitoring on normal smoking behavior. *Journal of Consulting and Clinical Psychology, 35,* 135–142.

McFall, R. M. (1977). Analogue methods in behavioral assessment: Issues and prospects. In J. D. Cone & R. P. Hawkins (Eds.), *Behavioral assessment: New direction in clinical psychology* (pp. 152–177). New York: Brunner/Mazel.

McFall, R. M., & Lillesand, D. B. (1971). Behavior rehearsal with modeling and coaching in assertion training. *Journal of Abnormal Psychology, 77,* 313–323.

McFarlain, R. A., & Hersen, M. (1974). Continuous measurement of activity level in psychiatric patients. *Journal of Clinical Psychology, 30,* 37–39.

McKnight, D. L., Nelson, R. O., Hayes, S. C., & Jarrett, R. B. (1983). Importance of treating individually assessed response classes in the amelioration of depression. *Behavior Therapy.*

McLaughlin, T. F., & Malaby, J. (1972). Intrinsic reinforcers in a classroom token economy. *Journal of Applied Behavior Analysis, 5,* 263–270.

McLean, A. P., & White, K. G. (1981). Undermatching and contrast within components of multiple schedules. *Journal of the Experimental Analysis of Behavior, 35,* 283–291.

McMahon, R. J., & Forehand, R. L. (1983). Consumer satisfaction in behavioral treatment of children: Types, issues, and recommendations. *Behavior Therapy, 14,* 209–225.

McNamara, J. R. (1972). The use of self-monitoring techniques to treat nailbiting. *Behaviour Research and Therapy, 10*, 193–194.

McNamara, J. R., & MacDonough, T. S. (1972). Some methodological considerations in the design and implementation of behavior therapy research. *Behavior Therapy, 3*, 361–378.

Melin, L., & Götestam, K. G. (1981). The effects of rearranging ward routines of communication and eating behaviors of psychogeriatric patients. *Journal of Applied Behavior Analysis, 14*, 47–51.

Metcalfe, M. (1956). Demonstration of a psychosomatic relationship. *British Journal of Medical Psychology, 29*, 63–66.

Michael, J. (1974). Statistical inference for individual organism research: Mixed blessing or curse? *Journal of Applied Behavior Analysis, 7*, 647–653.

Miller, P. M. (1973). An experimental analysis of retention control training in the treatment of nocturnal enuresis in two institutionalized adolescents. *Behavior Therapy, 4*, 288–294.

Miller, P. M., Hersen, M., Eisler, R. M., & Watts, J. G. (1974). Contingent reinforcement of lowered blood/alcohol levels in an outpatient chronic alcoholic. *Behaviour Research and Therapy, 12*, 261–263.

Mills, H. L., Agras, W. S., Barlow, D. H., & Mills, J. R. (1973). Compulsive rituals treated by response prevention: An experimental analysis. *Archives of General Psychiatry, 28*, 524–529.

Minkin, N., Braukmann, C. J., Minkin, B. L., Timbers, G. D., Timbers, B. J., Fixsen, D. L., Phillips, E. L., & Wolf, M. M. (1976). The social validation and training of conversational skills. *Journal of Applied Behavior Analysis, 9*, 127–139.

Mischel, W. (1968). *Personality and assessment.* New York: Wiley.

Mitchell, S. K. (1979). Interobserver agreement, reliability, and generalizability of data collected in observational studies. *Psychological Bulletin, 86*, 376–390.

Montague, J. D., & Coles, E. M. (1966). Mechanism and measurement of the galvanic skin response. *Psychological Bulletin, 65*, 261–279.

Monti, P. M., Corriveau, E. P., & Curran, J. P. (1982). Social skills training for psychiatric patients: Treatment and outcome. In J. P. Curran & P. M. Monti (Eds.), *Social skills training* (pp. 185–223). New York: Guilford Press.

Moses, L. E. (1952). Nonparametric statistics for psychological research. *Psychological Bulletin, 49*, 122–143.

Munford, P. R., & Liberman, R. P. (1978). Differential attention in the treatment of operant cough. *Journal of Behavioral Medicine, 1*, 280–289.

Nathan, P. E., Titler, N. A., Lowenstein, L. M., Solomon, P., & Rossi, A. M. (1970). Behavioral analysis of chronic alcoholism. *Archives of General Psychiatry, 22*, 419–430.

National Institute of Mental Health. (1980). *Behavior therapies in the treatment of anxiety disorders: Recommendations for strategies in treatment assessment research. (Final report of NIMH conference #RFP NIMH ER-79-003).* Unpublished manuscript.

Nay, W. R. (1977). Analogue measures. In A. R. Ciminero, K. S. Calhoun, & H. E. Adams (Eds)., *Handbook of behavioral assessment* (pp. 233–279). New York: Wiley.

Nay, W. R. (1979). *Multimethod clinical assessment.* New York: Gardner Press.

Neale, J. M., & Oltmanns, T. (1980). *Schizophrenia.* New York: Wiley.

Neef, N. A., Iwata, B. A., & Page, T. J. (1980). The effects of interspersal training versus high-density reinforcement on spelling acquisition and retention. *Journal of Applied Behavior Analysis, 13*, 153–158.

Nelson, R. O. (1977). Methodological issues in assessment via self-monitoring. In J. D. Cone & R. P. Hawkins (Eds.), *Behavioral assessment: New directions in clinical psychology* (pp. 217–254). New York: Brunner/Mazel.

Nelson, R. O., & Hayes, S. C. (1979). Some current dimensions of behavioral assessment. *Behavioral Assessment, 1*, 1–16.

Nelson, R. O., & Hayes, S. C. (1981). Nature of behavioral assessment. In M. Hersen & A. S. Bellack (Eds.), *Behavioral assessment: A practical handbook*, (2nd ed.) (pp. 3–37). Elmsford,

New York: Pergamon Press.

Nietzel, M. T., & Bernstein, D. A. (1981). Assessment of anxiety and fear. In M. Hersen & A. S. Bellack (Eds.), *Behavioral assessment: A practical handbook*, (2nd ed.) (pp. 215–245). Elmsford, New York: Pergamon Press.

Nordquist, V. M. (1971). The modification of a child's enuresis: Some response-response relationships. *Journal of Applied Behavior Analysis, 4*, 241–247.

Nunnally, J. (1978). *Psychometric theory*, (2nd ed.). New York: McGraw-Hill.

O'Brien, F., Azrin, N. H., & Henson, K. (1969). Increased communication of chronic mental patients by reinforcement and by response priming. *Journal of Applied Behavior Analysis, 2*, 23–29.

O'Brien, F., Bugle, C., & Azrin, N. H. (1977). Training and maintaining a retarded child's proper eating. *Journal of Applied Behavior Analysis, 10*, 465–478.

O'Leary, K. D. (1979). Behavioral assessment. *Behavioral Assessment, 1*, 31–36.

O'Leary, K. D., & Becker, W. C. (1967). Behavior modification of an adjustment class: A token reinforcement program. *Exceptional Children, 9*, 637–642.

O'Leary, K. D., Becker, W. C., Evans, M. B., & Saudargas, R. A. (1969). A token reinforcement program in a public school: A replication and systematic analysis. *Journal of Applied Behavior Analysis, 2*, 3–13.

O'Leary, K. D., Kent, R. N., & Kanowitz, J. (1975). Shaping data collection congruent with experimental hypotheses. *Journal of Applied Behavior Analysis, 8*, 43–51.

O'Leary, K. D., & Turkewitz, H. (1981). A comparative outcome study of behavioral marital therapy and communication therapy. *Journal of Marital and Family Therapy, 7*, 159–169.

Ollendick, T. H. (1981). Self-monitoring and self-administered overcorrection: *Behavior Modification, 5*, 75–84.

Ollendick, T. H., Matson, J. L., Esveldt-Dawson, K., & Shapiro, E. S. (1980). Increasing spelling achievement: An analysis of treatment procedures utilizing an alternating treatments design. *Journal of Applied Behavior Analysis, 13*, 645–654.

Ollendick, T. H., Shapiro, E. S., & Barrett, R. P. (1981). Reducing stereotypic behaviors: An analysis of treatment procedures utilizing an alternating treatments design. *Behavior Therapy, 12*, 570–577.

Orne, M. T. (1962). On the social psychology of the psychological experiment: With particular reference to demand characteristics and their implications. *American Psychologist, 17*, 776–783.

Paris, S. G., & Cairns, R. B. (1972). An experimental and ethological analysis of social reinforcement with retarded children. *Child Development, 43*, 717–719.

Parsonson, B. S., & Baer, D. M. (1978). The analysis and presentation of graphic data. In T. R. Kratochwill (Ed.), *Single-subject research: Strategies for evaluating change* (pp. 101–167). New York: Academic Press.

Patterson, G. R. (1982). *Coercive family process*. Eugene, OR: Castalia.

Paul, G. L. (1967). Strategy of outcome research in psychotherapy. *Journal of Consulting Psychology, 31*, 104–118.

Paul, G. L. (1969). Behavior modification research: Design and tactics. In C. M. Franks (Ed.), *Behavior therapy: Appraisal and status* (pp. 29–62). New York: McGraw-Hill.

Paul, G. L. (1979). New assessment systems for residential treatment, management, research, and evaluation: A symposium. *Journal of Behavioral Assessment, 1*, 181–184.

Paul, G. L., & Lentz, R. J. (1977). *Psychosocial treatment of chronic mental patients: Milieu versus social-learning programs*. Cambridge: Harvard University Press.

Pavlov, I. P. (1928). *Lectures on conditioned reflexes*. (W. H. Gantt, Trans.) New York: International.

Pendergrass, V. E. (1972). Timeout from positive reinforcement following persistent, high-rate behavior in retardates. *Journal of Applied Behavior Analysis, 5*, 85–91.

Pertschuk, M. J., Edwards, N., & Pomerleau, O. F. (1978). A multiple baseline approach to behavioral intervention in anorexia nervosa. *Behavior Therapy*, *9*, 368–376.

Peterson, L., Homer, A. L., & Wonderlich, S. A. (1982). The integrity of independent variables in behavior analysis. *Journal of Applied Behavior Analysis*, *15*, 477–492.

Peterson, R. F., & Peterson, L. (1968). The use of positive reinforcement in the self-control of self-destructive behavior in a retarded boy. *Journal of Experimental Child Psychology*, *6*, 351–360.

Pinkston, E. M., Reese, N. M., LeBlanc, J. M., & Baer, D. M. (1973). Independent control of a preschool child's aggression and peer interaction by contingent teacher attention. *Journal of Applied Behavior Analysis*, *6*, 115–124.

Poche, C., Brouwer, R., & Swearingen, M. (1981). Teaching self-protection to young children. *Journal of Applied Behavior Analysis*, *14*, 169–176.

Porterfield, J., Blunden, R., & Blewitt, E. (1980). Improving environments for profoundly handicapped adults: Using prompts and social attention to maintain high group engagement. *Behavior Modification*, *4*, 225–241.

Powell, J., & Hake, D. F. (1971). Positive vs. negative reinforcement: A direct comparison of effects on a complex human response. *Psychological Record*, *21*, 191–205.

Powell, J., Martindale, A., & Kulp, S. (1975). An evaluation of time-sample measures of behavior. *Journal of Applied Behavior Analysis*, *8*, 463–469.

Power, C. T. (1979). The Time-Sample Behavioral Checklist: Observational assessment of patient functioning. *Journal of Behavioral Assessment*, *1*, 199–210.

Powers, E., & Witmer, H. (1951). *An experiment in the prevention of delinquency.* New York: Columbia University Press.

Rachlin, H. (1973). Contrast and matching. *Psychological Review*, *80*, 297–308.

Rachman, S. J., & Hodgson, R. J. (1980). *Obsessions and compulsions.* Englewood Cliffs, NJ: Prentice-Hall.

Ramp, E., Ulrich, R., & Dulaney, S. (1971). Delayed timeout as a procedure for reducing disruptive classroom behavior: A case study. *Journal of Applied Behavior Analysis*, *4*, 235–239.

Rapport, M. D., Sonis, W. A., Fialkov, M. J., Matson, J. L., & Kazdin, A. E. (1983). Carbamazepine and behavior therapy for aggressive behavior: Treatment of a mentally retarded, postencephalic adolescent with seizure disorder. *Behavior Modification*, *7*, 255–265.

Ray, W. J., & Raczynski, J. M. (1981). Psychophysiological assessment. In M. Hersen & A. S. Bellack (Eds.), *Behavioral assessment: A practical handbook*, (2nd ed.) (pp. 175–211). Elmsford, New York: Pergamon Press.

Redd, W. H. (1980). Stimulus control and extinction of psychosomatic symptoms in cancer patients in protective isolation. *Journal of Consulting and Clinical Psychology*, *48*, 448–456.

Redd, W. H., & Birnbrauer, J. S. (1969). Adults as discriminative stimuli for different reinforcement contingencies with retarded children. *Journal of Experimental Child Psychology*, *7*, 440–447.

Redfield, J. P., & Paul, G. L. (1976). Bias in behavioral observation as a function of observer familiarity with subjects and typicality of behavior. *Journal of Consulting and Clinical Psychology*, *44*, 156.

Rees, L. (1953). Psychosomatic aspects of the prementrual tension system. *Journal of Mental Science*, *99*, 62–73.

Reid, J. B. (1978). The development of specialized observation systems. In J. B. Reid (Ed.), *A social learning approach to family intervention: Vol. 2. Observation in home settings* (pp. 43–49). Eugene, OR: Castalia.

Reid, J. B. (1982). Observer training in naturalistic research. In D. P. Hartmann (Ed.), *Using observers to study behavior: New directions for methodology of social and behavioral science* (pp. 37–50). San Francisco: Jossey-Bass.

Revusky, S. H. (1976). Some statistical treatments compatible with individual organism methodology. *Journal of the Experimental Analysis of Behavior*, *10*, 319–330.

Reynolds, G. S. (1968). *A primer of operant conditioning.* Glenview, IL: Scott, Foresman.

Reynolds, N. J., & Risley, T. R. (1968). The role of social and material reinforcers in increasing talking of a disadvantaged preschool child. *Journal of Applied Behavior Analysis, 1,* 253–262.

Rickard, H. C., Dignam, P. J., & Horner, R. F. (1960). Verbal manipulation in a psychotherapeutic relationship. *Journal of Clinical Psychology, 16,* 164–167.

Rickard, H. C., & Dinoff, M. (1962). A follow-up note on "Verbal manipulation in a psychotherapeutic relationship." *Psychologicl Reports, 11,* 506.

Rickard, H. C., & Saunders, T. R. (1971). Control of "clean-up" behavior in a summer camp. *Behavior Therapy, 2,* 340–344.

Risley, T. R. (1968). The effects and side-effects of punishing the autistic behaviors of a deviant child. *Journal of Applied Behavior Analysis, 1,* 21–34.

Risley, T. R. (1970). Behavior modification: An experimental-therapeutic endeavor. In L. A. Hamerlynck, P. O. Davidson, & L. E. Acker (Eds.), *Behavior modification and ideal health services* (pp. 103–127). Calgary, Alberta, Canada: University of Calgary Press.

Risley, T. R., & Wolf, M. M. (1972). Strategies for analyzing behavioral change over time. In J. Nesselroade & H. Reese (Eds.), *Life-span developmental psychology: Methodological issues* (pp. 175–183). New York: Academic Press.

Roberts, M. W., Hatzenbuehler, L. C., & Bean, A. W. (1981). The effects of differential attention and timeout on child noncompliance. *Behavior Therapy, 12,* 93–99.

Rogers, C. R., Gendlin, E. T., Kiesler, D. J., & Truax, C. B. (1967). *The therapeutic relationship and its impact: A study of psychotherapy with schizophrenics.* Madison: University of Wisconsin Press.

Rogers-Warren, A., & Warren, S. F. (1977). *Ecological perspectives in behavior analysis.* Baltimore: University Park Press.

Rojahn, J., Mulick, J. A., McCoy, D., & Schroeder, S. R. (1978). Setting effects, adaptive clothing, and the modification of head-banging and self-restraint in two profoundly retarded adults. *Behavioural Analysis and Modification, 2,* 185–196.

Rosen, J. C., & Leitenberg, H. (1982). Bulimia Nervosa: Treatment with exposure and response evaluation. *Behavior Therapy, 13,* 117–124.

Rosenblum, L. A. (1978). The creation of a behavioral taxonomy. In G. P. Sackett (Ed.), *Observing behavior: Vol. 2. Data collection and analysis methods* (pp. 15–24). Baltimore: University Park Press.

Rosenthal, R. (1976). *Experimenter effects in behavioral research* (enlarged ed.). New York: Irvington.

Rosenzweig, S. (1951). Idiodynamics in personality theory with special reference to projective methods. *Psychological Review, 58,* 213–223.

Ross, A. O. (1981). *Child behavior therapy: Principles, procedures, and empirical basis.* New York: Wiley.

Roxburgh, P. A. (1970). Treatment of persistent phenothiazine-induced oraldyskinesia. *British Journal of Psychiatry, 116,* 277–280.

Rubenstein, E. A., & Parloff, M. B. (1959). Research problems in psychotherapy. In E. A. Rubenstein & M. B. Parloff (Eds.), *Research in psychotherapy,* (Vol. 1) (pp. 276–293). Washington, DC: American Psychological Association.

Rugh, J. E., & Schwitzgebel, R. L. (1977). Instrumentation for behavioral assessment. In A. R. Ciminero, K. S. Calhoun, & H. E. Adams (Eds.), *Handbook of behavioral assessment* (pp. 79–113). New York: Wiley.

Rusch, F. R., & Kazdin, A. E. (1981). Toward a methodology of withdrawal designs for the assessment of response maintenance. *Journal of Applied Behavior Analysis, 14,* 131–140.

Rusch, F. R., Walker, H. M., & Greenwood, C. R. (1975). Experimenter calculation errors: A potential factor affecting interpretation of results. *Journal of Applied Behavior Analysis, 8,* 460.

Russell, M. B., & Bernal, M. E. (1977). Temporal and climatic variables in naturalistic observation. *Journal of Applied Behavior Analysis, 10,* 399–405.

Russo, D. C., & Koegel, R. L. (1977). A method for integrating an autistic child into a normal public school classroom. *Journal of Applied Behavior Analysis, 10,* 579–590.

Sackett, G. P. (1978). Measurement in observational research. In G. P. Sackett (Ed.), *Observing behavior: Vol. 2. Data collection and analysis methods* (pp. 25–43). Baltimore: University Park Press.

St. Lawrence, J. S., Bradlyn, A. S., & Kelly, J. A. (1983). Interpersonal adjustment of a homosexual adult: Enhancement via social skills training. *Behavior Modification, 7,* 41–55.

Sajwaj, T. E., & Dillon, A. (1977). Complexities of an "elementary" behavior modification procedure: Differential adult attention used for children's behavior disorders. In B. C. Etzel, J. M. LeBlanc, & D. M. Baer (Eds.), *New developments in behavioral research: Theory, methods and application* (pp. 303–315). Hillsdale, NJ: Erlbaum.

Sajwaj, T. E., & Hedges, D. (1971). Functions of parental attention in an oppositional retarded boy. In *Proceedings of the 79th Annual Convention of the American Psychological Association* (pp. 697–698). Washington, DC: American Psychological Association.

Sajwaj, T. E., Twardosz, S., & Burke, M. (1972). Side effects of extinction procedures in a remedial preschool. *Journal of Applied Behavior Analysis, 5,* 163–175.

Sanson-Fisher, R. W., Poole, A. D., Small, G. A., & Fleming, I. R. (1979). Data acquisition in real time: An improved system for naturalistic observations. *Behavior Therapy, 10,* 543–554.

Scheffé, H. (1959). *The analysis of variance.* New York: Wiley.

Schindele, R. (1981). Methodological problems in rehabilitation research. *International Journal of Rehabilitation Research, 4,* 233–248.

Schleien, S. J., Weyman, P., & Kiernan, J. (1981). Teaching leisure skills to severely handicapped adults: An age appropriate darts game. *Journal of Applied Behavior Analysis, 14,* 513–519.

Schreibman, L., Koegel, R. L., Mills, D. L., & Burke, J. C. (in press). Training parent child interactions. In E. Scholper & G. Mesibov (Eds.), *Issues in autism:* Vol. III. The effects of autism on the family. New York: Plenum.

Schumaker, J., & Sherman, J. A. (1970). Training generative verb usage by imitation and reinforcement procedure. *Journal of Applied Behavior Analysis, 3,* 273–287.

Schutte, R. C., & Hopkins, B. L. (1970). The effects of teacher attention following instructions in a kindergarten class. *Journal of Applied Behavior Analysis, 3,* 117–122.

Sechrest, L. (Ed.). (1979). *Unobtrusive measurement today: New directions for methodology of behavioral science.* San Francisco: Jossey-Bass.

Shapiro, D. A., & Shapiro, D. (1983). Comparative therapy outcome research: Methodological implications of meta-analysis. *Journal of Consulting and Clinical Psychology, 51,* 42–53.

Shapiro, E. S., Barrett, R. P., & Ollendick, T. H. (1980). A comparison of physical restraint and positive practice overcorrection in treating stereotypic behavior. *Behavior Therapy, 11,* 227–233.

Shapiro, E. S., Kazdin, A. E., & McGonigle, J. J. (1982). Multiple-treatment interference in the simultaneous- or alternating-treatments design. *Behavioral Assessment, 4,* 105–115.

Shapiro, M. B. (1961). The single case in fundamental clincial psychological research. *British Journal of Medical Psychology, 34,* 255–263.

Shapiro, M. B. (1966). The single case in clinical-psychological research. *Journal of General Psychology, 74,* 3–23.

Shapiro, M. B. (1970). Intensive assessment of the single case: An inductive-deductive approach. In P. Mittler (Ed.), *Psychological assessment of mental and physical handicaps.* London: Methuen.

Shapiro, M. B., & Ravenette, A. T. (1959). A preliminary experiment on paranoid delusions. *Journal of Mental Science, 105,* 295–312.

Shine, L. C., & Bower, S. M. (1971). A one-way analysis of variance for single-subject designs. *Educational and Psychological Measurement, 31,* 105–113.

Shontz, F. C. (1965). *Research methods in personality.* New York: Appleton-Century-Crofts.

Shrout, P. E., & Fleiss, J. H. (1979). Intraclass correlations: Uses in assessing rater reliability. *Psychological Bulletin, 86,* 420–428.

Shuller, D. Y., & McNamara, J. R. (1976). Expectancy factors in behavioral observation. *Behavior Therapy, 7,* 519–527.

Sidman, M. (1960. *Tactics of scientific research: Evaluating experimental data in psychology.* New York: Basic Books.

Simon, A., & Boyer, E. G. (1974). *Mirrors for behavior: Vol. 3. An anthology of observation instruments.* Eyncote, PA: Communication Materials Center.

Simpson, M. J. A. (1979). Problems of recording behavioral data by keyboard. In M. E. Lamb, S. J. Suomi, & G. R. Stephenson (Eds.), *Social interaction analysis: Methodological issues* (pp. 137–156). Madison: University of Wisconsin Press.

Singh, N. N., Dawson, J. H., & Gergory, P. R. (1980). Suppression of chronic hyperventilation using response contingent aromatic ammonia. *Behavior Therapy, 11,* 561–566.

Singh, N. N., Manning, P. J., & Angell, M. J. (1982). Effects of an oral hygiene punishment procedure on chronic schizophrenic rumination and collateral behaviors in monozygous twins. *Journal of Applied Behavior Analysis, 15,* 309–314.

Singh, N. N., Winton, A. S., & Dawson, M. H. (1982). Suppression of antisocial behavior by facial screening using multiple baseline and alternating treatments designs. *Behavior Therapy, 13,* 511–520.

Skiba, E. A., Pettigrew, E., & Alden, S. E. (1971). A behavioral approach to the control of thumbsucking in the classroom. *Journal of Applied Behavior Analysis, 4,* 121–125.

Skinner, B. F. (1938). *The behavior of organisms.* New York: Appleton-Century-Crofts.

Skinner, B. F. (1953). *Science and human behavior.* New York: Macmillan.

Skinner, B. F. (1966a). Invited address to the Pavlovian Society of America, Boston.

Skinner, B. F. (1966b). Operant behavior. In W. K. Honig (Ed.), *Operant behavior: Areas of research and application* (pp. 12–32). New York: Appleton-Century-Crofts.

Slavon, R. E., Wodarski, J. S., & Blackburn, B. L. (1981). A group contingency for electricity conservation in master-metered apartments. *Journal of Applied Behavior Analysis, 14,* 357–363.

Sloane, H. N., Johnston, M. K., & Bijou, S. W. (1967). Successive modification of aggressive behavior and aggressive fantasy play by management of contingencies. *Journal of Child Psychology and Psychiatry, 8,* 217–226.

Smeets, P. M. (1970). Withdrawal of social reinforcers as a means of controlling rumination and regurgitation in a profoundly retarded person. *Training School Bulletin, 67,* 158–163.

Smith, C. M. (1963). Controlled observations on the single case. *Canadian Medical Association Journal, 88,* 410–412.

Smith, M. L., & Glass, G. V. (1977). Meta-analysis of psychotherapy outcome studies. *American psychologist, 32,* 752–760.

Smith, P. C., & Kendall, L. M. (1963). Retranslation of expectations: An approach to the construction of unambiguous anchors for rating scales. *Journal of Applied Psychology, 47,* 149–155.

Sowers, J., Rusch, F. R., Connis, R. T., & Cummings, L. T. (1980). Teaching mentally retarded adults to time-manage in a vocational setting. *Journal of Applied Behavior Analysis, 13,* 119–128.

Spitzer, R. L., Forman, J. B. W., & Nee, J. (1979). DSM-III field trials: I. Initial interrater diagnostic reliability. *American Journal of Psychiatry, 136,* 815–817.

Steinman, W. M. (1970). The social control of generalized imitation. *Journal of Applied Behavior Analysis, 3,* 159–167.

Steketee, G., & Foa, E. B. (in press). Obsessive-compulsive disorders. In D. H. Barlow (Ed.), *Behavioral treatment of adult disorders.* New York: Guilford Press.

Steketee, G., Foa, E. B., & Grayson, J. B. (1982). Recent advances in the behavioral treatment of obsessive compulsives. *Archives of General Psychiatry, 39,* 1365–1371.

Stern, R. M., Ray, W. J., & Davis, C. M. (1980). *Psychophysiological recording.* New York: Oxford University Press.

Stilson, D. W. (1966). *Probability and statistics in psychological research and theory.* San Francisco: Holden-Day.

Stokes, T. F., & Baer, D. M. (1977). An implicit technology of generalization. *Journal of Applied Behavior Analysis, 10,* 349–367.

Stokes, T. F., & Kennedy, S. H. (1980). Reducing child uncooperative behavior during dental treatment through modeling and reinforcement. *Journal of Applied Behavior Analysis, 13,* 41–49.

Stoline, M. R., Huitema, B. E., & Mitchell, B. T. (1980). Intervention time-series model with different pre- and postintervention first-order autoregressive parameters. *Psychological Bulletin, 88,* 46–53.

Stravynski, A., Marks, I. M., & Yule, W. (1982). The sleep of patients with obsessive-compulsive disorder. *Archives of General Psychiatry, 39,* 1378–1385.

Striefel, S., Bryan, K. S., & Aikens, D. A. (1974). Transfer of stimulus control from motor to verbal stimuli. *Journal of Applied Behavior Analysis, 7,* 123–135.

Striefel, S., & Wetherby, B. (1973). Instruction following behavior of a retarded child and its controlling stimuli. *Journal of Applied Behavior Analysis, 6,* 663–670.

Strupp, H. H., & Hadley, S. W. (1979). Specific vs. nonspecific factors in psychotherapy. *Archives of General Psychiatry, 36,* 1125–1137.

Strupp, H. H., & Luborsky, L. (Eds.) (1962). *Research in psychotherapy* (Vol. 2). Washington, DC: American Psychological Association.

Stuart, R. B. (1971). A three-dimensional program for the treatment of obesity. *Behaviour Research and Therapy, 9,* 177–186.

Sulzer-Azaroff, B., & deSantamaria, M. C. (1980). Industrial safety hazard reduction through performance feedback. *Journal of Applied Behavior Analysis, 13,* 287–295.

Sulzer-Azaroff, B., & Mayer, R. G. (1977). *Applying behavior-analysis procedures with children and youth.* New York: Holt, Rinehart and Winston.

Swan, G. E., & MacDonald, M. L. (1978). Behavior therapy in practice: A national survey of behavior therapists. *Behavior Therapy, 9,* 799–807.

Taplin, P. S., & Reid, J. B. (1973). Effects of instructional set and experimental influences on observer reliability. *Child Development, 44,* 547–554.

Tate, B. B., & Baroff, G. S. (1966). Aversive control of self-injurious behavior in a psychotic boy. *Behaviour Research and Therapy, 4,* 281–287.

Taylor, C. B., & Agras, W. S. (1981). Assessment of phobia. In D. H. Barlow (Ed.), *Behavioral assessment of adult disorders* (pp. 181–209). New York: Guilford Press.

Thomas, D. R., Becker, W. C., & Armstrong, M. (1968). Production and elimination of disruptive classroom behavior by systematically varying teachers' behavior. *Journal of Applied Behavior Analysis, 1,* 35–45.

Thomas, D. R., Nielsen, T. J., Kuypers, D. S., & Becker, W. C. (1968). Social reinforcement and remedial instruction in the elimination of a classroom behavior problem. *Journal of Special Education, 2,* 291–305.

Thomas, J. D., & Adams, M. A. (1971). Problems in teacher use of selected behaviour modification techniques in the classroom. *New Zealand Journal of Educational Studies, 6,* 151–165.

Thomson, C., Holmberg, M., & Baer, D. M. (1974). A brief report on a comparison of time-sampling procedures. *Journal of Applied Behavior Analysis, 7,* 623–626.

Thoreson, C. E., & Elashoff, J. D. (1974). Some comments on "An analysis-of-variance model for the instrasubject replication design." *Journal of Applied Behavior Analysis, 7,* 639–641.

Thorne, F. C. (1947). The clinical method in science. *American Psychologist, 2,* 161–166.

Tinsley, H. E. A., & Weiss, D. J. (1975). Interrater reliability and agreement of subjective judgments. *Journal of Counseling Psychology, 22*, 358–376.

Truax, C. B. (1966). Reinforcement and non-reinforcement in Rogerian psychotherapy. *Journal of Abnormal Psychology, 71*, 1–9.

Truax, C. B., & Carkhuff, R. R. (1965). Experimental manipulation of therapeutic conditions. *Journal of Consulting Psychology, 29*, 119–124.

Tyron, W. W. (1982). A simplified time-series analysis for evaluating treatment interventions. *Journal of Applied Behavior Analysis, 15*, 423–429.

Turkat, I. D., & Maisto, S. (in press). Personality disorders. In D. H. Barlow (Ed.), *Behavioral treatment of adult disorders*. New York: Guilford Press.

Turner, S. M., Hersen, M., & Alford, H. (1974). Effects of massed practice and meprobamate on spasmodic torticollis: An experimental analysis. *Behaviour Research and Therapy, 12*, 259–260.

Turner, S. M., Hersen, M., & Bellack, A. S. (1978). Social skills training to teach prosocial behaviors in an organically impaired and retarded patient. *Journal of Behavior Therapy and Experimental Psychiatry, 9*, 253–258.

Turner, S. M., Hersen, M., Bellack, A. S., Andrasik, F., & Capparell, H. V. (1980). Behavioral and pharmacological treatment of obsessive-compulsive disorders. *Journal of Nervous & Mental Disease, 168*, 651–657.

Twardosz, S., & Sajwaj, T. E. (1972). Multiple effects of a procedure to increase sitting in a hyperactive, retarded boy. *Journal of Applied Behavior Analysis, 5*, 73–78.

Ullmann, L. P., & Krasner, L. (Eds.) (1965). *Case studies in behavior modfication*. New York: Holt, Rinehart and Winston.

Ulman, J. D., & Sulzer-Azaroff, B. (1973, August). *Multielement baseline design in applied behavior analysis*. Symposium conducted at the annual meeting of the American Psychological Association, Montreal.

Ulman, J. D., & Sulzer-Azaroff, B. (1975). Multielement baseline design in educational research. In E. Ramp & G. Semb (Eds.), *Behavior analysis: Areas of research and application* (pp. 377–391). Englewood Cliffs, NJ: Prentice-Hall, 1975.

Underwood, B. J. (1957). *Psychological research*. New York: Appleton-Century-Crofts.

VanBierliet, A., Spangler, P. F., & Marshall, A. M. (1981). An ecobehavioral examination of a simple strategy for increasing mealtime language in residential facilities. *Journal of Applied Behavior Analysis, 14*, 295–305.

Van Hasselt, V. B., & Hersen, M. (1981). Applications of single-case designs to research with visually impaired individuals. *Journal of Visual Impairment and Blindness, 75*, 359–362.

Van Hasselt, V. B., Hersen, M., Kazdin, A. E., Simon, J., & Mastantuono, A. K. (1983). Social skills training for blind adolescents. *Journal of Visual Impairment and Blindness, 75*, 199–203.

Van Houten, R., Nau, P. A., MacKenzie-Keating, S. E., Sameoto, D., & Colavecchia, B. (1982). An analysis of some variables influencing the effectiveness of reprimands. *Journal of Applied Behavior Analysis, 15*, 65–83.

Varni, J. W., Russo, D. C., & Cataldo, M. F. (1978). Assessment and modification of delusional speech in an 11-year-old child: A comparative analysis of behavior therapy and stimulant drug effects. *Journal of Behavior Therapy and Experimental Psychiatry, 9*, 377–380.

Veenstra, M. (1971). Behavior modification in the home with the mother as the experimenter: The effect of differential reinforcement on sibling negative response rates. *Child Development, 42*, 2079–2083.

Venables, P. H., & Christie, M. H. (1973). Mechanism, instrumentation, recording techniques and quantification of responses. In W. F. Prokasy & D. C. Raskin (Eds.), *Ectodermal activity in psychological research* (pp. 1–124). New York: Academic Press.

Venables, P. H., & Martin, I. (1967). *A manual of psychophysiological methods*. Amsterdam: North-Holland.

Vermilyea, J. A., Boice, R., & Barlow, D. H. (in press). Rachman and Hodgson (1974) a decade later: How do desynchronous response systems relate to the treatment of agoraphobia?

Behaviour Research and Therapy.

Vukelich, R., & Hake, D. F. (1971). Reduction of dangerously aggressive behavior in a severely retarded resident through a continuation of positive reinforcement procedures. *Journal of Applied Behavior Analysis, 4,* 215–225.

Wade, T. C., Baker, T. B., & Hartmann, D. P. (1979). Behavior therapists' self-reported views and practices. *Behavior Therapist, 2,* 3–6.

Wahler, R. G. (1968, April). *Behavior therapy for oppositional children: Love is not enough.* Paper presented at the meeting of the Eastern Psychological Association, Washington, DC.

Wahler, R. G. (1969a). Oppositional children: A quest for parental reinforcement control. *Journal of Applied Behavior Analysis, 2,* 159–170.

Wahler, R. G. (1969b). Setting generality: Some specific and general effects of child behavior therapy. *Journal of Applied Behavior Analysis, 2,* 239–246.

Wahler, R. G., Berland, R. M., & Coe, T. D. (1979). Generalization processes in child behavior change. In B. B. Lahey & A. E. Kazdin (Eds.), *Advances in clinical child psychology* (pp. 36–72). New York: Plenum.

Wahler, R. G., & Pollio, H. R. (1968). Behavior and insight: A case study in behavior therapy. *Journal of Experimental Research in Personality, 3,* 45–56.

Wahler, R. G., Sperling, K. A., Thomas, M. R., Teeter, N. C., & Luper, H. L. (1970). Modification of childhood stuttering: Some response-response relationships. *Journal of Experimental Child Psychology, 9,* 411–428.

Wahler, R. G., Winkel, G. H., Peterson, R. F., & Morrison, D. C. (1965). Mothers as behavior therapists for their own children. *Behaviour Research and Therapy, 3,* 113–124.

Waite, W. W., & Osborne, J. G. (1972). Sustained behavioral contrast in children. *Journal of the Experimental Analysis of Behavior, 18,* 113–117.

Walker, H. M., & Buckely, N. K. (1968). The use of positive reinforcement in conditioning attending behavior. *Journal of Applied Behavior Analysis, 1,* 245–250.

Walker, H. M., & Lev, J. (1953). *Statistical inference.* New York: Holt, Rinehart and Winston.

Wallace, C. J., Boone, S. E., Donahoe, C. P., & Foy, D. W. (in press). Chronic mental disabilities. In D. H. Barlow (Ed.), *Behavioral treatment of adult disorders.* New York: Guilford Press.

Wallace, C. J. (1982). The social skills training project of the Mental Health Clinical Research Center for the Study of Schizophrenia. In J. P. Curran & P. M. Monti (Eds.), *Social skills training* (pp. 57–89). New York: Guilford Press.

Wallace, C. J., & Elder, J. P. (1980). Statistics to evaluate measurement accuracy and treatment effects in single subject research designs. In M. Hersen, R. M. Eisler, & P. M. Monti (Eds.), *Progress in behavior modification,* (Vol. 10, pp. 40–82). New York: Academic Press.

Wampold, B. E., & Furlong, M. J. (1981a). The heuristics of visual inference. *Behavioral Assessment, 3,* 79–82.

Wampold, B. E., & Furlong, M. J. (1981b). Randomization tests in single-subject designs: Illustrative examples. *Journal of Behavioral Assessment, 3,* 329–341.

Ward, M. H., & Baker, B. L. (1968). Reinforcement therapy in the classroom. *Journal of Applied Behavior Analysis, 1,* 323–328.

Warren, V. L., & Cairns, R. B. (1972). Social reinforcement satiation: An outcome of frequency or ambiguity. *Journal of Experimental Child Psychology, 13,* 249–260.

Watson, J. B., & Rayner, R. (1920). Conditioned emotional reactions. *Journal of Experimental Psychology, 3,* 1–14.

Watson, P. J., & Workman, E. A. (1981). The non-concurrent multiple baseline across-individuals design: An extension of the traditional multiple baseline design. *Journal of Behavior Therapy and Experimental Psychiatry, 12,* 257–259.

Webb, E. J., Campbell, D. T., Schwartz, R. D., & Sechrest, L. (1966). *Unobtrusive measures: Nonreactive research in the social sciences.* Chicago: Rand McNally.

Webb, E. J., Campbell, D. T., Schwartz, R. D., Sechrest, L., & Grove, J. B. (1981). *Nonreactive measures in the social sciences,* (2nd ed.). Boston: Houghton Mifflin.

Weick, K. E. (1968). Systematic observational methods. In G. Lindzey & E. Aronson (Eds.)., *The handbook of social psychology*, (Vol. 2, 2nd ed.). (pp. 357–451). Menlo Park, CA: Addison-Wesley.

Weinrott, M. R., Garrett, B., & Todd, N. (1978). The influence of observer presence on classroom behavior. *Behavior Therapy*, *9*, 900–911.

Weinrott, M. R., Jones, R. R., & Boler, G. R. (1981). Convergent and discriminant validity of five classroom observation systems: A secondary analysis. *Journal of Educational Psychology*, *73*, 671–679.

Wells, K. C., Hersen, M., Bellack, A. S., & Himmelhock, J. M., (1979). Social skills training in unipolar nonpsychotic depression. *American Journal of Psychiatry*, *136*, 1331–1332.

Werner, J. S., Minkin, N., Minkin, B. L., Fixsen, D. L., Phillips, E. L., & Wolf, M. M. (1975). "Intervention package": An analysis to prepare juvenile delinquents for encounters with police officers. *Criminal Justice and Behavior*, *2*, 55–83.

Whang, P. L., Fletcher, R. K., & Fawcett, S. B. (1982). Training counseling skills: An experimental analysis and social validation. *Journal of Applied Behavior Analysis*, *15*, 325–334.

Wheeler, A. J., & Sulzer, B. (1970). Operant training and generalization of a verbal response form in a speech-deficient child. *Journal of Applied Behavior Analysis*, *3*, 139–147.

White, O. R. (1971). *A glossary of behavioral terminology.* Champaign, IL: Research Press.

White, O. R. (1972). *A manual for the calculation and use of the median slope: A technique of progress estimation and prediction in the single case.* Eugene, OR: University of Oregon, Regional Resource Center for Handicapped Children.

White, O. R. (1974). *The "split middle": A "quickie" method of trend estimation.* Seattle, WA: University of Washington, Experimental Education Unit, Child Development and Mental Retardation Center.

Wildman, B. G., & Erickson, M. T. (1977). Methodological problems in behavioral observation. In J. D. Cone & R. P. Hawkins (Eds.), *Behavior assessment: New directions in clinical psychology* (pp. 255–273). New York: Brunner/Mazel.

Williams, C. D. (1959). Case report: The elimination of tantrum behavior by extinction procedures. *Journal of Abnormal and Social Psychology*, *59*, 269.

Williams, J. G., Barlow, D. H., & Agras, W. S. (1972). Behavioral measurement of severe depression. *Archives of General Psychiatry*, *27*, 330–334.

Wilson, C. W., & Hopkins, B. L. (1973). The effects of contingent music on the intensity of noise in junior high home economics classes. *Journal of Applied Behavior Analysis*, *6*, 269–275.

Wilson, G. T., & Rachman, S. J. (1983). Meta-analysis and the evaluation of psychotherapy outcome limitations and liabilities. *Journal of Consulting and Clinical Psychology*, *51*, 54–64.

Wincze, J. P. (1982). Assessment of sexual disorders. *Behavioral Assessment*, *4*, 257–271.

Wincze, J. P., & Lange, J. D. (1981). Assessment of sexual behavior. In D. H. Barlow (Ed.), *Behavioral assessment of adult disorders* (pp. 301–329). New York: Guilford Press.

Wincze, J. P., Leitenberg, H., & Agras, W. S. (1972). The effects of token reinforcement and feedback on the delusional verbal behavior of chronic paranoid schizophrenics. *Journal of Applied Behavior Analysis*, *5*, 247–262.

Winkler, R. C. (1977). What types of sex-role behavior should behavior modifiers promote? *Journal of Applied Behavior Analysis*, *10*, 549–552.

Winett, R. A., & Winkler, R. C. (1972). Current behavior modification in the classroom: Be still, be quiet, be docile. *Journal of Applied Behavior Analysis*, *5*, 499–504.

Wittlieb, E., Eifert, G., Wilson, F. E., & Evans, I. M. (1979). Target behavior selection in recent child case reports in behavior therapy. *Behavior Therapist*, *1*, 15–16.

Wolery, M. & Billingsley, F. F. (1982). The application of Revusky's R_n test to slope and level changes. *Behavioral Assessment*, *4*, 93–103.

Wolf, M. M. (1978). Social validity: The case for subjective measurement or how applied behavior analysis is finding its heart. *Journal of Applied Behavioral Anslysis*, *11*, 203–215.

Wolf, M. M., Brinbrauer, J. S., Williams, T., & Lawler, J. (1965). A note on apparent extinction of the vomiting behavior of a retarded child. In L. P. Ullmann & L. Krasner (Eds.), *Case studies in behavior modification* (pp. 364–366). New York: Holt, Rinehart and Winston.

Wolf, M. M., & Risley, T. R. (1971). Reinforcement: Applied research. In R. Glaser (Ed.), *The nature of reinforcement* (pp. 310–325). New York: Academic Press.

Wolfe, J. L., & Fodor, I. G. (1977). Modifying assertive behavior in women: A comparison of three approaches. *Behavior Therapy, 8*, 567–574.

Wolpe, J. (1958). *Psychotherapy by reciprocal inhibition.* Stanford: Stanford University Press.

Wolpe, J. (1976). *Theme and variations: A behavior therapy casebook.* Elmsford, New York: Pergamon Press.

Wolstein, B. (1954). *Transference: Its meaning and function in psychoanalytic therapy.* New York: Grune & Stratton.

Wong, S. E., Gaydos, G. R., & Fuqua, R. W. (1982). Operant control of pedophilia. *Behavior Modification, 6*, 73–84.

Wood, D. D., Callahan, E. J., Alevizos, P. N., & Teigen, J. R. (1979). Inpatient behavioral assessment with a problem-oriented psychiatric logbook. *Journal of Behavior Therapy and Experimental Psychiatry, 10*, 229–235.

Wood, L. F., & Jacobson, N. S. (in press). Marital disorders. In D. H. Barlow (Ed.), *Behavioral treatment of adult disorders.* New York: Guilford Press.

Wright, H. F. (1960). Observational child study. In P. Mussen (Ed.), *Handbook of research methods in child development* (pp. 71–139). New York: Wiley.

Wright, J., Clayton, J., & Edgar, C. L. (1970). Behavior modification with low-level mental retardates. *Psychological Record, 20*, 465–471.

Yarrow, M. R., & Waxler, C. Z. (1979). Dimensions and correlates of prosocial behavior in young children. *Child Development, 47*, 118–125.

Yates, A. J. (1970). *Behavior therapy.* New York: Wiley.

Yates, A. J. (1975). *Theory and practice in behavior therapy.* New York: Wiley.

Yawkey, T. D. (1971). Conditioning independent work behavior in reading with seven-year-old children in a regular early childhood classroom. *Child Study Journal, 2*, 23–34.

Yelton, A. R., Wildman, B. G., & Erickson, M. T. (1977). A probability-based formula for calculating interobserver agreement. *Journal of Applied Behavior Analysis, 10*, 127–131.

Zeilberger, J., Sampen, S. E., & Sloane, H. N. (1968). Modification of a child's problem behaviors in the home with the mother as therapist. *Journal of Applied Behavior Analysis, 1*, 47–53.

Zilbergeld, B., & Evans, M. B. (1980). The inadequacy of Masters and Johnson. *Psychology Today, 14*, 28–43.

Zimmerman, E. H., & Zimmerman, J. (1962). The alteration of behavior in a special classroom situation. *Journal of the Experimental Analysis of Behavior, 5*, 59–60.

Zimmerman, J. Overpeck, C., Eisenberg, H., & Garlick, B. (1969). Operant conditioning in a sheltered workshop. *Rehabilitation Literature, 30*, 326–334.

Subject Index

Name Index

About the Authors

DAVID H. BARLOW received his Ph.D from the University of Vermont in 1969 and has published over 150 articles and chapters and seven books, mostly in the areas of anxiety disorders, sexual problems, and clinical research methodology. He is formerly Professor of Psychiatry at the University of Mississippi Medical Center and Professor of Psychiatry and Psychology at Brown University, and founded clinical psychology internships in both settings. Currently he is Professor in the Department of Psychology at the State University of New York at Albany and has been a consultant to the National Institute of Mental Health and the National Institutes of Health since 1973. He is Past President of the Association for Advancement of Behavior Therapy, past Associate Editor of the *Journal of Consulting and Clinical Psychology*, past Editor of the *Journal of Applied Behavior Analysis*, and currently Editor of *Behavior Therapy*. At the present he is also Director of the Phobia and Anxiety Disorders Clinic and the Sexuality Research Program at SUNY at Albany. He is a Diplomate in Clinical Psychology of the American Board of Professional Psychology and maintains a private practice.

MICHAEL HERSEN (Ph.D., State University of New York at Buffalo, 1966) is Professor of Psychiatry and Psychology at the University of Pittsburgh. He is the Past President of the Assocation for Advancement of Behavior Therapy. He has co-authored and co-edited 33 books including: *Single-Case Experimental Designs: Strategies for Studying Behavior Change (1st edition), Behavior Therapy in the Psychiatric Setting, Behavior Modification: An Introductory Textbook, Introduction to Clinical Psychology, International Handbook of Behavior Modification and Therapy, Outpatient Behavior Therapy: A Clinical Guide, Issues in Psychotherapy Research, Handbook of Child Psychopathology, The Clinical Psychology Handbook,* and *Adult Psychopathology and Diagnosis*. With Alan S. Bellack, he is editor and founder of *Behavior Modification* and *Clinical Psychology Review*. He is Associate Editor of *Addictive Behaviors* and Editor of *Progress in Behavior Modification*. Dr. Hersen is the recipient of several grants from the National Institute of Mental Health, the National Institute of Handicapped Research, and the March of Dimes Birth Defects Foundation.